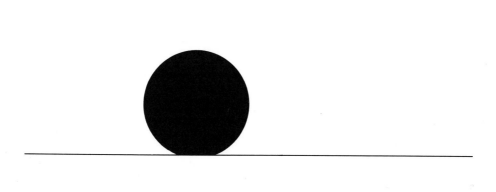

Energy, Resources, & Policy

Richard C. Dorf
University of California at Davis

Addison-Wesley Publishing Company
Reading, Massachusetts
Menlo Park, California • Don Mills, Ontario
London • Amsterdam • Sydney

Second printing, July 1978

Copyright © 1978 by Addison-Wesley Publishing Company, Inc. Philippines copyright 1978 by Addison-Wesley Publishing Company, Inc.

ISBN 0-201-01673-7
ABCDEFGHIJK-MA-798

 # Preface

"Now is the winter of our discontent." The winter of 1977 was the winter of discontent for a large proportion of the U.S. population. There was frost in Florida, drought in California, and flooding in the Midwest, and a continuing cold wave kept natural gas in short supply in scores of cities east of the Mississippi. Crop losses ran into the billions of dollars, energy costs increased by several billion dollars, and more than a million workers were laid off for several weeks. Suddenly, people became aware of their great dependence on the ready supply of energy. The lights that illuminate the monuments and memorials in Washington, D.C., were turned off to conserve energy, and store owners were given the choice of restricting their hours of business to 40 a week or lowering their thermostats to 50° F.

President Jimmy Carter presented two televised speeches during the week of April 18th, 1977, calling for a new energy policy for the nation. He said, "With the exception of preventing war, this is the greatest challenge our nation will face during our lifetime." The energy crisis had become real! Mr. Carter proposed new legislation for taxes on "gas-guzzling" cars, a standby tax on gasoline, tax credit for home insulation, and an increase in the price of natural gas, among other recommended measures. Mr. Carter described the approach needed as "the moral equivalent of war."

The U.S., Canada, Japan, and Europe are greatly dependent upon fossil fuels to maintain their economies. Pres. Carter has proposed that the U.S. launch a new policy based on energy conservation, a reduction of waste, and the development of solar-energy conversion methods and devices. Certainly, it has become clear to all that energy is a critical commodity in the world today.

The energy crisis is the concern of many nations throughout the world. The reader will gain an understanding of the situation that faces the world today by examining national policies and their economic impact. Yet, for all its timeliness, the book attempts to state the facts clearly and objectively. The reader is encouraged to formulate his or her own analysis of the energy question.

The purpose of this book is to present an introduction to the uses of energy, supplies of fossil fuels, alternative energy sources, and various policy alternatives for the U.S. and the world. The history of energy use and projections for future consumption are discussed. The fossil fuels are considered, and electric power, a carrier of energy, is discussed in a separate chapter. The uses of energy in transportation and agriculture are explored. Hydroelectric, wind, tidal, geothermal, and nuclear power generating methods are considered. Solar-energy methods are considered, and alternative conversion and storage systems are explored. The conservation of energy and the relationship of energy and the environment are discussed. Energy economics, policy, and international factors are considered.

The text is written in an integrated form so that the reader should proceed from the first to the last chapter. However, it is not necessary to include all the chapters of the book in any given course. The book is designed for an introductory course on the subject of energy, resources, economics, and policy. It is written to be used in a second- or third-year college course open to students of the sciences, engineering, or economics.

With the knowledge gained from a course on energy, one should be able to face the future with an ability to choose energy policies, institute conservation measures, make economic judgments concerning energy use, and assist in the development of new alternative energy technologies. These tasks will be the most exciting opportunities of the 1980's. It is hoped that this book will assist the reader in encountering this possibility. Hopefully, you will be able to participate, on however small the scale, in the forming of the new society.

This book has been developed with the assistance of many individuals to whom I owe a debt of gratitude. I wish particularly to acknowledge the assistance of Mrs. Phyllis Needle who typed the manuscript. Finally, I can only partially acknowledge the encouragement, patience, and good humor of my wife, Joy MacDonald Dorf, who helped to make this book possible.

Davis, California
November, 1977   Richard C. Dorf

Contents

Introduction

1.1 ENERGY AND SOCIETY

The operation of our technological society depends upon the production and use of large amounts of energy. Many of the world's present problems are closely related to problems of energy distribution, dwindling fossil-fuel supplies, and environmental effects of various methods of energy production and utilization. Energy is not only a commodity; it is also an idea, an intellectual concept that stands out in the history of modern scientific and engineering thought.

The problem of energy use and availability is common, to a greater or lesser extent, throughout the whole world. While the industrialized nations of Europe and North America depend heavily upon fossil fuels for their industrial processes, the developing nations also desire to increase their technological capabilities and thus their use of energy in its various forms. In addition, because deposits of fossil fuels are unequally distributed throughout the world, profound economic and political issues are associated with energy use.

Energy, the ability to accomplish physical work, comes to our attention principally as an input for economic processes and as an intermediate good. It is rare to find energy demanded for its own sake. Energy is usually valued as an input in some process of production or utilization which results in a final product. Economic development has gone hand in hand with increased energy used per capita, beginning in history with solar energy embodied in plants and animals, and continuing through draft animals, wind energy, and fossil fuels. Energy input always produces some measure of pollution, since waste products occur along with the desired goods.

There is a strong correlation between the standard of living as measured by the per capita gross national product and the per capita energy consumption. In industrialized nations such as the United States, energy used per capita has increased steadily over the past century. Table 1.1 shows the growth in population and the energy consumed per capita over the past years. The graph in Fig. 1.1

Table 1.1
ENERGY CONSUMPTION IN THE UNITED STATES*

Year	Total energy consumed $(10^{12}\ Btu)$	Population (thousands)	Index of industrialized production	Total energy per capita (millions Btu†)
1850	2,500	24,000	—	105
1900	8,300	75,000	—	110
1920	19,782	106,466	16.6	186
1925	20,809	115,832	19.9	180
1930	22,288	123,077	20.2	181
1935	19,107	127,250	19.4	150
1940	23,908	131,954	27.8	181
1945	31,541	132,481	44.6	238
1950	34,153	151,241	47.4	226
1955	39,956	164,309	61.1	243
1960	44,816	179,992	68.8	249
1965	53,969	193,815	90.7	278
1970	67,444	203,185	106.7	330
1972	72,108	206,073	106.4	345
1973	75,561	210,400	—	359
1974	73,941	213,500	—	346
1976	74,500	217,700		342

* SOURCE: Reference 1.
† A British Thermal Unit (Btu) is a common unit of energy. Energy units are discussed fully in Chapter 3.

shows the growth of energy consumption and population in the United States during the period 1850 to 1972. Figure 1.2 shows the growth of energy use per capita and an index of industrial productivity during the period 1920 to 1972. Note that energy use per capita dropped significantly during the depression in the 1930's and the recession in 1950. Energy use per capita is tied to the general affluence of the population and thus tends to drop during economic slowdowns.

1.2 ENERGY CRISES

In recent years we have experienced what is often called an "energy crisis." A crisis is a turning point in the course of history, and recent events clearly fit this definition. Since it is a basic economic commodity, energy is available to users to the extent that they are willing to pay the price that is asked by the seller. The relevant questions are how much customers will demand at a given price at a given time and how much business can afford to spend to supply the energy. At some point the answers to these questions match, and the market of supply and demand meet. This theoretical economic action is altered by regulatory and other government controls. During the past few years we have experienced a period of rising prices, shifting relationships among energy suppliers, uncertain supply, and political influences on fuel supply. Each of these factors constitutes an important element of change.

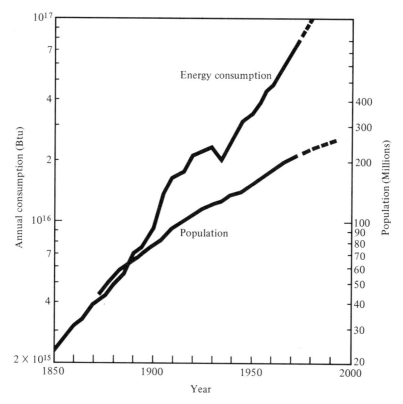

Fig. 1.1   *Energy consumption and population growth in the United States from 1850 to 1972.*

In addition, there is the problem of overcoming the threat of excessive damage to the environment due to extensive use of fossil fuels in transportation and industry. The process of compensating for these environmental effects will consume more energy and will require new approaches. Recycling of materials, rapid transit, sewage treatment, and air-pollution control equipment often consume additional energy and thus increase our total use of energy.

1.3   ENERGY CONSUMPTION

An imbalance of energy consumption exists within the world. Less than 50 percent of the world's population consumes close to 90 percent of its commercial energy; this is a major reason for the great chasm between the industrialized and the underdeveloped nations. The United States itself consumes approximately one-third of the world's energy, although it has only about six percent of the world's population.

The economy of the United States is highly energy-intensive, and per capita use continues to grow. During the period 1945 to 1965, energy consumption

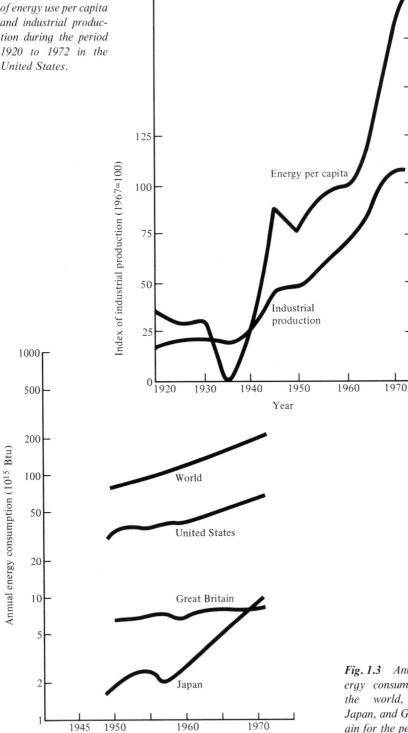

Fig. 1.2 *The growth of energy use per capita and industrial production during the period 1920 to 1972 in the United States.*

Energy per capita

Industrial production

Index of industrial production (1967=100)

Total energy per capita (Millions of Btu)

Year

Annual energy consumption (10^{15} Btu)

World

United States

Great Britain

Japan

Year

Fig. 1.3 *Annual energy consumption for the world, U.S.A., Japan, and Great Britain for the period 1949 to 1971.*

increased at an annual rate of 3 percent. During the period 1965 to 1971 the rate accelerated to 4.8 percent, then slowed again to 3 percent following the increases in gasoline and heating oil prices in late 1973 and subsequent years.

World energy consumption has also grown rapidly during the period 1950 to the present. Figure 1.3 shows the growth of world energy consumption during the period 1949 to 1971, during which a rate of growth of 5 percent per year was experienced. The figure also shows that Japan has expanded its use of energy in recent years at an annual rate of approximately 11 percent.

Affluence, as measured by the Gross National Product of a nation, is closely related to the energy consumption of that nation. Figure 1.4 shows the energy consumption per capita versus the gross national product per capita of several nations. There is a clear trend toward higher consumption of energy as a nation industrializes and increases its gross national product. Note how increasing use of energy per capita tends to produce an increase in gross national product per capita, but that the effect is less pronounced once a nation passes into a state of

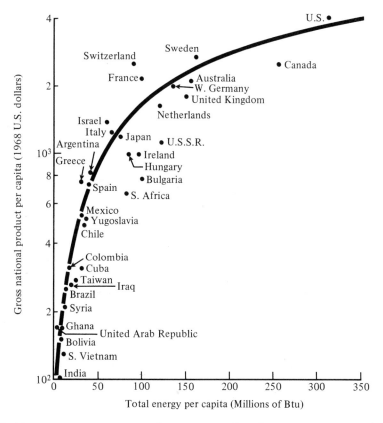

Fig. 1.4 *Energy use per capita versus the gross national product per capita in 1968 for several nations.*

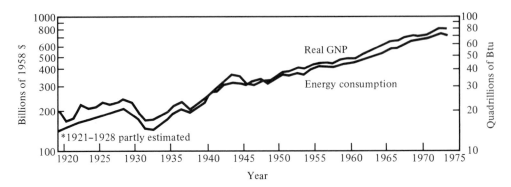

Fig. 1.5 *U.S. energy consumption and real GNP as measured in 1958 dollars for the period 1920–1974.*

relative affluence (approximately beyond $1000 per capita GNP). (See Fig. 1.4.) The very close relationship of energy consumption and real gross national product, in constant 1958 dollars, is shown in Fig. 1.5. [3] Whether GNP pushes up energy consumption or energy use causes increased GNP is an interesting and important question. We may compare U.S. energy use with consumption in other countries. Our workers commute longer distances than in foreign countries (usually by auto); do our centralized, efficient industrial processes make up for the added expenditure of energy for transportation? In Fig. 1.6, the energy consumed per unit of economic output is shown. [3] Clearly, by this measure, U.S. industry is an efficient user of energy.

The increased standard of living in the U.S.A. over the period 1920 to 1970 can be visualized in Fig. 1.7, where energy consumption per capita is shown versus real GNP per capita. Energy consumption continues to grow throughout the

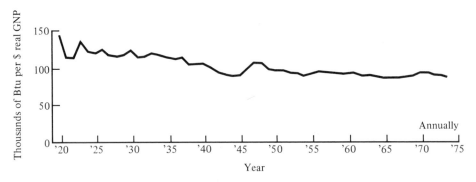

Fig. 1.6 *Energy consumed per unit of economic output (real GNP in 1958 dollars) in the U.S.A.*

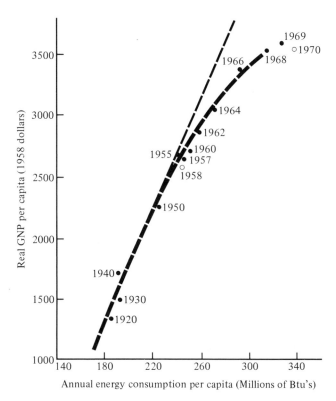

Fig. 1.7 *Energy consumption versus real GNP per capita for the period 1920 to 1970, in the U.S.A.*

world, as shown in Table 1.2 (for the period 1968–1971). Following the Arab oil embargo and oil price increase of late 1973, the growth in energy use has slowed somewhat, but has not plateaued. As energy costs have risen, energy has taken a greater share of a nation's budgeted allocations.

Table 1.2
ENERGY CONSUMPTION PER CAPITA IN
SELECTED COUNTRIES IN 1968 AND 1971

Country	1968	1971	Increase (%)
USA	312*	337	8.1
France	98	117	19.7
W. Germany	134.5	156.5	16.4
Great Britain	148.7	165	11.0
USSR	121.4	136	12.0
Japan	75.5	98	29.7
World	52	58	11.1

* All per capita figures in millions of Btu's.

1.4 ENERGY ECONOMICS

The profound effect of the recent increase in the price of energy has been felt throughout the world's economy. A simple illustration is the cost of energy to U.S. colleges and universities. In the five years from 1969 to 1974, U.S. colleges and universities experienced a 246-percent average cost increase. [4] For example, the energy cost for the University of California, Davis, went from $812,000 to $1,553,000 over the five-year period even after a 10-percent reduction in energy consumption due to energy conservation. Michigan State University experienced a 160-percent increase (from $2,432,000 to $6,332,000) in the five-year period. These are significant portions of the budget of a university. For example, the percentage of the total budget allocated to energy at Michigan State increased from 1.44 percent to 2.81 percent over the five-year period from 1969 to 1974.

Energy is used for four major sectors of the U.S. economy: (1) residential and commercial; (2) transportation; (3) industrial; and (4) electrical power generation. The portion of the total U.S. energy use attributed to each sector is shown in Fig. 1.8. Transportation and electrical generation have grown proportionately as significant users of the nation's energy (as shown graphically in Fig. 1.8 and listed as percentage use in Table 1.3). Electrical power generation is projected to increase its share of the market over the next twenty years as the nation utilizes more nuclear energy.

The flow of energy inputs to the four market sectors and then to useful and rejected (or waste) energy is shown in Fig. 1.9 for 1970. The supply of fuels is shown on the left, and each fuel is shown as a flow of percentage use by each sector. In 1970, the total input of energy was 71.6×10^{15} Btu's. Examining the output side we note that the useful energy, 31.8Q, is only 50.5 percent of the total output

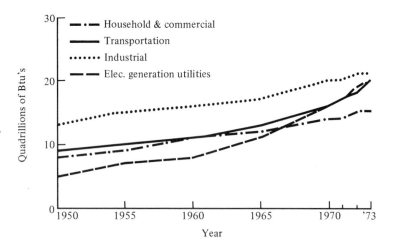

Fig. 1.8   *The use of energy in the four sectors of the U.S. market.*

Table 1.3

CONSUMPTION OF FUEL RESOURCES BY
MAJOR CONSUMER GROUP IN THE U.S.A.

	Percent distribution		
Consumer sector	*1965*	*1970*	*1973*
Household and commercial	22.2	20.8	20.3
Industrial	32.3	30.2	28.4
Transportation	23.8	24.5	24.8
Electrical generation	20.7	24.2	26.2
Other	1.0	0.3	0.3
Totals	100.0	100.0	100.0

63Q. In other words, approximately half of the energy is lost as rejected or waste energy. In Chapter 3 we will examine further this high proportion of waste energy, in terms of thermodynamic principles.

Energy has been a key to the increased industrialization of Europe and the United States. The availability of inexpensive and abundant energy is important

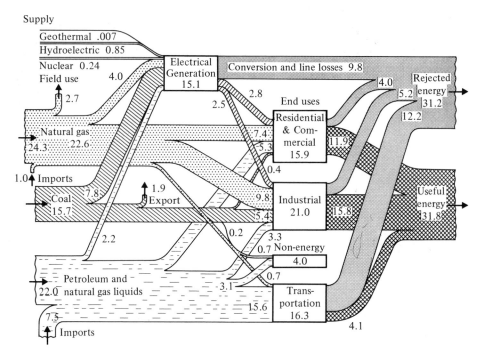

Fig. 1.9   *The flow of energy in the U.S. in 1970. All values are given in 10^{15} Btu $= 1Q$ ($Q = $ quadrillion Btu's). Total input is 71.6Q.*

to mechanized agriculture, industrial manufacture, modern transportation, and modern physical comfort. The world has entered an era of profound alteration in traditional patterns and trends in the field of energy. Price relationships, sources of supply, and in its broadest sense, national security, all have become fraught with uncertainty and conflict. The current division of the uses of energy is shown in Fig. 1.10.

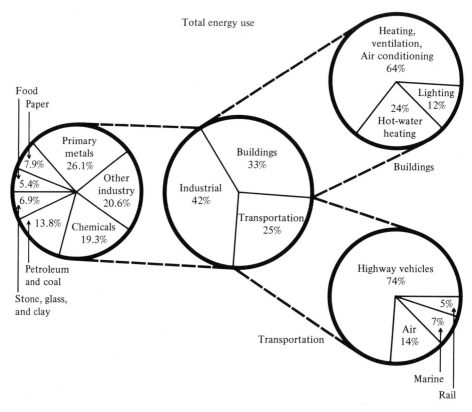

Fig. 1.10   *The uses for energy in the U.S.*

1.5 ENERGY PROBLEMS

Energy problems did not originate in 1973 with the Arab oil embargo; they were evident much earlier: gas utilities refusing to connect service for new residential customers, enforced electrical voltage reductions, and the electrical blackout in northeastern U.S. in 1965. Recently, the closing of gasoline stations, the rises in gasoline prices, and the shortage of natural gas have brought the energy issue to most Americans. Spurts in the sales of compact autos over those of larger autos may be one of the harbingers of change.

Some Americans are searching for a villain who can be blamed for energy shortages and higher prices. Some blame the oil companies, others the federal government, while others blame the Arab governments or the environmentalists. The three most convenient targets are: the energy industry, the federal government, and the environmentalists. As seen by their adversaries, the first conspires, the second bungles, and the third obstructs. The first is a knave, the second a fool, and the third a dreamer.

The winter of 1976–1977 brought another energy crisis pressing the nation. The harshest winter in decades gripped more than half of the U.S. Factories, businesses, and schools were forced to close because energy was not available to run the machines or heat the buildings. The combination of frigid temperatures and the shortage of natural gas, which provides energy for half of the nation's homes and 40 percent of its industries, was devastating. Schools and factories in Ohio, Pennsylvania, New York, and New Jersey were forced to close for lack of natural gas. Frozen rivers and icy roads blocked deliveries of oil and coal. Trains delivering coal were stalled due to snow and ice. The winter of 1977 exemplified to many the dependence of the nation on the ready availability of energy sources.

Industry is accused of withholding supplies—and information on the magnitude of its resources, which are alleged to be much greater than reported. Thus it is alleged the energy industry creates shortages, resulting in higher prices, and, as a bonus, is able to squeeze out independent refiners and service stations. Government is charged by the energy industry with holding down prices (or, by consumers, with not holding them down), reducing incentives, cutting subsidies, and enmeshing industry in a thicket of agencies, standards, and regulations, causing profits to decline, capital to become scarce, and initiative to evaporate. Environmentalists are held guilty of halting the wheels of progress, of following elitist aspirations, and of shunning reality. [5] In a recent survey in the U.S.A., over 40 percent of the people felt the federal government was at fault, while about 30 percent blamed the energy industry. About 15 percent blamed the Arab countries for the energy shortages and price increases.

During the past two decades there has been an increasing tendency for industry and transportation to become more energy-intensive. Most industries responded to increased demand for production by becoming more energy-intensive and less labor-intensive. This is readily seen in the auto industry, which has become increasingly automated, thus replacing manual labor with energy-using machines. Perhaps during the next decades, there will be a decline in the use of automation and a resultant shift back to the use of labor. Nevertheless, fossil-fuel energy has been useful in multiplying man's ability to accomplish work, to move himself and his goods, and to provide physical comfort. Energy will be needed in the future to run the new sewage-treatment plant, for the recycling of materials, and for the creation of new jobs. The challenge remains to supply the energy in a way compatible with a wise use of the environment and the earth's resources.

1.6 THE ENERGY ISSUE

The three dimensions of the energy issue, as shown in Fig. 1.11, are: (1) energy sources, (2) energy processing, and (3) energy policy. In the following chapters we will examine the energy sources available now and those still to be developed, the processing and conversion of energy, energy conservation, and finally the economic, environmental, and political policy issues involving energy use.

In the next chapter we will explore the history of the use of energy throughout the world, and thus enhance our understanding of the current energy dilemma.

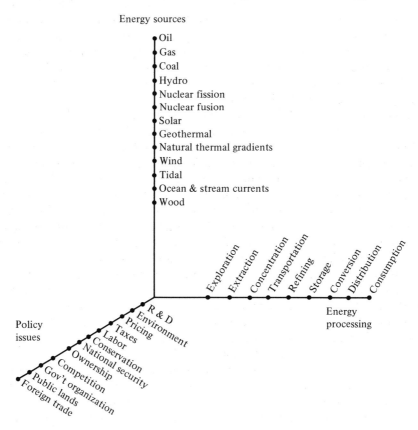

Fig. 1.11   *The three dimensions of the energy issue.*

REFERENCES

1. *Energy Facts*, Science Policy Research Div., Library of Congress, U.S. Government Printing Office, Washington, D.C., 1973.

2. *Social Indicators*, U.S. Government Printing Office, Washington, D.C., 1973.

3. "Pages with the editor," *Public Utilities Fortnightly*, Jan. 30, 1975, p. 6.

4. "Less energy, more dollars," *Chronicle of Higher Education*, July 15, 1975, p. 7.

5. H. H. LANDSBERG, "Low-cost, abundant energy—Paradise lost?," *Science*, 19 April 1974, pp. 247–253.

6. J. R. MURRAY, *et. al.*, "Evolution of public response to the energy crisis," *Science*, 19 April 1974, pp. 257–263.

7. J. O'TOOLE, *Energy and Social Change*, The M.I.T. Press, Cambridge, 1977.

8. H. S. STOCKER, S. L. SEAGER, and R. L. CAPENER, *Energy: From Source to Use*, Scott Foresman and Co., Glenview, Illinois, 1976.

EXERCISES

1.1 The most fundamental energy source, without which there would be no life on earth, is _____.

1.2 Of the annual use of commercial energy worldwide, the United States accounts for:

    a) almost all of it,
    b) about 6 percent of it,
    c) about a third of it,
    d) about three quarters of it.

1.3 The use of wood as a fuel is usually not included in the statistics of commercial energy use in the world. Is this because its contribution to meeting human energy needs is unimportant or insignificant? Why is it not included in the statistics?

1.4 a) Suppose energy use is growing at a geometric rate such that it doubles every 15 years. Assuming this rate continues, how big would energy use be in 2030 compared to the 1970 use?

    b) Suppose that, by means of powerful measures to reduce energy waste, the growth of energy use had been slowed down as of 1970 such that the doubling time was 30 years. If this rate continued, how big would energy use be in 2030 compared to the 1970 use?

***1.5** Obtain the population and energy consumption figures for your state during the period 1900 to 1975, and plot a pair of curves similar to Fig. 1.1. Calculate the energy per capita in your state and compare it with the national figures given in Fig. 1.2 for 1940 and 1970.

1.6 Determine the increase in the cost of energy for your college or university for the five-year period 1969 to 1974. Compare the increase at your college with the national average increase experienced by colleges during that period.

---

* An asterisk indicates a relatively difficult or advanced exercise.

2

The History of Energy Use

2.1 INTRODUCTION

The artful manipulation of energy has been an essential component of human ability to survive and to develop socially. When people first learned to use fuel for warmth, they were taking the first step in the use of an energy resource.

The use of energy has been important in the development of a supply of fuel to provide physical comfort and to improve the quality of life beyond the rudiments of survival. The utilization of energy depends upon the availability of resources and development of the technological skill to use them. Energy resources have always been available to man since recorded time. Figure 2.1 shows the solar input to earth, the basic source over time, and some related energy rates.

Most of the energy utilized by mankind until the advent of nuclear power originated in the sun. This solar energy is absorbed and stored in plants by photosynthesis and provides the energy found in foodstuffs. The energy of coal and oil is also derived from sunlight. These fuels are derived from decomposed plants that lived millions of years ago. Energy from the sun also gives rise to the winds in our atmosphere, which have powered windmills over the ages, and the rain that fills our rivers and drives waterwheels.

It is only in recent times that we have begun to fully utilize the resource of fossil fuels. People often depended upon solar heating, draft animals, and wind energy. The development of power devices to convert energy into useful work has been a recent historical development. The prehistoric domestication of animals represented a multiplication of the power resources then available to man. Still, the use of the horse and the ox limited the power available for irrigation, cultivation, and transportation.

The horizontal waterwheel appeared in the first century B.C.; it normally had a power capacity of about 300 watts. By about the fourth century the vertical

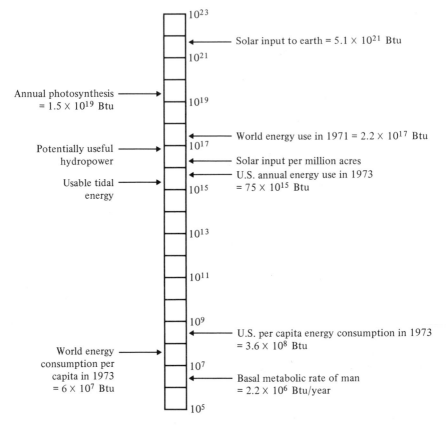

Fig. 2.1 *The solar input to the earth and related energy rates. The unit of energy is the British Thermal Unit. (See Chapter 3 for a full discussion of energy units.)*

waterwheel had been developed, and this yielded about 2000 watts of power. These wheels, used primarily to grind grain, could be used as well for other mechanical tasks. By the 16th century the waterwheel was the prime mover of early machinery. The famous Versailles waterworks at Marly-la-Machine, France, is said to have had a power of about 50,000 watts.

The first windmill appeared in western Europe in the 12th century and was used for grinding grain, hoisting materials, and pumping water. The power of windmills ranged from several thousand watts to over ten thousand watts. Their biggest disadvantage was the intermittent nature of their operation.

Throughout most of man's history the population of a region grew as control and consumption of energy increased. Population growth has taken place in surges that could be said to reflect quantum jumps in the use of energy. One

illustration is the history of China during the past 1000 years. During this mil-
lennium, each sharp rise in Chinese population can be correlated either with
extensive cultivation of a new or improved food plant, or with intensive use of
fertilizer and increased use of fossil fuels; each peak in population can be correlated with saturation of the land area adaptable to the new crop; and each
subsequent decline in population can be correlated with declining agricultural
production brought about by overplanting, soil depletion, erosion, flooding, and
subsequent civil war or invasion.

The ability of government to make and implement decisions to preserve a
society is also a function of the energy supply available to that society. With a
limited amount of energy, cultural and economic development can progress only
to the limits of the efficiencies of the tools and machines used by a civilization.
Past civilizations progressed rapidly after learning to harness a new energy source;
then they reached a plateau where they marked time until some new way was
found to harness additional amounts of energy per capita per year.

For the first million years of more of man's existence, the energy required to
activate his cultures was the energy of his body—principally his muscles. The
amount of energy obtainable from this source is quite small—about one-tenth
of a horsepower per adult male. Consequently, all human cultures were simple,
primitive, and crude until energy was harnessed and utilized in the form of cultivated plants and domesticated animals.

Food surpluses and relative economic security gave rise to large populations
and the consequent development of complex social and political structures. Great
public works, such as irrigation systems and pyramids, were undertaken, and
great armies were formed. Tools, most of which were never needed before, were
devised and improved, and technology moved forward rapidly.

Waterwheel power and windpower technologies were developed intensively
in the late Middle Ages. The great power sources in Europe were the horse, the
windmill, and the watermill. Teams of horses drawing heavy plows powered an
agricultural revolution in the northern plains. The windmill helped the Dutch
reclaim their country from the sea. Water power was applied to all sorts of
industrial processes, from crushing ore to making mash for beer.

By the late 13th century, the forests of England had become so depleted that
a new fuel was required and the nation turned to burning coal in spite of its noxious
smoke.

2.2 COAL TECHNOLOGY

Before the Industrial Revolution began, coal had already become almost the sole
fuel of British industry. The principal sources of industrial power before 1700
came not from fossil fuels, but rather from the horse and from wind and watermills.
In the 18th and 19th centuries, coal replaced wood in many industrial processes,
among them glassmaking and the smelting of nonferrous metals. By the middle of

the 18th century, coal was the primary fuel for industrial use in Britain. [1] By the end of the 17th century in Britain, coal production was between two and three million tons per year. Coal was used by blacksmiths and other metal workers such as coppersmiths and gunsmiths. It was subsequently used in wire drawing. Coal entered into the manufacture of brick, tile, and earthenware. In the chemical industry it was employed in the refining of saltpeter, alum, and copper (as in the making of gunpowder). The production of glass with coal was particularly important and occurred during the 17th century.

By the end of the 17th century steel was being made in covered containers in a closed furnace using coal as a fuel. By 1740 the crucible was involved in the steelmaking process, and coal was often used in blast furnaces.

As might be expected, the expanding industrial utilization of coal led to increased demand for the fuel, and therefore to improvements and innovations in the coal-mining industry itself. Perhaps the most remarkable development in mining coal was the invention of a machine to pump ground water from the mine shaft.

The first use of coal in a blast furnace was in 1709, by Abraham Darby in England. In the blast furnace the fuel and ore are in contact. Air forced through the furnace burns the fuel at high temperature; the ore is reduced and the molten iron is tapped off at the bottom. The product is called pig iron, which is easily cast later into innumerable useful forms.

By the end of the 18th century coal was being used in England in a broad spectrum of methods, processes, devices, and tools. Coal was the fuel that facilitated the birth and growth of the Industrial Revolution. It remains today an important fuel for industrial processes.

2.3 THE STEAM ENGINE

The heat engine was first demonstrated by Hero in the Greek era, but the first practical steam engines were used in the early 18th century. Thomas Savery's steam pump was demonstrated in 1698, although it proved to have only limited possibilities. The engine invented by Thomas Newcomen was installed at Dudley near Birmingham in 1722, and first used to pump water from the mines.

The Newcomen steam engine provided power for a vertical reciprocating lift pump. The whole assemblage looked like a familiar well pump, greatly enlarged. The pump rod was hung on one end of a heavy, horizontal working-beam, pivoted at the center; the steam cylinder of the engine was located beneath the other end; a movable piston within the cylinder was attached by a chain to the beam end. When at rest, the pump end of the beam was down and the engine end was up. The steam cylinder was then filled with steam. Next, a water spray, injected into the cylinder, condensed the steam, forming a vacuum within the cylinder. The atmosphere, acting on the upper side of the piston, pushed it downward into the vacuum, dragging with it one end of the beam; thus the pump end of the beam was raised and a pumping stroke took place. The cycle could be repeated as often as 14 times per minute. The Newcomen engine could be built with sufficient volume to handle large quantities of water, and the engine became very successful.

James Watt turned his curiosity to the steam engine in 1763, and subsequently invented the separate condenser in 1765. In the Newcomen engine the cylinder itself was heated and cooled during each cycle. Watt devised a separate condenser (or *sink*) for condensing the exhaust steam from the power cylinder. This innovation reduced the waste of heat and increased the efficiency of the engine to five percent, twice that of the Newcomen engine. [2] The second of Watt's innovations was the double-acting feature which made practical the rotative engine. By introducing steam into the cylinder first at one end and then at the other, force could be exerted alternately on each side of the piston. Then the working beam could be used to push as well as to pull. Watt's third invention was the speed governor. The fly-ball governor, whose weighted arms swing farther away from the vertical axis of a rotating shaft as the speed increases, was used to regulate steam flow into the cylinder. (See Fig. 2.2.)

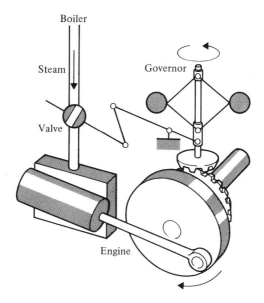

Fig. 2.2   *Watt flyball governor.*

Jonathan Hornblower saw the possibility of using the same steam in more than one cylinder, and patented the compound engine in 1781. The first practical application of the compound engine occurred after 1804, when Arthur Woolf and others made improvements in the engine.

The compound engine found its greatest use in ocean-going steamships, where economy in fuel consumption meant increased cargo capacity. These engines materially hastened the triumph of the steamship over the sailing vessel. By the

1860's, for instance, before the opening of the Suez Canal, steamers equipped with compound engines were able to compete with sailing ships in the Australian trade. By the 1880's marine engines built on this principle were capable of delivering one horsepower (746 watts) for each $2\frac{1}{2}$ pounds (1.13 kilograms) of coal. The *triple* expansion engine, which appeared in 1906, had an efficiency of 23 percent. [3]

The spread of steam engines was rapid and widespread. In Birmingham, for example, the number of engines in use in 1820 was 60, developing an estimated 1000 horsepower (746 kilowatts). By 1840 there were 240 engines developing 3436 horsepower (2.6 million watts). In the United States in 1838, the Secretary of the Treasury reported to the Congress that there were 3010 steam engines at work in the nation, of which 800 were used on board steamboats, 350 in locomotives, and 1860 in various manufacturing establishments and public works. It was an impressive record. [2]

The factory system of manufacture, in which machines and their operators were brought together in a single building around a single power source, was soon well established in both Europe and the U.S. The steam engine also gave transportation systems a power source of tremendous importance. The railway steam locomotive, used extensively in the 19th century, was called the "Iron Horse." Man had harnessed a valuable new source of energy.

Thermodynamics

It can be said that science owes more to the steam engine than the steam engine does to science. The science of thermodynamics grew out of the need to find a method of calculating the efficiency of a steam engine. Sadi Carnot (1796–1832) was the founder of engineering thermodynamics. Carnot concluded that the efficiency of a heat engine was measured by the difference in temperature between the heat source, the input steam, and the heat sink or the steam output temperature as it leaves the engine.

James Joule (1818–1889) of Manchester, England, provided the scientific proof of the principle of the conservation of energy: Energy can neither be created nor destroyed. This principle is called the First Law of Thermodynamics. Joule also worked out the mechanical equivalent of heat: 778 foot-pounds of work is equivalent to the heat required to raise the temperature of one pound of water one degree Fahrenheit.*

The British physicist William Thompson (Lord Kelvin), along with the German physicist Rudolph Clausius, completed the theory of heat engines by defining entropy as the amount of energy (in a heat engine) that cannot be transformed into mechanical work and the natural tendency of this to increase with use. This finding, the Second Law of Thermodynamics, accounted for the observed loss of heat in engines, and enjoyed general consensus by 1865.

---

* Based on the *specific heat* of water at 39° F.

The practical application of thermodynamics to the design of steam engines was promoted by William Rankine (1820–1872), who was Professor of Engineering at Glasgow University. The subject of thermodynamics will be more fully discussed in Chapter 3.

2.4 COAL AND NATURAL GAS

The idea of having energy immediately available, like water on tap, led to development of the idea of using fuel gas. Active experimentation with coal gas began about 1760, and it was used as a fuel for tar ovens in Scotland. In the late 18th century it was used for illumination in a laboratory.

The first public illumination of a building by gas was achieved in Paris in 1801 by Philippe Lebon. Lebon used a fuel obtained by distilling gas from wood. However, it proved to be more efficient to distill gas from coal than from wood.

William Murdock (1754–1839) began work with gas in the 1790's and lighted a room with it in 1792. Murdock conceived the idea of pumping gas throughout a town to illuminate the streets and homes. In 1806 he provided the lights for a textile factory at Salford, England. Gas remained the best solution for the problem of illumination, until it was superseded by electricity in the 20th century.

In 1825 Thomas Drummond produced a brilliant light by using a stick of limestone (CaO) in a gas flame, a technique from which we have derived the term "limelight." Gas lighting was introduced on Pall Mall in London in 1807 and in Baltimore, U.S.A., in 1816. The U.S. developed supplies of natural gas and began using this in place of coal gas at the end of the 19th century. The 19th century was the era of gas lighting. Gas lighting permitted industrial plants to work at night, and allowed people to study and read at night. Good artificial light was important to the development of community activities, adult education, and recreation.

2.5 WIND ENERGY

Civilization's use of the wind dates back to the beginning of history, with the earliest use for sailing ships. Windmills were used as early as 250 B.C. for pumping and milling. The primitive windmill retained the same basic structure until the 12th century, when the horizontal-axis (or Dutch) windmill made its appearance in Europe. Dutch settlers brought the windmill to the United States in the mid-18th century. The Dutch windmill used four to eight or more sails made of cloth stretched over a wooden frame, which could be rotated to face into the wind. The efficiency of these windmills was about 5 percent.

The wind is actually sun-driven due to atmospheric changes, and thus operates on the original fuel of the sun. To pioneers on the windswept Great Plains, the windmills that tapped these air currents substituted for the millraces they had left behind in the East. Between 1880 and 1930, over six million windmills pumped water, sawed wood, ground grain, and generated electric power in the American west. The utilization of windmills continues in the U.S., where it is estimated that over 300,000 pumping windmills are in use today. In addition, over 50,000 wind-mills are used in isolated areas of the country to drive electric generators. In 1941 a large windmill was installed in Vermont by Palmer Putnam. This windmill delivered 1250 kilowatts for a short period, until one of the blades failed; it was not put back into service because of the limited availability of materials during World War II. Wind-driven machines are again of great interest today, as we shall discuss more fully in Chapter 14.

2.6 ELECTRICITY

A very useful form of energy, electrical energy, can be converted into heat, light, and mechanical energy. The earliest forms of electrical machinery were the static electricity devices of the 18th century, used to generate sparks and arcs. The electric battery was developed by Humphrey Davy, and the arc lamp followed in the 1800's.

In 1831 Michael Faraday discovered the principle of electromagnetic induc-tion. He inserted a magnet into a coil of wire and generated, as he described it, "a wave of electricity." The greatest of these discoveries was "that an electric current could be generated in a wire or conductor merely by moving it in a strong magnetic field." Thus, the first rudimentary dynamo was born.

The dynamo or electric generator provided electric power that could be transmitted over wires to users of the energy. The dynamo could be rotated by a steam engine, a water wheel, or a windmill. By 1900, the U.S. census reported that there was 300,000 electrical horsepower (223 million watts) in use in industry. The use of electrical motors at the end of distribution lines enabled these industries to tap power generated at a power station elsewhere. By 1914 electrical machines in American industry were using 8.85 million horsepower (6.6×10^9 watts). As electricity became readily available throughout cities, streetcars, elevators, and other devices became available by the late 1800's.

Thomas A. Edison (1847–1931) was a prolific inventor of devices using electric current. He invented the stockmarket ticker, the multiplex telegraph, and the incandescent lamp. The Edison light, invented in 1882, used a carbon filament and gave the same light as a gas lamp. Edison pursued the development of a complete electric generation and distribution system, and the Edison Electric Light Company was incorporated in 1878. By 1882 the company was supplying 2323 customers from the central Pearl Street power station in New York City. By 1884 it was supplying power for 11,272 lamps in 500 homes. [4]

2.7 PETROLEUM

The medicinal use of natural oil was known by the Indians early in the history of America. In 1854 Abraham Gesner patented a process for distilling petroleum to produce illuminating gas and kerosene. By 1859, others had distilled kerosene from coal, which they called coal oil. This coal oil was used for illumination. Since it was known that kerosene could also be produced from petroleum, it became important to locate sources of petroleum for refining into illuminating kerosene.

After 1850, much effort was devoted to obtaining petroleum from underground wells by pumping. Edwin L. Drake followed the idea of drilling for oil in Pennsylvania, and the first drilling began in June 1859. By August 27, 1859, oil had come to the surface through the drilled shaft, and oil was being pumped out by October, 1859. The drill shaft was $69\frac{1}{2}$ feet deep, and from this well Drake was able to produce about 10 to 35 barrels a day. Total output of Drake's well at Titusville, Pennsylvania, was 2,000 barrels in 1859. By 1863 that figure had climbed to three million barrels.

John D. Rockefeller saw a future in oil and developed ways of obtaining oil, transporting it, refining it, and marketing it. By 1865 Rockefeller had a refinery with a capacity of 505 barrels a day of kerosene. By 1872 he had increased production to 10,000 barrels of kerosene a day. Then he went into the pipeline business, a step towards achieving his sought-after monopoly in the oil business through his Standard Oil Company.

The age of oil at the end of the 19th century led to the practical development of the internal combustion engine.

2.8   THE INTERNAL COMBUSTION ENGINE

The thirst for mechanical power could not be satisfied by the bulky steam engine. What was desired was a light, powerful engine to provide motive power for transport. The decisive breakthrough came in 1876 when Nicolaus A. Otto (1832–1891) of Germany found a way of compressing a combustible mixture inside the working cylinder before ignition. In the following 25 years a reliable electric ignition and a light and powerful engine operating on fuel oil was devised.

The great advantages of the internal combustion engine over the steam engine were its smaller size and its adaptability to intermittent operation. The vision of an engine without a boiler and smaller than a steam engine led to the idea of powering a carriage or autocar with the internal combustion engine.

The Lenoir engine of 1860 was the first internal-combustion engine to be sold in some quantity. It was a one-cylinder, double-acting affair, like a steam engine, that developed one or two horsepower at 100 revolutions per minute. The piston sucked in a mixture of illuminating gas and air for the first half of its stroke. An electric spark then ignited the mixture, and the expanding gases drove the piston through the second half of its stroke. A similar process on the other side of the piston then drove it back.

By 1900 the proven practicality of the internal combustion engine led to the ready production of Ford motor cars, among others. The Wright Brothers also used this engine in their first airplane flight in 1903.

2.9 SOLAR ENERGY

Solar energy has been used by people over the centuries first to grow their crops and later to dry them for storage. In addition, solar energy was used to dry bricks and other building materials and to provide heat for homes in suitable climes.

The idea of solar engines and other solar-powered devices has been with civilization for a long time. However, solar energy has always represented only a small percentage of the total energy used by civilization. Nevertheless, in 1913 an English company named the Sun Power Company, Ltd., opened an irrigation plant near Cairo, Egypt. The steam engine, fueled by the sun's rays, generated 65 horsepower (4.8 kilowatts) and pumped 6,000 gallons an hour (23×10^3 liters/hour). The developer was Frank Shuman of Tacony, Pennsylvania. Shuman used five large heat absorbers that covered 14,000 square feet. Each absorber was a long, curved mirror of inexpensive window glass silvered on one side. The mirror focused the sun's rays upon a black pipe, which acted as the steam engine's boiler. Mounted on a light steel frame, the mirror could be turned to catch the sun's rays throughout the day. Shuman could increase the power of the engine by increasing the size of the absorbers.

As we will discuss more fully in Chapter 17, the disadvantages of solar power are (1) the diffuseness of solar energy, and (2) the intermittency of the availability of sunlight.

Another interesting solar-energy experiment took place in Pasadena, California, in 1901. A huge disk, made up of 1700 small pieces of glass, focused the sun's rays on a 100-gallon boiler; a block mechanism turned the disk to follow the sun. The steam power was used to drive an irrigation pump. This project and the larger Shuman project never became practical operating systems because of materials limitations and high cost. Nevertheless, they led the way, along with many others, toward providing the foundation for solar-energy power systems.

2.10 COMPARISON OF POWER MACHINES

From 1700 on, the power output of energy-conversion devices increased 10,000-fold. Figure 2.3 shows the maximum power output of several devices for the period 1700 to 1970. As can be seen, the steam engine developed over the period 1700 to 1850 up to a maximum power level of about 1000 kilowatts. The internal combustion engine, developed at a later date, achieved a higher level of maximum power output.

The introduction of the turbine, which was a wheel propelled by the force of water or steam coming out through blades or vanes on its circumference, was, in effect, the application of the jet principle. This more efficient device, the reaction turbine, was developed over the period 1830 to the present. By 1837, a water turbine was available that was capable of 45 kilowatts at 80 percent efficiency. Steam turbines, introduced into practical use in the early 1900's, operated on the same principle, and contemporary models are capable of up to several million kilowatts output.

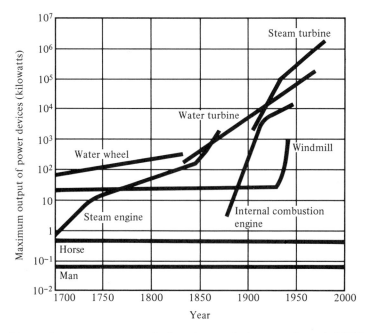

Fig. 2.3   *The maximum power output of selected power devices over the period 1700 to 1970.*

We have come a long way from the simple power available from the horse and man's muscle. As an example of this transition, suppose that there were only horses available in California and there were one horse for every two persons, to be used for transport, farming, and industry. Then for these 10 million horses, we might need 10 million acres of land just to feed the horses. However, California has only nine million acres, which it now farms to feed a large portion of the nation. Clearly, man could not be where he is today if he depended only upon the horse and his own muscles. The machines he has developed are part of his heritage and his future hopes.

2.11   THE HISTORY OF FUEL USE

As man has changed his use of machines to provide power for his needs, he has also changed the use of different fuels. Figure 2.4 shows the history of U.S. energy sources and consumption since 1850. While wood was the fuel used by the early Americans, coal became an important fuel by 1870. By 1910, petroleum and natural gas had become an important fuel for American processes and transportation. Hydroelectric power has been an important part of the history of American energy resources since 1900, but its contributions proportionately have only approached 5 percent, and it is limited by the number of rivers and waterways that can be utilized to drive generators. (See summary in Table 2.1.)

Table 2.1

HISTORICAL SUMMARY OF ENERGY

	Technology	Supply	Life style changes
1776	James Watt's steam engine	Fuel sources are wood, hay, coal	
1807	Gas street lights first installed in London Clermont—first commercially successful steamship	Gas	Gas lamp illumination
1814	First practical steam locomotive		
1830	Best Friend locomotive		
1831	Henry's electric motor	Coal	
1859		Drake well hits oil at 69 feet	Kerosene lamps
1863	Open-hearth furnace	Coal	
1867	Refrigerated rail car patented		
1880	Edison patents electric light bulb		U.S. railroad mileage is 87,800
1882	Rockefeller forms Standard Oil Trust	U.S. is producing 85% of world's crude oil	
1888	Tesla's first electric motor		London subway opens
1890	Suez Canel opens		
1893	Benz and Ford construct their first car		
1895	Los Angeles oil field discovered		
1900		Standard Oil trust said to control 80% of U.S. oil	
1903	Airplane—Wright's first flight		
1908			Model-T assembly line; improvement in salaries for workers
1911	Kettering develops electric self-starter for auto	Standard Oil trust broken up	

Table 2.1 (*Continued*)

	Technology	Supply	Life style changes
1914	Ford develops first tractor	Crude oil worth 88¢/barrel	
1921		Gasoline production in U.S. = 472 million barrels	
1927	Lindberg flies (N.Y. to Paris) in 33.5 hours nonstop		
1929		World's largest refinery built in Arabia	Stock market collapse
1931	E. O. Lawrence invents cyclotron	Natural gas from Texas reaches Chicago through 24″ pipeline	
1937	First jet engine built by Frank Whittle		
1942	Enrico Fermi splits the atom		
1945	First atomic bomb		
1949	U.S.S.R. tests its first atomic bomb	Texas producing 85% of world's petrochemicals	British gas industry nationalized
1953	First hydrogen bomb		
1957	U.S.S.R. launches Sputnik	U.S. crude oil production at 915 million barrels/day	
1962		200 atomic reactors in US (military and civilian)	
1963	First oil drilling off California Coast		
1965		Gas discovered in Alaskan North Slope	Northeast U.S. loses electricity (blackout) for several hours
1968		Aswan Dam completed	
1973		Middle East oil prices doubled	Gasoline rationing or allocation
1975		Oil depletion allowance eliminated	China becomes oil exporter

SOURCE. *Two Hundred Years of Energy*, Chilton's Oil and Gas Energy, January, 1976.

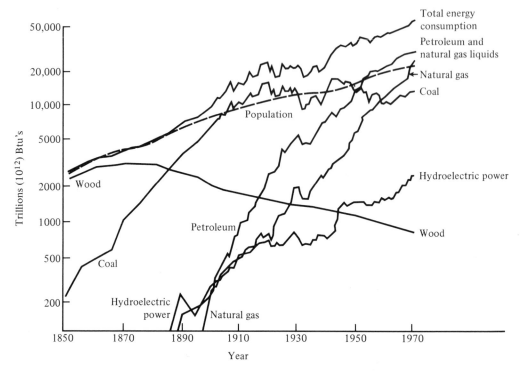

Fig. 2.4 *The history of U.S. energy resources since 1850. The dotted line shows the population growth through that period (ordinate value × 10). Note that the vertical scale is logarithmic.* [*Adapted from Hottel and Howard,* New Energy Technology, *MIT Press, 1971, with permission.*]

Energy Shortages

World Wars I and II imposed enormous military demands for fuel, and resulted in a shortage of fuels for industry and private use. During World War II, rationing of gasoline to the public and industry was instituted, in order to conserve fuel for use by the military. Immediately following World War II, an energy shortage occurred because of a rapid increase of nonmilitary demand. In Northern California an electric power shortage occurred in 1948 that resulted in compulsory rationing aimed at 20 percent curtailment of use. In Great Britain, rationing of gasoline remained in force for several years after World War II. Shortages of energy are not new—they are an omnipresent fact of life. Still, most people would prefer to live *without* shortages, and with an unlimited supply of fuel. In the following chapters we will examine the potential for meeting our demands for energy, as well as the possibility of reducing the demand through energy conservation.

REFERENCES

1. J. R. HARRIS, "The rise of coal technology," *Scientific American*, August 1974, pp. 92–97.

2. E. S. FERGUSON, "The steam engine before 1830," in *Technology in Western Civilization*, M. KRANZBERG and C. W. PURSELL, eds. Oxford University Press, New York, 1967, Volume 1.

3. J. B. RAE, "Energy conversion," in *Technology in Western Civilization, ibid.*

4. H. I. SHAELIN, "Applications of electricity," in *Technology in Western Civilization, ibid.*, pp. 563–648.

EXERCISES

2.1 While Americans worry about the availability of gasoline and fuel oil, millions throughout the world face a shortage of wood for heating homes and cooking food. In undeveloped nations without resources, called the "Fourth World," 90 percent of the people depend on this subsistence fuel, and their forests are rapidly disappearing. So severe is the shortage, *Smithsonian* magazine reports, that many families are spending up to 30 percent of their income on firewood. Like the energy shortage in industrialized nations, this crisis is creating a number of side effects. In South America, India, and central Africa, where the problem is acute, depleted forests tend to hasten soil erosion. In some areas, flooding results; rivers and reservoirs silt up in others, and deserts encroach on formerly verdant land. Where wood is unavailable, people burn dried cow dung as fuel and deprive their crops of fertilizer.

Explore the use of wood in the undeveloped nations and provide a plan for the wise use of a forest resource. Note, that, in this case, the resource is renewable within a person's lifetime, while fossil fuels are nonrenewable.

2.2 Examine the history of the petroleum industry in the U.S., with special attention to the breakup of the Standard Oil trust.

2.3 Prepare a history of the energy consumption and supply for your city or state. Determine what was the primary energy source in 1925 for your city. Determine the date of operation of the first electric generating plant in your city.

2.4 Gas lamps were commonly used as street lamps in many cities in the 1800's. Determine whether the largest city in your state used gas street lamps and when they were replaced by electric lamps. Also determine the source of gas for the lamps.

3

Principles of Energy and Thermo- dynamics

After the development of the steam engine and other means of obtaining increased force and power for human needs, many began to examine the underlying principles of the science of heat engines. In this case, the science of thermodynamics followed from the practice of steam engines. The harnessing of energy stored in fossil fuels such as coal, as well as the energy of falling water, led to attempts to define and understand the principles of energy flow and use. Some of the questions that arise are: What is energy? Is it limitless? What is a measure of goodness for an engine? In this chapter we will review the principles of energy and thermodynamics, and attempt to provide an underlying basis for the examination of various energy systems and fuels in the following chapters.

3.1 ENERGY CONVERSION

We can define *energy* as anything that makes it possible to do work, anything capable of bringing about movement against resistance. All matter and all things have energy. Therefore, energy is a property of matter and takes many forms. Table 3.1 lists recognizable forms of energy; you are probably familiar with most, if not all, of them. Mechanical energy, as evidenced in a rotating shaft, can be

Table 3.1
FORMS OF ENERGY

Kinetic	Potential
Electrical	Magnetic
Heat	Chemical
Nuclear	Sound
Light	Mass

converted to electrical energy through an electrical generator attached to the rotating shaft. This process of converting from one form of energy to another is logically called *energy conversion.* Table 3.2 lists several energy-conversion devices or processes. We are quite familiar with devices for burning fuels and thus converting chemical energy to thermal energy, for example.

Table 3.2
CONVERSION OF FORMS OF ENERGY

From	*To*				
	Mechanical	*Electrical*	*Heat*	*Chemical*	*Nuclear*
Mechanical	Gears, piston	Microphone, electric generator	Friction		
Electrical	Loudspeaker, electric relay		Electric heater	Electrolysis, battery	Particle accelerator
Heat	Turbine	Thermocouple	Heat exchanger		
Chemical	Rocket, combustion engines	Battery, fuel cell	Fires, boilers	Chemical processes	
Nuclear			Nuclear reactor		
Solar		Conversion cells	Collector	Photosynthesis	

Let us trace, for example, the flow of energy from the sun to earth to man. When protons in the sun unite to form helium nuclei, nuclear energy is released. This energy may go first to the kinetic energy of motion of nuclei. Some of the energy is then carried away from the sun by photons, bundles of electromagnetic energy, which is a combination of electrical and magnetic energy. The energy content of the photons may be transformed, by the complicated and not yet fully understood process of photosynthesis, into chemical energy stored in plants. Either by eating plants or by eating animals that have eaten plants, people acquire this chemical energy and convert it into chemical energy to power brain and muscles and into heat energy to keep themselves warm.

That mass is one of the forms of energy was first realized at the beginning of this century. Mass energy can be thought of as the "energy of being," matter possessing energy just by virtue of existing. A material particle is nothing more than a highly concentrated and localized bundle of energy. The amount of concentrated energy for a motionless particle is proportional to its mass. If the particle is moving, it has still more energy, its kinetic energy. A massless particle, such as a photon, has only energy of motion and no energy of being (mass).

The relation between the mass, m, of a particle and its energy, E, is given by Einstein's equation,

$$E = mc^2 \qquad \text{joules,} \qquad (3.1)$$

where $c = 3 \times 10^8$ meters/second (the speed of light), mass is given in kilograms, and energy has the unit joule.

Another common form of energy is *kinetic energy*, which is a measure of how much work is required to set an object into motion (if it is at rest) or to bring it to rest in a specified distance (if it is in motion). For speeds much smaller than the speed of light c, kinetic energy is

$$K = \tfrac{1}{2}mv^2 \qquad \text{joules,} \qquad (3.2)$$

where $v = $ velocity. Similarly, the *potential energy* of a mass, m, at a height above a reference plane h, is

$$P = mgh \qquad \text{joules,} \qquad (3.3)$$

where g is the acceleration due to gravity (9.8 m/sec).

Energy is the capacity to accomplish work, and therefore may be defined for a mass experiencing a constant force **F** and moving through a displacement **x**. The work done by the force is the scalar product of **F** and **x** and is written as

$$\text{Work} = \mathbf{F} \cdot \mathbf{x}$$
$$= Fx \cos \phi \qquad \text{joules,} \qquad (3.4)$$

where ϕ is the angle between the force and the line of movement. Using the metric (or SI) units, we have force in newtons and distance in meters, and therefore work or energy in joules.

For example, a mass with a potential energy $P = mgh$ will convert the energy to kinetic energy in falling through the height h. The work done is then the force $F = mg$ times the distance $x = h$, which is then equal to the original potential energy, or

$$\text{Work} = Fx$$
$$= (mg)h \qquad \text{joules.} \qquad (3.5)$$

The transformation of energy from potential energy to kinetic energy is shown in graphic form in Fig. 3.1. Heat energy, resulting from a body moving over a surface with friction, for example, is a form of kinetic energy of particles. As the body travels over the surface it heats up. (Check your auto tires after a trip!)

The conversion of energy from one form to another occurs in many forms. Heat will make a thread of mercury rise in a thermometer against the pull of gravity; infrared radiation will turn the vanes of a radiometer against the slowing effect of friction, and magnetism will raise a nail. Heat, light, electricity, magnetism, motion, sound, chemical bonds, nuclear forces—all represent manifestations of energy and all are different forms of essentially the same thing, for one form can be freely turned into another. Charges moving through a wire can produce

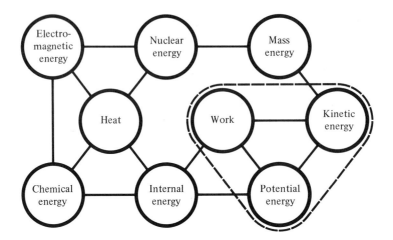

Fig. 3.1 *Forms of energy and their interconnections. Potential energy can be converted to kinetic energy, for example. The dashed lines incorporate the forms of energy relevant to macroscopic mechanics or the mechanics of large masses as contrasted to particles.*

light, and a paddle rotating rapidly in water can produce heat. Magnetism can be turned into motion of electrons, chemical explosions into motion, nuclear reactions into sound, and so on.

3.2 CONSERVATION OF ENERGY

Energy can be converted from one form into another, but it cannot be created or destroyed. When energy is used it is only changed into another form. The hot water in a kettle may cool down, but the energy does not disappear; it is transferred to the outside world.

The law of the conservation of energy states that: The total quantity of energy in the universe is constant. In other words, energy can neither be created nor destroyed. The study of changes of energy from one form to another and the transport of energy from one place to another is called "thermodynamics" (from Greek words meaning heat and motion). For that reason, the law of the conservation of energy is often called the *First Law of Thermodynamics*.

The law of conservation of energy may be written as

$$E_{in} = E_{out} + \Delta E_{stored} \tag{3.6}$$

where

$$\Delta E_{stored} = E_{\substack{final \\ stored}} - E_{\substack{initial \\ stored}}. \tag{3.7}$$

The *change* in the stored energy is denoted as ΔE_{stored}. In other words, all the energy is conserved within the boundary of the energy system.

Consider, for example, the case of a person holding a rock in his hand, as in Fig. 3.2(a). The rock has a potential energy P at that time. In Fig. 3.2(b) the stone has been released and has fallen to a point just above the ground. In Fig. 3.2(c) the stone has hit the ground and come to rest. At time t_1, when the rock is being held, the total energy is $E_{in} = 0$ since

$$E_{out} = 0 \quad \text{and} \quad \Delta E_{stored} = E_{initial \atop stored} = P.$$

When the stone reaches the point above the ground, we have

$$E_{out} = E_{in} = 0 \quad \text{and} \quad E_{final \atop stored} = E_{initial \atop stored},$$

or

$$K = P. \tag{3.8}$$

In Fig. 3.2(c) the stone has hit the ground and come to rest. The kinetic energy has been dissipated in heat, and we can assume that the heat has moved into the air (out of the system). Then we have, at time t_3,

$$E_{out} = K \quad \text{joules,} \tag{3.9}$$

and the air has carried away the energy originally present as potential energy. Of course, *within the total universe* the energy is still conserved, since the heat in the air remains within the universe.

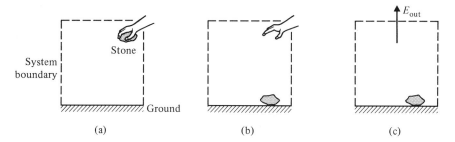

Fig. 3.2 *A person holds a rock with potential energy P in (a). In (b) the rock falls to a point just above the ground where it has a kinetic energy K which is equal to P. In (c) the rock has come to rest and the heat energy equal to K has risen through the air.*

Now let us consider a steam power plant with the boundaries as shown in Fig. 3.3. Fuel and air enter the boiler, and steam is used to drive a turbine, which in turn drives an electric generator. A condenser is used to return the waste steam to water to be used again in the boiler. The cooling water enters from a river (or lake) and is returned to that river with an increase in energy resulting from the condensation of the waste steam. Assuming the plant has been operating for some time and is in its steady-state operation, we have $\Delta E_{stored} = 0$. Therefore,

$$E_{in} = E_{out} \quad \text{joules.} \tag{3.10}$$

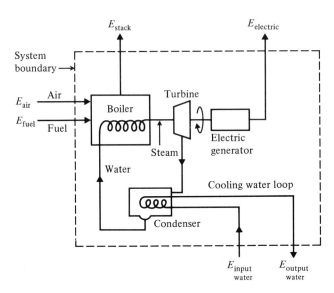

Fig. 3.3 *An electrical generating plant uses a steam turbine to drive the generator and a condenser to return the waste steam to water.*

The input of energy is

$$E_{in} = E_{fuel} + E_{air} + E_{input \atop water} \qquad \text{joules.} \qquad (3.11)$$

The output of energy is

$$E_{out} = E_{electric} + E_{stack} + E_{output \atop water} \qquad \text{joules.} \qquad (3.12)$$

Therefore, Eq. (3.10) becomes

$$E_{fuel} + E_{air} + E_{input \atop water} = E_{electric} + E_{stack} + E_{output.\atop water} \qquad (3.13)$$

We may define the change in the energy of the cooling water as:

$$\Delta E_{water} = E_{output \atop water} - E_{input \atop water} \qquad \text{joules,} \qquad (3.14)$$

and neglect E_{air} since it is significantly less than E_{fuel}. Then, Eq. (3.13) may be written as

$$E_{fuel} = E_{electric} + E_{stack} + \Delta E_{water} \qquad \text{joules.} \qquad (3.15)$$

The electric energy is the desired output, and the stack loss and the cooling water net energy are waste outputs. The overall plant efficiency (η) is thus written as:

$$\eta = \frac{E_{electric}}{E_{fuel}} \qquad \text{joules.} \qquad (3.16)$$

The efficiency of a typical electric power plant is approximately 33 percent. Therefore, two-thirds of the fuel energy is lost in the stack output or cooling-water output. This energy is not lost to the universe, but is dissipated throughout the air and water. The heat in the air and water is spread out and is relatively unavailable. Can we recover the energy from the air or water and use it again? This question leads to the subject of entropy.

3.3 ENTROPY

If we want to use the dissipated heat to do work again, we have to collect it from the surroundings and concentrate it again. However, it takes more energy to collect and concentrate the dissipated heat energy than would be obtained from the collected energy. Therefore, you might as well use the new energy itself rather than try to collect the dissipated energy. In other words, you cannot get something for nothing and you cannot invent a perpetual-motion machine.

The French engineer Sadi Carnot demonstrated in 1824 that the steam engine did work because part of its system was quite hot, the steam, and part of it was cold, the condenser. Essentially the engine worked by taking some heat from the hot part, transforming some of it into work, and dumping some of the heat into the cold part. The fraction of energy that could be turned into work depended upon the temperature difference between the hot and cold parts of the system. The work obtained depends upon the difference in energy concentration within the system.

Therefore, no device can deliver work unless there is a difference in energy concentration within the system, no matter how much total energy is used. Since there is no way of putting all the energy into one part and none in the other, we can never turn every bit of energy into work. As the energy flows from the high concentration to low concentration, the energy at the higher point is also dissipating to the universe. Also, the low-energy point is increasing in energy from the universe, as well as from the higher point. In terms of temperatures, the *difference* in temperature between the hot and cold points is shrinking faster than you would expect. A German physicist, Rudolf Clausius, pointed this out in 1865 and defined a quantity consisting of the change in heat divided by its absolute temperature; he called this a change in *entropy, S.* The change in entropy may be written as

$$dS = \frac{dQ}{T} \quad \frac{\text{joules}}{\text{kelvin}}, \tag{3.17}$$

where dQ is the differential change in heat energy. The change of entropy of any part of the system is equal to the increment of heat added to that part of the system, divided by its absolute temperature at the moment the heat is added, provided the change is from one equilibrium state to another. For heat gain, dQ is positive and entropy increases. The temperature is given in its absolute scale of kelvin (degrees Kelvin) (K) when heat is given in joules. Heat always flows from a hotter to a cooler body. Entropy always increases with time. For an engine or heat device it may be stated that, for any device operating in a repetitive cycle, heat flow *out* of one part of the system during one cycle cannot be transformed wholly into mechanical energy (via work) but must be accompanied by heat flow *into* a cooler part of the system. In brief, heat cannot be completely transformed to work by any cyclical machine.

Every time a moving object is brought to rest by friction, all of its ordered energy of motion is converted to disordered energy of molecular motion, and entropy increases rapidly. As entropy increases, energy is transformed from available to unavailable (or less available) form.

The *Second Law of Thermodynamics* may be stated as: Every system left to itself changes in such a way as to approach a definite final state of rest or equilibrium. The diffusion of materials from regions of high energy concentration to regions of low energy concentrations, the passage of heat from hot to cold bodies are illustrative of the process.

Many scientists have defined entropy as the practical measure of the random element (or *chaos*), which can increase but never decrease in the universe. This means the Second Law of Thermodynamics can be stated as: The entropy of an isolated system increases or remains the same. Entropy is a measure of the extent of disorder in a system or of the probability of the arrangement of the parts of a system. For greater probability, which means greater disorder, the entropy is higher. An arrangement of less probability (greater order) has less entropy.

The entropy is defined as Boltzmann's constant k multiplied by the natural logarithm of the probability, p, of any particular state of the system:

$$S = k \ln p \qquad \text{(joules/K)}, \tag{3.18}$$

where $k = 1.38 \times 10^{-23}$ J/K. Again, entropy always increases or stays the same, so that

$$\Delta S \geq 0. \tag{3.19}$$

In the case of a heat engine operating in a cyclical manner and with all changes taking place through equilibrium states, let us define the heat input at temperature T_1 and the output heat flow at T_2. The amount of entropy flow with heat across a system boundary is written as:

$$\Delta S_{in} = \frac{\Delta Q_{in}}{T_{boundary}}.$$

Figure 3.4 shows a simple schematic diagram of a heat engine. (For notational simplicity we shall omit the symbol Δ.) The First Law states that energy is conserved, or

$$Q_1 = Q_2 + W. \tag{3.20}$$

Entropy flows from the heat reservoir in the amount

$$S_{in} = \frac{Q_1}{T_1}$$

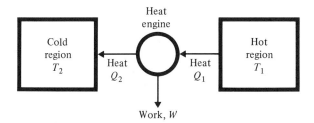

Fig. 3.4 *The schematic of the functioning of a heat engine operating in steady state.*

and out into the cold region as

$$S_{\text{out}} = \frac{Q_2}{T_2}.$$

There is no storage of entropy within the system in steady-state operation, so the entropy increase is

$$\Delta S = S_{\text{out}} - S_{\text{in}}$$

$$= \frac{Q_2}{T_2} - \frac{Q_1}{T_1}. \tag{3.21}$$

Since $\Delta S \geq 0$ we note that

$$\frac{Q_2}{Q_1} \geq \frac{T_2}{T_1}. \tag{3.22}$$

From Eq. (3.20) we have

$$\frac{W}{Q_1} = 1 - \frac{Q_2}{Q_1}, \tag{3.23}$$

where the efficiency, η, is W/Q_1. Substituting Eq. (3.22) into Eq. (3.23), we have:

$$\eta \leq 1 - \frac{T_2}{T_1}. \tag{3.24}$$

The maximum efficiency, the Carnot efficiency, is obtained when the equality is used and

$$\eta_{\text{Carnot}} = 1 - \frac{T_2}{T_1}. \tag{3.25}$$

Clearly, it is most desirable to use a very high-temperature source and a very low-temperature cooling reservoir. For example, when $T_2 = 300$ K and $T_1 = 1000$ K, we have:

$$\eta_{\text{Carnot}} = 1 - \frac{3}{10} = 0.70. \tag{3.26}$$

Typically, an actual plant can yield about half of the ideal efficiency, or about 35 percent.

The heat, Q_2, exhausted into the environment is sometimes called *thermal waste* or *thermal pollution*. Thermal pollution is the inevitable concomitant of power generation and cannot be eliminated. Heat exhausted into water is a serious problem because of its effect on aquatic life. In larger urban areas, heat exhausted into the atmosphere may also be of significance because of its effect on local weather.

In general, the trend of nature toward greater disorder is a trend towards less available energy. In summary, a way of stating the First and Second Laws of Thermodynamics is: "The total energy content of the universe is constant, and the total entropy is continually increasing."* This means that although the universe never loses any energy, less and less of that energy can be converted into work as time goes on.

3.4 AVAILABLE WORK

Available work, or *availability*, is the thermodynamic property that measures the potential of a system to do work. Unlike energy, availability is not conserved, and any degradation by friction or by heat transfer through a temperature difference results in destruction of availability and thus an increase in entropy. [7] Available work is consumed in a process.

The concept of available work makes possible a useful definition of Second Law efficiency called ε, which is

$$\varepsilon = \frac{\text{Least available work required for a task}}{\text{Available work actually consumed in the task}}.$$

As an example, consider a typical home furnace, which may be delivering 60 percent of the fuel energy as heat at 43°C from the register. However, the actual flame temperature is much higher, and this high temperature heat is "wasted" in delivering low-temperature heating so that the Second Law efficiency is only about 7 percent. The use of the Second Law efficiency may allow the user to select an appropriate heating device matched to the task. For example, based on the Second Law efficiency, one might select a heat pump rather than an oil furnace.

3.5 POWER

An important aspect of systems that use or produce energy is the *rate* at which this occurs. The *power* of a device is the time rate of delivery of energy from the device, written as

$$P = \frac{dE}{dt}. \tag{3.27}$$

---

* When everything (including the dying sun) has reached the same temperature, there is no temperature difference to work with; and this is known as the "heat death of the universe."

The unit of power in the International System of Units is joules/second, or watts. The power used by a bread toaster is about 1000 watts, while the power used by a typical reading lamp is 200 watts. A large electric-power generating plant will provide 1000 megawatts. Some relative power levels of devices and of human effort are given in Fig. 3.5. For example, a resting person consumes about 40 watts metabolically. The installed electrical generating capacity per capita in the U.S.A. in 1975 was 2200 watts. A person can generate a large power for a short period of time only. A golfer during the fraction of a second required for a full stroke generates about 2500 watts, and can store about half of that power in the club head. This level of power generation is similar to that of a high jumper.

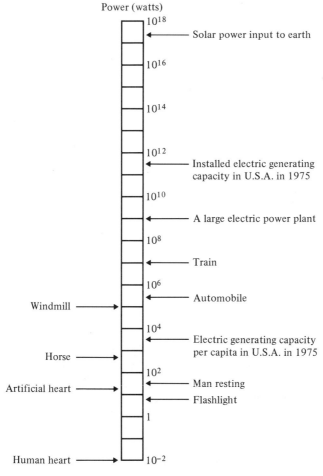

Fig. 3.5 *The power of various devices and of animals and humans (in watts).*

Units of Energy and Power

Since there are several systems of measurement used today, we have to account for several sets of units. Nevertheless, we prefer to use the International System of Units (SI) in contrast to the British system of units. The International System of Units is given in Table 3.3. The unit of energy is joules and the unit of power is watts in SI. (Prefixes for the powers of ten are given in Table 3.4.) Therefore, we call 10,000,000 watts by the measure ten megawatts (or 10 MW).

Table 3.3

THE INTERNATIONAL SYSTEM OF UNITS (SI)

	Unit	Symbol
Basic units		
Length	meter	m
Mass	kilogram	kg
Time	second	s
Temperature	kelvin	K
Electric current	ampere	A
Derived units		
Velocity	meters per second	m/s
Area	square meter	m^2
Force	newton	$N = kgm/s^2$
Energy	joule	$J = Nm$
Power	watt	$W = J/s$

Table 3.4

PREFIXES FOR POWERS
OF TEN

Power	Prefix	Symbol
10^{12}	tera	T
10^{9}	giga	G
10^{6}	mega	M
10^{3}	kilo	k
10^{-2}	centi	c
10^{-3}	milli	m

The conversion of other systems of units to SI units is facilitated by the conversion factors given in Table 3.5. When possible, throughout this book, measures will be given in SI units alongside the familiar British units (if the latter are used at all). Conversion factors and definitions of energy units are given in Table 3.6 for conversion to SI units. A triangle of energy and power unit conversion factors is given in Fig. 3.6. The three systems of units shown in the figure are commonly used in North America. With these factors, you will be able to readily convert from one set of units to another. Also, in many cases, the unit $1Q = 10^{15}$ Btu is used to denote large quantities of energy.

Table 3.5

CONVERSION FACTORS FOR CONVERTING TO SI UNITS

From	Multiply by	To obtain
Length		
inches	25.4	millimeters
feet	30.48	centimeters
miles	1.609	kilometers
Area		
square feet	0.09290	square meters
square miles	2.590	square kilometers
acres	0.4047	hectares (equiv. to 10^4 sq. meters)
Volume		
gallons	3.785	liters
cubic feet	0.02832	cubic meters
barrels (U.S. petroleum)	0.1590	cubic meters
Speed		
miles per hour	0.4470	meters per second
Mass		
pounds	0.4536	kilograms
ton (2000 lbs.)	0.9072	metric tons (1000 kg, also called tonne)
Force		
pounds—force	4.448	newtons
Work–Energy–Power		
foot-pounds—force	1.356	joules
British thermal unit	1055	joules
horsepower	746	watts
calorie	4.184	joules

Table 3.6

ENERGY CONVERSION FACTORS AND DEFINITIONS

Unit	Definition	Multiply by	To obtain
kilowatt-hour	One thousand watts for one hour	3.6×10^6	joules
British thermal unit	Heat needed to raise the temperature of 1 lb of water by 1 degree Fahrenheit	1055	joules
therm	100,000 Btu	29.3	kilowatt-hour
		1.055×10^8	joules
calorie	Heat needed to raise the temperature of 1 cc of water 1 degree centigrade (at 20°C)	4.184	joules
Calorie = one kilocalorie	1000 calories (used in dietetics)	4184	joules
Electron volt	Energy acquired by one electron passing through a potential of one volt	1.602×10^{-19}	joules

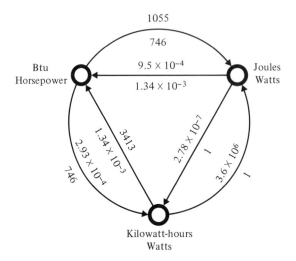

Fig. 3.6 *A triangle of energy unit conversion. The number given above the line converts the energy units and the number below the line converts the power units.*

3.6 ENERGY FLOW

When we consider the earth and its atmosphere as a total system, it is of interest to analyze the energy flow within that system and across the boundary of the atmosphere. The overall flow of energy on and to the earth is shown in Fig. 3.7. The energy inputs to the earth are the energy coming from the sun by means of solar radiation and the energy derived from the mechanical, kinetic, and potential energy of the earth–sun–moon system, which is manifested in the ocean tides and currents.

The rate of energy flow from the sun can be calculated to be equal to 1.73×10^{17} watts. The tidal-energy input is 3×10^{12} watts. Therefore, the sun supplies *essentially all* of the energy flowing to the earth. Terrestrial energy escaping from the earth is about 3.2×10^{13} watts, ten times larger than the tidal energy.

When the solar power enters the earth, a large part of it, 30 percent, is reradiated or reflected back into space as short-wavelength radiation (flow ①). The direct conversion of part of the remaining energy is reradiated as long-wavelength radiation in flow ②. Another part, in flow ③, sets up differences of temperature in the atmosphere and oceans in such a manner that the convective currents produce the winds, ocean currents, and waves. This mechanical energy is eventually dissipated into heat and radiated into space.

Still another part, flow ④, follows the evaporation, precipitation, and surface-runoff channel of the hydrologic cycle. Heat energy is absorbed during the evaporation of water, but it is again released when the water is precipitated. However, the water vapor, being a part of the atmosphere, is convected to high elevations by means of the convective energy already discussed. When precipitation occurs at these elevations, the water possesses potential energy, which again is dissipated back to low-temperature heat on the descent to sea level. It is this

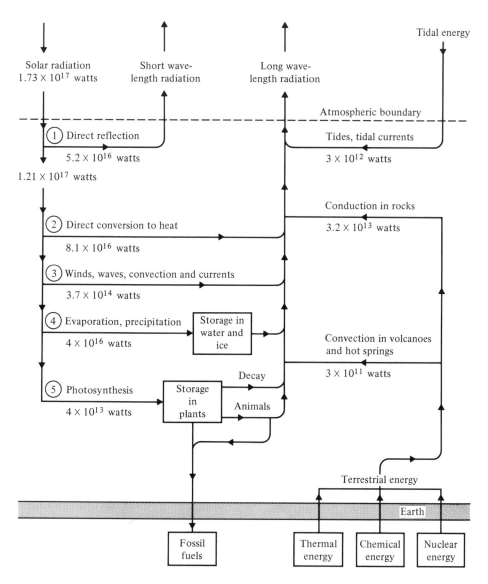

Fig. 3.7   *Energy-flow diagram for the earth.*

energy that is responsible for all precipitation on the land, and for the potential and kinetic energy of surface lakes and streams.

A final fraction of the incident solar energy is captured by the leaves of plants by the process of photosynthesis. In this process, solar energy is stored as chemical energy. If energy could be stored in plants and retained indefinitely, as in firewood, the aggregate amount of stored energy could increase rapidly. In nature, however, eventual decay occurs; and the release of stored energy is nearly equal to the rate of photosynthesis. In a few special locations, such as swamps and peat bogs, vegetable material becomes submerged in a reducing environment so that the rate of decay is greatly retarded and a storage of a small fraction of the photosynthesized energy becomes possible. Over the past 500 million years, this storage and decay process has resulted in our present stores of fossil fuels: coal, petroleum and natural gas, and related materials.

Flow diagrams for transfers of energy can be profitably used in the analysis of energy production and consumption. For example, consider the diagram shown in Fig. 3.8, illustrating energy use on a farm. [4, 9] The external inputs are the solar energy, the rain energy, and the machine and human energy. Nutrients stored in the soil also provide energy to the food production process. As a result of the production process, energy is degraded to heat and waste. The farm product is then equal to 2×10^8 joules, or 47,800 kilocalories. Since the average person consumes about 3000 kilocalories per day, this farm can supply food for about

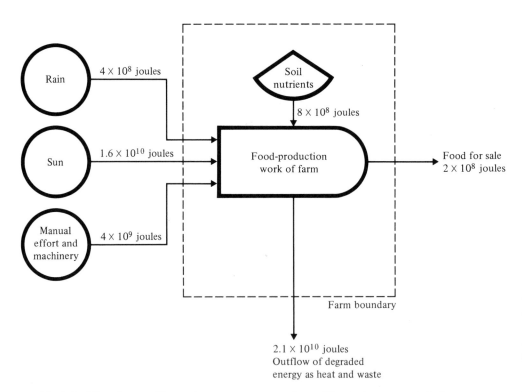

Fig. 3.8   *Flow diagram illustrating energy use on a farm. The circles are symbols for energy sources. The symbol appearing as a tank with nutrients is a storage symbol. The flows are given for one day. The farm's salable output is 2×10^8 joules $= 47,800$ kilocalories.*

16 persons. The proportion of energy that is stored in the food, relative to that lost in heat and wastes, is about one percent.

Typical energy expenditures, in kilowatt-hours, are given in Fig. 3.9. The flow of energy through our environment is dependent upon a myriad of complex processes that are linked together. Flow diagrams can assist in the understanding of energy flow and the expenditure of energy for various processes.

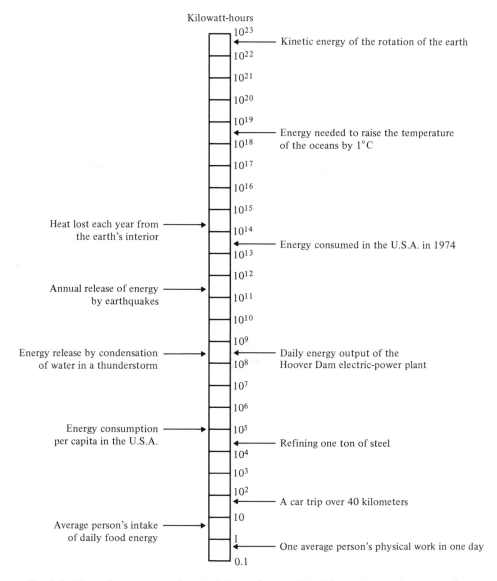

Fig. 3.9 *Typical energy expenditure in kilowatt-hours. (One kilowatt-hour = 3.6 megajoules.)*

REFERENCES

1. K. W. Ford, *Classical and Modern Physics*, Xerox College Publishers, Lexington, Mass, 1972, Chapter 14.

2. W. C. Reynolds, *Energy: From Nature to Man*, McGraw-Hill, Inc., New York, 1974, Chapter 3.

3. M. K. Hubbert, "The Energy Resources of the Earth," in *Energy and Power*, W. H. Freeman and Co., San Francisco, 1971, Chapter 3.

4. H. T. Odum, *Environment, Power and Society*. Wiley and Sons, Inc., New York, 1971.

5. G. M. Reistad, "Available energy conversion and utilization in the United States," *Journal of Engineering for Power*, July 1975, pp. 429–434.

6. J. R. Dixon, *Thermodynamics: An Introduction to Energy*, Prentice-Hall, Inc., Englewood Cliffs, N.J., 1975.

7. *Efficient Use of Energy: A Physics Perspective*, American Physical Society, January, 1975.

8. J. Priest, *Problems of Our Physical Environment*, Addison-Wesley Publishing Co., Reading, Mass., 1973.

9. H. T. Odum and E. C. Odum, *Energy Basis for Man and Nature*, McGraw-Hill, New York, 1976.

10. R. H. Romer, *Energy: An Introduction to Physics*, W. H. Freeman, San Francisco, 1976.

EXERCISES

3.1 Which of the following global energy flows is biggest?

a) sunlight
b) the winds
c) ocean currents
d) the flow of energy in civilization

It is not wholly impossible that a heat engine could be made of plastic. Estimate the upper limit of efficiency of such an engine.

3.2 The steam engines of Newcomen and Watt ran between reservoirs at temperatures of 100°C and 10°C. Find the maximum efficiency of these engines. What was the actual efficiency?

*3.3** The refrigerator transfers heat from a cooler to a warmer region and uses electric energy to do this. Consider the case where T_1 is the temperature inside the refrigerator and T_2 is the room temperature. Heat is removed from the refrigerator at the rate dQ_1/dt and added to the room at the rate dQ_2/dt. Using the two Laws of Thermodynamics, show that the required electric power is

$$P \geq \frac{dQ_1}{dt}\left[\frac{T_2}{T_1} - 1\right].$$

---

* An asterisk indicates a relatively difficult or advanced exercise.

3.4 According to the Second Law of Thermodynamics:

a) Heat cannot be converted completely into useful work;

b) In the universe as a whole, the availability of energy to do useful work is continuously decreasing;

c) A drop of ink placed in a glass of water will spread out by itself, but it will not reconcentrate into a drop again by itself;

d) All of the above;

e) None of the above.

3.5 A typical electric power plant using coal as a fuel is shown in Fig. E3.5. [8] Calculate the overall efficiency of this generating plant. Calculate separately the efficiency of the boiler, the turbine and the generator.

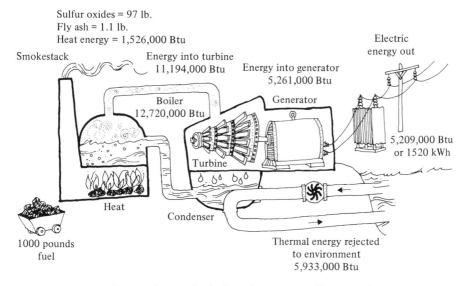

Sulfur oxides = 97 lb.
Fly ash = 1.1 lb.
Heat energy = 1,526,000 Btu

Smokestack

Energy into turbine
11,194,000 Btu

Energy into generator
5,261,000 Btu

Electric
energy out

Boiler
12,720,000 Btu

Generator

5,209,000 Btu
or 1520 kWh

Turbine

Heat

Condenser

1000 pounds
fuel

Thermal energy rejected
to environment
5,933,000 Btu

Fig. E3.5   From Problems of Our Physical Environment, *Addison-Wesley Publishing Co., 1973.*

3.6 A gallon of gasoline has an energy content of 125,000 Btu's. Dry organic material (for example, wood, dried grass, dried buffalo dung, paper) has an energy content of around 4 kilocalories per gram. How many grams of such material does it take to make the energy equivalent of a gallon of gasoline? How many pounds of dry organic material is this?

3.7 Producing a ton of raw steel in the U.S. today takes about 10 million kilocalories of energy.

a) How many gallons of gasoline does this represent? How many pounds of wood?

About 10 tons of steel are in use in the U.S. for every member of the population. (This is the total stock of steel tied up in cars, buildings, bridges, etc., *not* the annual production of steel, which is much smaller.)

b) If wood had been the source of the energy used to build up this stock of steel for the U.S. population of 215 million, how much wood would have been required?

*3.8 It would be wise to utilize the waste heat from an internal combustion (I.C.) engine in an automobile and thus raise the overall efficiency of the auto. An additional engine using the exhaust of the internal combustion engine uses a working fluid heated by the exhaust gases and waste heat. The output of this second engine is coupled to the output of the I.C. engine, thus increasing the overall efficiency of the total power plant in the car. One estimate states that the efficiency of a diesel truck could be increased from 25% to 40%. Explore the value of the use of waste heat for a second engine and the progress in this field by contacting Thermo Electron of Waltham, Mass., or Chapman Engines of Reseda, Calif.

3.9 The power output of machines has increased significantly over the past ten thousand years. The power output of an ox is about 0.2 kilowatt and of a horse about 0.5 kilowatt. The power output of a windmill, over five hundred years ago, was about 15 kilowatts. Estimate the power output of the following devices: (1) waterwheel, (2) steam engine, (3) internal combustion engine, (4) water turbine, (5) steam turbine, (6) liquid fuel rocket. Prepare a table displaying this set of power figures.

4

Projections of Energy Consumption

Continued growth in the use of fossil fuels can occur only for a number of years before we exhaust these resources. The implications of compound growth often escape our attention until it is too late. For example, in the U.S. we have used about 100 billion barrels of our domestic oil resources, and there may be 100 billion barrels remaining as recoverable resources. Since we have used oil for about 100 years, it may be tempting to assume that the remaining oil should last for another 100 years, but the fact is that the consumption of oil has been increasing at a rate of 5 percent per year, and the remaining oil could be depleted in another 14 years.

4.1 COMPOUND GROWTH

If an amount of consumption grows from an initial value A_0 at a rate r (a decimal), then we have, after one year,

$$A_1 = A_0(1 + r), \tag{4.1}$$

where A_1 = the consumption value after one year. After the second year, we have

$$\begin{aligned} A_2 &= A_1(1 + r) \\ &= A_0(1 + r)^2. \end{aligned} \tag{4.2}$$

In general, after n years, we have

$$A_n = A_0(1 + r)^n. \tag{4.3}$$

Thus, if we consumed one billion barrels of oil in 1933 and the growth rate was 5 percent per year, we calculate the consumption, at a time 20 years later (in

1953), as follows:

$$\text{Consumption in 1953} = A_0(1 + 0.05)^{20}$$
$$= 1(2.65) \text{ billion barrels.} \tag{4.4}$$

One can calculate $(1.05)^{20}$ using logarithms, a table, or a calculator.

If we ask the question, "How long will it take for a doubling of a value A_0 at a specified rate r?", we can use logarithms to write:

$$\log A_n = \log A_0 + n \log (1 + r). \tag{4.5}$$

Solving for n, we have

$$n = \frac{\log A_n - \log A_0}{\log (1 + r)}$$
$$= \frac{\log (A_n/A_0)}{\log (1 + r)}. \tag{4.6}$$

Since we want to determine the number of years for doubling, we have

$$n_{\text{double}} = \frac{\log 2}{\log (1 + r)} = \frac{0.3010}{\log (1 + r)}. \tag{4.7}$$

The doubling times for several values of r are given in Table 4.1. Examining Table 4.1, we note that, as an approximation,

$$n_{\text{double}} \approx \frac{72}{r} \tag{4.8}$$

over the range $0.01 \leq r \leq 0.12$.

Table 4.1
DOUBLING PERIODS FOR SPECIFIED GROWTH RATES

r	0.01	0.03	0.05	0.06	0.07	0.09	0.10	0.12
n_{double}	69.6	23.4	14.2	11.9	10.2	8.0	7.3	6.1

The use of petroleum in the world is growing at the rate of 5 percent per year, and the consumption in 1975 was 20 billion barrels per year. Thus we may expect to be using 40 billion barrels 14 years later (in 1989) if this growth rate is sustained over the period.

4.2 EXPONENTIAL GROWTH

The growth in the consumption of oil occurs continuously, and we state that the growth is *exponential* and follows the equation

$$A(t) = A_0 e^{rt}. \tag{4.9}$$

If we replace t by n periods of time of length Δ, we have

$$A_n = A_0 e^{rn\Delta}; \tag{4.10}$$

and if, for example, $\Delta = 1$ year, we then write $A_n = A_0 e^{rn}$. Recalling that

$$e^x = 1 + x + \frac{x^2}{2} + \cdots + \frac{x^m}{m!},$$

we may use only the first two terms when r is small, so that

$$A_n = A_0 (1 + r)^n. \tag{4.11}$$

Therefore, for r less than 0.10, compound growth and exponential growth yield the same result over a limited period of time.

Another approach is to reconsider Eq. (4.3). We may rewrite this equation as follows:

$$A_0(1 + r)^n = A_0 e[\ln (1 + r)]t$$
$$= A_0 e^{\alpha t}, \tag{4.12}$$

where $\alpha = \ln (1 + r)$. Again we note that α is approximately equal to r if $r \ll 1$, and therefore, we obtain Eq. (4.9).

Some Consequences of Growth

When consumption of a resource proceeds exponentially, we can show that the amount of a product consumed during one doubling period is equal to the total used over *all preceding time*. The total used up to time t_1, C_1, is

$$C_1 = \int_{-\infty}^{t_1} A_0 e^{rt} \, dt = \frac{A_0}{r} e^{rt_1}. \tag{4.13}$$

The total used during the doubling time t_1 to t_2, C_2, is

$$C_2 = \int_{t_1}^{t_2} A_0 e^{rt} \, dt = \frac{A_0}{r} (e^{rt_2} - e^{-rt_1}). \tag{4.14}$$

Since it is a doubling period, we have $e^{rt_2} = 2e^{rt_1}$; therefore Eq. (4.14) becomes

$$C_2 = \frac{A_0}{r} (2e^{rt_1} - e^{rt_1})$$

$$= \frac{A_0}{r} e^{rt_1}, \tag{4.15}$$

and thus C_2 equals C_1; that is, the consumption over the doubling period is equal to all prior consumption. Thus, given that oil is being consumed at an exponentially growing rate, when half of that resource has been consumed, then only one doubling period is left before the resource is totally depleted.

Let us obtain an expression for the period of consumption time remaining if a resource has a finite value A_f and we have consumed C already. The remaining period of time, t, for using this resource is then

$$A_f - C = \int_0^t A_0 e^{rt}\, dt, \tag{4.16}$$

where A_0 is the rate of consumption at the beginning of the period (when $t = 0$). Therefore,

$$A_f - C = \frac{A_0}{r}(e^{rt} - 1) \tag{4.17}$$

and

$$e^{rt} = \frac{(A_f - C)r}{A_0} + 1. \tag{4.18}$$

Solving for the period t we have

$$t = \left(\frac{1}{r}\right)\ln\left[\frac{(A_f - C)r}{A_0} + 1\right]. \tag{4.19}$$

As an example, let us assume that 100 billion barrels of oil remain as U.S. resources and we continue to increase our consumption at 5 percent per year. We use our resources at a rate of 3.5 billion barrels in 1975, and we wish to calculate the number of years that the resource will last. Using Eq. (4.19) we have

$$
\begin{aligned}
t &= \left(\frac{1}{0.05}\right)\ln\left[\left(\frac{100}{3.5}\right)0.05 + 1\right] \\
&= 20\ln(2.43) \\
&= 20(0.888) \\
&= 17.76 \text{ years.} \tag{4.20}
\end{aligned}
$$

Limits to Growth

Exponential growth cannot continue indefinitely since some natural limitation sets in. In the case of a resource such as uranium or natural gas, as consumption continues and the resource becomes scarcer, classical economic theory of supply and demand suggests that the price will increase. This price increase should dampen demand and thus lengthen the lifetime of the resource. Other factors such as environmental controls or conservation measures may also limit or alter the growth pattern of the consumption of a fossil fuel.

As an example, consider the growth in the consumption of coal in the U.S. since 1860. During the period 1860 to 1910, the consumption of coal grew at a rate of 6.6 percent per year, thus doubling every 11 years. After 1910, however, as supplies of natural gas and petroleum became available, the rate of growth in the consumption of coal slowed until the average yearly consumption became essentially constant.

The production of an exhaustible fossil fuel may follow a curve, as shown in Fig. 4.1. The production increases initially along an exponential curve, then passes through a maximum and declines eventually to a very small quantity as the fuel source approaches exhaustion.

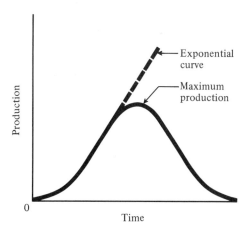

Fig. 4.1   *The production of a fossil fuel in finite supply.*

The cumulative production up to a time t_1 is

$$Q = \int_{-\infty}^{t_1} P \, dt, \tag{4.21}$$

where P = production rate. For a finite quantity of fuel Q_F, we have

$$Q_F = \int_{-\infty}^{t_F} P \, dt, \tag{4.22}$$

where t_F = time when the fuel is exhausted.

If we define Q_R as the proven reserves and Q_P as the total accumulated quantity of fuel removed from the ground, the cumulative discoveries Q_D represents the quantities removed from the ground plus the recoverable quantities remaining in the ground. Therefore,

$$Q_D = Q_P + Q_R. \tag{4.23}$$

Since Q_D is eventually limited to Q_F, we may represent the changes in these quantities as shown in Fig. 4.2. As the cumulative discoveries approach the finite quantity of a fuel, proven reserves drop and the price of fuel rises, thus limiting production.

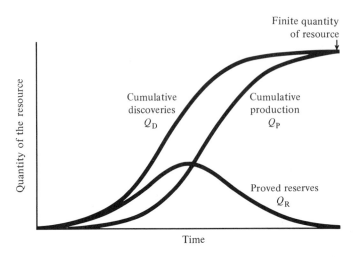

Finite quantity
of resource

Cumulative
discoveries
Q_D

Cumulative
production
Q_P

Proved reserves
Q_R

Quantity of the resource

Time

Fig. 4.2 *The consumption of a limited fuel over the lifetime of the fuel.*

4.3 USE OF ENERGY IN THE U.S.

As an example of a shift from exponential growth, consider the growth in per capita consumption of energy in the U.S. Historically, energy consumption has grown at a rate of about 3 percent over the 70-year period from 1900 to 1970. If we were to extend this rate of growth over the next 30-year period, we could expect a doubling of per capita consumption every 24 years. A more likely event is a gradual deceleration of the rate, so that the growth rate might reach zero by the year 2000.

For example, the rate of increase might decelerate following the equation

$$\frac{dr}{dt} = -a, \tag{4.24}$$

where $(-a)$ is the constant deceleration. Then we have

$$\int_{t_0}^{t_1} dr = \int_{t_0}^{t_1} (-a)\, dt \tag{4.25}$$

or

$$r(t_1) = r(t_0) - a(t_1 - t_0). \tag{4.26}$$

Thus, if $r(t_0) = 0.03$ and $a = 0.0015$ per year, we would expect the rate of growth to be equal to zero after 20 years.

In Fig. 4.3 the trend of the growth in energy use in the U.S. is shown for the period 1900 to 1975. The curve of the historical exponential growth* is extrapolated to the year 2020, yielding a value of 300×10^{15} Btu for that year. A

---

* When exponential growth is plotted against a logarithmic scale (on the vertical axis), the curve becomes a straight line.

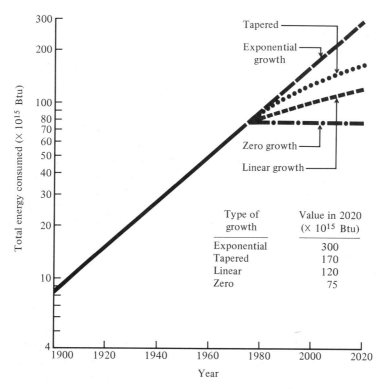

Fig. 4.3 *The historical growth of energy consumed in the U.S. and several projections for the period 1975 to 2020.*

curve for zero growth is also shown. The curve for linear growth following an increase of 1×10^{15} Btu's per year is also shown. We can also calculate a curve for a tapered growth resulting from a deceleration of the use of energy in the U.S. In Fig. 4.3, one curve shows a linear deceleration of the growth rate to a zero growth rate by 2025 ($a = 0.0006$ per year).

While the exponential projection yields a value of total energy consumed in 2020 equal to 300×10^{15} Btu, the value in 2020 is 170 and 120×10^{15} Btu, for the tapered and linear models respectively. Of course, the zero-growth projection remains at 75×10^{15} Btu. We can expect that many of the curves related to energy consumption in the world will experience a tapered growth as prices of fossil fuels increase and supply falls behind demand.

Forecasting

Forecasts of future energy consumption in the U.S. have traditionally turned out to be low. However, recent events have caused forecasts of energy and power trends to be highly questionable. Various forecasts of U.S. energy consumption in 1980 which were prepared in 1970 or 1971 varied from 95 to 110 $\times$ 10^{15} Btu's. [2] As we now rapidly approach 1980, we can more accurately estimate energy consumption for that year as lying in the range 80 to 90 $\times$ 10^{15} Btu's. Forecasting is an art, not a science, and variations in forecasts are to be expected.

Per Capita Use of Energy

Energy use per capita has remained a relatively low-growth item in our U.S. energy history. Considering the use of wood, and agricultural products for work-animal feed, as well as the use of coal and oil, energy per capita has not increased continuously. [3] Two jumps in our energy history occurred: early in the 20th century when agriculture became energy-intensive, and again recently because of a marked increase in the number of Americans employed. However, as the population of the U.S. levels off and the per-capita use of energy stabilizes, the nation's energy curve may flatten out over the next 40 years.

The average standard of living, L, for a nation may be expressed by the following equation: [4]

$$L = \frac{R \times E \times I}{P},$$
(4.27)

where R represents the consumption of raw materials, E represents the consumption of energy, and I the consumption of all forms of ingenuity. The population of the nation is represented by P. As population stabilizes, the average level of affluence and well-being may stabilize also as long as a stable supply of energy and raw materials is available.

4.4 FORECASTS OF U.S. ENERGY CONSUMPTION

There have been many forecasts of the consumption of energy in the U.S. for the period 1975 to 2025. The projections provided in Fig. 4.3 are relatively useful and accurate. A recent study sponsored by the Ford Foundation proposed three possible growth patterns for the period 1975 to 2000, a relatively intermediate-term period. [5]

The first forecast is based on the assumption that energy use in the U.S. will continue to grow until the end of the century at about 3.4 percent annually, the average rate of growth from 1950 to 1970. Thus it assumes that no deliberate effort would be made to alter our habitual patterns of energy use, nor would higher energy prices cause a decreasing rate of growth in consumption. The curve depicting this growth is shown in Fig. 4.4, and the value of energy consumption is listed for 1975, 1985, and 2000 in Table 4.2. While the trends of energy prices cast doubt upon the likelihood that historical growth trends will persist, it is nevertheless the one trend that is often assumed by many government and industry leaders.

The second forecast is based on the assumption of a *conscious national effort* to use energy more efficiently through the introduction of energy-conserving technology, and is called the *technical fix* scenario. The Ford Foundation project

Table 4.2
THREE FORECASTS OF THE FORD ENERGY PROJECT

	Energy consumption ($\times 10^{15}$ Btu)		
	1975	1985	2000
Historical growth	78	116	187
Technical fix	78	91	124
"Zero growth"	78	93	100
Tapered growth of Fig. 4.3	78	100	132

report states that if these technical approaches were consistently applied, an energy growth of 1.9 percent per year would be adequate to satisfy our national needs. The growth of energy use in the technical-fix scenario is shown in Fig. 4.4 and the value of energy consumption is given for 1975, 1985, and 2000 in Table 4.2. Note that the energy consumption of the technical-fix response is one-third less than that of the historical response. This savings of 63×10^{15} Btu in 2000 is four-fifths as large as the total energy consumption in 1975.

The Ford project proposes a *zero energy-growth* scenario. While called zero growth, it actually has decreasing growth rates, eventually reaching zero after 2000. The zero-growth scenario assumes a growth rate of 1.76 percent from 1975 to 1985, and 0.47 percent over the period 1985 to 2000. The growth of energy in

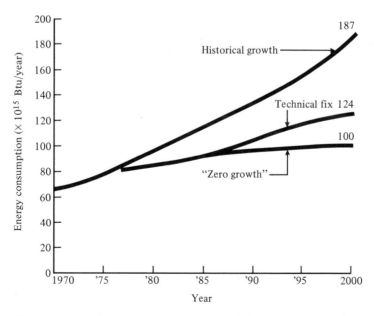

Fig. 4.4 *Three forecasts of energy consumption in the U.S. during the period 1975 to 2000. These forecasts are a result of the Ford Foundation Energy Policy Project.*

this forecast is shown in Fig. 4.4, and reflected values are listed in Table 4.2. The zero-growth response estimates a value of 100×10^{12} Btu for 2000, only 28 percent larger than the value of consumption in 1975, and 46 percent less than the value forecast for the historical response.

For comparative purposes, the *tapered growth* of energy consumption shown in Fig. 4.3 is listed for 1975, 1985, and 2000 in Table 4.2. This tapered response is only 6 percent larger in 2000 than that forecast for the technical fix case.

Only time will tell what the energy consumption will be between 1976 and 2000. In fact, whether we experience faster, lower, or zero energy growth depends upon still conjectural factors such as fuel prices, environmental controls, lifestyles, and the value attached to energy uses. [6] A conservative but unsubstantiated approach might be to assume that a response similar to the technical fix of the Ford project or the tapered growth of Fig. 4.3 is a reasonable forecast. The actual growth rate, however, will depend upon national and international policy issues yet to be settled.

One Projection for the Period 1970 to 2000

A reasonable projection for the growth of energy use in the U.S. for the period 1970 to 2000 might call for a value of energy consumption in 2000 equal to 150×10^{15} Btu. A pattern of growth yielding this value in the year 2000 is shown in Fig. 4.5, which also indicates the proportions of each of the sources of supply that would be required in order to achieve this growth.

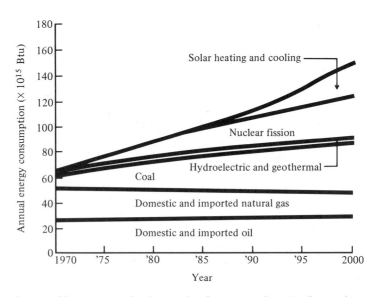

Fig. 4.5 *One possible projection for the supply of energy in the U.S. during the period 1970 to 2000.*

It is assumed that petroleum will supply 29×10^{15} Btu's in 2000, an increase of only 15 percent during the 30-year period. This could be achieved by increasing imports of foreign oil as well as increasing domestic exploitation of U.S. resources. It is assumed that the supply of natural gas will decrease over the period 1970 to 2000 as domestic sources decline and imports prove to be unable to balance this decline. Coal, in this projection, is assumed to supply an increasing percentage of the nation's energy. Furthermore, it is assumed that hydroelectric and geothermal energy sources can be developed to yield a steadily increasing amount of energy.

Within this projection, it is assumed that energy supplied by nuclear fission reactors generating electricity will grow from 0.25×10^{15} Btu's (7×10^{10} kilowatt-hours) in 1970 to 30×10^{15} Btu's (9×10^{12} kilowatt-hours) in 2000. This use of nuclear energy over the period 1970 to 2000 assumes a rapid growth that may be unrealizable. After 1982, it is assumed that the use of solar heating and cooling systems will yield an important portion of the increasing energy consumption in the U.S.

This one projection is illustrative of the various scenarios that may be proposed. In later chapters we will consider each source of supply and determine what reasonable amounts may be obtained and what policy decisions would need to be taken in order to reach a stated level of supply.

Table 4.3

ONE POSSIBLE PROJECTION FOR THE SUPPLY OF ENERGY IN THE U.S. IN 1970 AND 1985

	1970		1985	
Supply	Consumption ($\times 10^{15}$ Btu)	Percent of total	Consumption ($\times 10^{15}$ Btu)	Percent of total
Oil	27	39.7	28	28.0
Coal	16	23.5	26.5	26.5
Gas	23.8	35.0	23.5	23.5
Hydro and geothermal	0.9	1.3	1.6	1.6
Nuclear	0.25	0.4	17.5	17.5
Solar	0	0	3	3.0
Total	68	100	100	100

The changes that might be achieved by 1985 are summarized in Table 4.3. Examining this table, we can see that the projected growth will only be achieved if nuclear power increases at a rapid rate and becomes 17.5 percent of the total supply by 1985. Furthermore, this projection assumes that solar heating and cooling will account for 3 percent of the total energy used in 1985. Perhaps it is more reasonable to expect nuclear power to supply only 10×10^{15} Btu in 1985, in which case we might expect the total energy consumed in 1985 to amount to 92.5×10^{15} Btu instead of 100×10^{15} Btu as previously projected for the U.S. The exploitation of nuclear fission power over the next decade will largely determine whether the projected supply (as shown in Fig. 4.5) will be achieved. An unattractive alternative would be to increase the amount of imported oil by 7.5×10^{15} Btu in 1985, in order to yield a total consumption of 100 quads. (One quad is equal to 10^{15} Btu's.) This alternative is summarized in Table 4.4.

Table 4.4

AN ALTERNATIVE PROJECTION FOR THE SUPPLY
OF ENERGY IN THE U.S. IN 1985 WITH AN
INCREASED DEPENDENCE UPON IMPORTED OIL

Supply	*(Quads)*	*Percent of total*
Oil	35.5	35.5
Coal	26.5	26.5
Gas	23.5	23.5
Hydro and geothermal	1.6	1.6
Nuclear	10.0	10
Solar	3.0	3
Total	100	100

4.5 ENERGY SUPPLY FOR THE WORLD

The growth of energy use by the world has followed an exponential rate of 4.5 percent over the past several decades and many expect this growth rate to continue at least until the 21st century, barring a worldwide economic depression. The growth rate for Eastern Europe and Russia may approach 5 percent over the rest of this century as these nations reach out for affluence and high productivity.

It is expected that oil production will grow dramatically in mainland China, the Middle East, and Africa, in order to supply their increasing demands for energy. One intriguing possibility is the projection of rapid industrialization of the Middle East to the point of consuming the major part of its own oil production by the year 2000.

Europe and many nations of the world are basing their future expansion of energy consumption on the availability of nuclear-fission reactor power plants by the 1980's. In 1973, Western Europe was 67 percent dependent on imported energy sources and Japan 85 percent dependent. Russia is essentially self-sufficient, while the U.S. imports about 12 percent of its energy.

The world consumed approximately 260×10^{15} Btu (78×10^{12} kilowatt-hours) in 1975, and we may expect this to grow to 800×10^{15} Btu by the year 2000 if the rate of growth is maintained at 4.5 percent per year (a questionable assumption).

The unsettled political situation in the Middle East and the rising costs of oil have been major factors in the eager development of new energy sources in all industrialized countries. The emphasis has, up to now, been put on two sources, nuclear energy and domestic oil and gas production, but a third source is also receiving more and more interest—the use of coal directly and for the production of synthetic gas. [7] The development of the Alaskan and North Sea oil fields can supply only a fraction of the U.S. and European demands.

A decade ago, nuclear fission power was presented as the answer to all future energy problems. Now, however, serious concern with environmental and safety factors is delaying full and rapid exploitation of this energy source.

Table 4.5
INCREASE IN ENERGY USE FOR 1951 TO 1969

	Energy increase ($\times 10^{12}$ kilowatt-hours)	Percent increase
World total	37.2	100%
Third World	4.1	11%
U.S.S.R. and Eastern Europe	13.0	35%
U.S.A. and Western Europe	20.1	54%

The world doubled its use of energy over the period 1951 to 1969, as shown in Table 4.5. During the remaining years of the 20th century, we can expect the rate of growth of energy consumption to rise in the Third World while decreasing somewhat in the U.S. and Western Europe. As industrialized nations increase in affluence, their per capita energy use is increased. Table 4.6 shows the increase in per capita energy use for selected regions in the world for the period 1951 to 1969. [7] It is reasonable to expect that the increase in per capita consumption will be slowed down over the next decade for the industrialized nations, while the per capita consumption in less developed nations will continue to increase at a constant or growing rate. Energy consumption per capita in the U.S. has essentially leveled off since 1968 (see Table 1.2), and we can expect this effect to continue. Energy use per capita in the U.S. increased from 225×10^6 Btu in 1951 to 325×10^6 Btu in 1969—an increase of 44 percent over the period. We might project a modest growth in per capita consumption in the U.S., from 337×10^6 Btu in 1971 to 350×10^6 Btu in 1980.

Table 4.6
INCREASE IN PER CAPITA ENERGY USE FOR
CERTAIN REGIONS FOR 1951–1969

	Energy increase (megawatt-hours)
North America	23.6
South America	2.5
Western Europe	10.1
Africa	0.8
U.S.S.R. and Eastern Europe	8.0

While recognizing the increasing worldwide demand for energy, we must also work towards constraining this demand by means of energy conservation and lifestyle changes. Reduction in the rate of growth of demand in the transportation sector is discussed in Chapter 9. The conservation of energy is discussed in Chapter 21.

REFERENCES

1. R. C. DORF, *Technology, Society and Man*, Boyd and Fraser Pub. Co., San Francisco, 1974, Chapter 19.
2. J. MARTINO, "What do you do with 35 conflicting forecasts?," *The Futurist*, June 1973, pp. 134–135.
3. "Energy demand: Constant per capita," *Technology Review*, April 1974, pp. 67–68.
4. V. E. McKELVEY, "Mineral resource estimates and public policy," *American Scientist*, Jan. 1972, pp. 32–40.
5. *A Time to Choose*, Report of the Energy Policy Project of the Ford Foundation, Ballinger Pub. Co., Cambridge, Mass., 1974.
6. *No Time To Confuse: A Critique of the Report of the Energy Policy Project*, Institute for Contemporary Studies, San Francisco, 1975.
7. L. KRISTOFERSON, "Energy in society," *Ambio*, **2**, No. 6., 1973, pp. 178–185.
8. C. WILSON, *Energy Demand Studies: Major Consuming Countries*, M.I.T. Press, Cambridge, 1977.
9. G. LEACH, "Energy futures—wide open to change and choice," *Ambio*, **5**, No. 3, 1976, pp. 108–116.
10. C. WILSON, *Energy Demand to the Year 2000 and Energy Supply–Demand Integrations*, The M.I.T. Press, Cambridge, Mass., 1977.

EXERCISES

4.1 If the enrollment of your college grows at a rate of 2% per year for the next decade, by what factor is the enrollment multiplied? What is the total percentage increase over that ten-year period?

4.2 The U.S. has approximately 80×10^{18} Btu of energy in coal remaining to be exploited, and currently uses 16×10^{15} Btu of coal each year. If the growth rate of coal use is 5% over the next several centuries, how long would it take to completely exhaust this resource?

4.3 Draw a curve of energy consumption in the U.S., for the period 1970 to the year 2000, that assumes a growth rate of 2% per year. What are the values of consumption for the years 1985 and 2000?

4.4 Draw a curve of energy consumption in the U.S. for a tapered growth of 3% initially in 1970, decreasing to a zero-growth rate in 2010. What are the values of consumption for the years 1985 and 2000?

4.5 Using natural logarithms and the approximation that $\ln(1 + x) \simeq x$ for small x, show that

$$n_{\text{double}} \simeq \frac{69.3}{r}.$$

Compare this result with Eq. (4.8) and explain the discrepancy.

4.6 For the curve obtained in Exercise 4.3, sketch the supply elements in a manner similar to that shown in Fig. 4.5. Decrease the reliance on imported oil and nuclear energy in your projection.

4.7 Early hunting man, with fire available but no agriculture (1610), used 5000 kilocalories per day, while early agricultural man (1700) used 12,000 kilocalories per day. Advanced agricultural man (1780) used tools and domesticated animals and consumed about 30,000 kilocalories per day. Industrial man (about 1870) used about 70,000 kilocalories per day and contemporary man uses about 230,000 kilocalories per day. Complete a plot with logarithmic coordinates for energy per capita per day as the vertical scale and years (linear) on the horizontal. Determine the rate of growth over the period 1600 to 1975. Estimate, by extrapolating the curve, what the per capita use would be in the year 2100. Is it reasonable to extrapolate this curve over this many years? Do you expect that this per capita figure will be attained?

*4.8 When a resource is first discovered, its production may rise from zero at an exponential rate. Eventually, as the resource is depleted, it is more difficult and expensive to exploit, and the rate of growth of the consumption of the resource slows down. Let the value of the ultimate resource be Q_u and the cumulative consumption of the resource be Q. The equation for the use of the resource will be

$$\frac{dQ}{dt} = cQ(Q_u - Q),$$

where c is constant. Solve the ratio $Q(t)/Q_u$ which yields a curve called the logistics curve. Note that the rate of change approaches zero when Q approaches the ultimate value Q_u.

---

* An asterisk indicates a relatively difficult or advanced exercise.

Natural and Synthetic Gas

Natural gas and gas manufactured from coal or oil have been used as fuels for lighting, cooling, and heating for over a century. In 1880 approximately 200 billion cubic feet of fuel gas was used in the world. Today, about one century later, the world is using *three hundred times* that amount of gaseous fuel.

The use of natural gas and gaseous fuel generated from coal developed in the U.S. after 1890. In 1900 the U.S. used approximately 200 billion cubic feet of gas as energy fuel. Today, about 22 trillion cubic feet of gas is used as fuel in the U.S., an increase of over one hundred times the amount used 75 years ago.

Natural gas was discovered in the U.S. along with petroleum and became available as petroleum wells were developed. Natural gas, which is CH_4 (methane), was not generally used at first, since it was not as easily stored or transported as liquid petroleum. However, as natural gas was recognized as a valuable fossil fuel, methods of storage and transportation were developed.

Natural gas can be found together with oil in associated gas fields or independently from oil in unassociated gas fields. Some gas has been generated in the past from oil and some from coal. Gas is the cleanest (in terms of pollutants) and most flexible natural fuel, but it is more difficult to store and transport than liquid or solid fuels, especially when the distribution system must extend from one continent to another. In associated gas fields, the production of gas is directly linked to the production of oil and its flow cannot be modified separately from that of oil. Gas that is not used can be reinjected, destroyed, or liquefied. Reinjection helps to maintain pressure in the oilfield and preserves the gas for future use. However, much natural gas that is produced together with oil, especially in the Middle East, is still destroyed through burning (often called *flaring*).*

---

* It is estimated that a billion cubic feet per day are burned at the wellhead. (From Time, Inc., "Bechtel's Pipe Dream," reproduced in *Fortune*, Nov. 1952, and N. de Nevers, *Technology and Society*, Addison-Wesley, 1972.)

5.1   NATURAL GAS IN THE U.S.

Natural gas supplies 33 percent of the total energy consumed in the U.S. today, as shown in Fig. 5.1. Oil, gas, and coal supply essentially 95 percent of all the energy consumed in the U.S., and gas is an important second in order of priority of use.

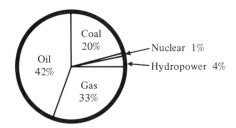

Fig. 5.1   *The sources of energy in the U.S. in 1974.*

The annual consumption of natural gas in the U.S. has grown from 812 billion cubic feet in 1920, to 22 trillion cubic feet in 1974, an increase by a factor of 27, or an annual increase of 6 percent, as shown in Fig. 5.2. The use of natural gas grew dramatically after 1945 as is illustrated by Fig. 5.2 and can be noted in Table 5.1. The per capita consumption of natural gas grew during the period 1925 to 1972

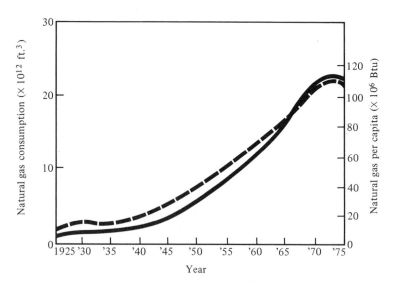

Fig. 5.2   *Annual natural gas consumption (solid line) in trillions of cubic feet, and per capita natural gas consumption in millions of Btu (dashed line) for the U.S.A.*

Table 5.1

USE OF NATURAL GAS IN THE U.S. FOR THE PERIOD 1920 TO 1974

Year	Annual gas consumption (Billions of ft³)	Gas consumption per capita (Millions of Btu)	Average wellhead price (Cents/1000 ft³)	Index of real price 1967 = 100
1920	812	8		
1925	1,209	10		
1930	1,979	16		
1935	1,969	16		
1940	2,734	21		
1945	4,042	30		
1950	6,282	41		
1955	9,405	56	10.4	75
1960	12,771	71	14.0	92
1965	16,040	83	15.6	101
1970	21,921	108	17.1	97
1972	22,532	110	18.6	98.6
1973	22,648	109	21.0	103
1974	21,712	108		
1975	20,000	96		
1976	19,000	89		

and has only recently leveled off, as can be seen by examining Fig. 5.2 and Table 5.1. The actual consumption of natural gas has declined in the U.S. since reaching a peak in 1973, as can be noted from Table 5.1. Convenient conversion factors for relating a cubic foot of gas to energy in Btu (or joules) is given in Table 5.2.

Table 5.2

CONVERSION FACTORS FOR NATURAL GAS

1 cubic foot of natural gas = 1031 Btu's = 1.088×10^6 joules;
1 cubic foot of natural gas = 0.0283 cubic meters = 28.3 liters.

Because natural gas has been an inexpensive fuel for the past 25 years, it became an important fuel for industrial and commercial uses. The average well-head price of natural gas for 1955 to 1974 is shown in Table 5.1, which also gives an index of the real price of natural gas, using the wholesale price index. The price of natural gas at the point of production was some 20 percent less than that of coal and 60 percent less than that of oil during most of that period (per 10^6 Btu's of energy), and this led to the wide use of gas.

Natural gas presently accounts for one-third of the total energy use in the U.S. In 1920, as shown in Table 5.3, gas accounted for only 4 percent of the total energy use. Natural gas became an important fuel in the U.S. because it was relatively inexpensive, relatively nonpolluting, and widely available throughout the country.

Table 5.3
PER CAPITA CONSUMPTION OF NATURAL GAS AND
TOTAL ENERGY FOR 1920 TO 1970 (IN MILLIONS OF BTU)

Year	Natural gas	Total energy	Natural gas as a percent of the total energy used
1920	8	186	4
1930	16	181	9
1940	21	181	12
1950	41	226	18
1960	71	249	28
1970	110	338	33

Transmission and Storage

The development of high-pressure pipelines for transporting natural gas changed gas from a fuel applied principally in local use, into a major industry. After 1945, the pipeline system in the U.S. was expanded from 77,000 miles in 1945 to 265,000 miles in 1974. This development permitted gas to be delivered to states that did not produce natural gas; and as a result, approximately 68 percent of the natural gas was sold outside the producing states by 1974, as shown in Table 5.4. In addition, more than 550,000 miles of lower-pressure distribution lines reach out to individual customers.

Table 5.4
THE DEVELOPMENT OF A U.S. INTERSTATE TRANSMISSION
PIPELINE SYSTEM

Year	Gas consumption (10^{12} cubic ft)	High-pressure pipelines (Thousands of miles)	Percent gas sold outside producing states
1945	4.0	77	48
1950	6.3	109	52
1955	9.4	142	55
1960	12.7	181	60
1965	16.0	210	64
1970	21.9	253	67
1974	21.7	265	68

The transportation of gas by pipeline is a relatively inexpensive means of transmitting energy. Natural gas can be transported via pipeline for 3 cents/10^6 Btu over a one-hundred mile distance. By contrast, it costs 21 cents/10^6 Btu to transmit electricity over the same 100 miles.

A typical pipeline is 48 inches in diameter; gas is pumped along the line using compressors spaced at 100-mile intervals. Pipelines operate at pressures of 600 psi

(40 atmospheres) and are capable of power capacities of 10,000 megawatts, several times the capacity of an overhead electric transmission line.

An aesthetic benefit of the gas pipeline system is the fact that the entire network lies underground at a depth of 24 inches, regardless of the terrain; this allows the land to be used in the same manner as was customary prior to the introduction of the pipeline.

Because natural-gas pipelines are a costly investment, sizable fixed charges are levied for their service. They have to be used at a high rate of capacity to keep unit transportation costs down. Maintaining this high load factor while meeting extra seasonal demands, especially for space heating, has instigated a search for large, natural, underground storage facilities, such as caverns or abandoned gas wells, near major marketing areas. The total capacity of such facilities in the U.S. is some 5 trillion cubic feet.

The Natural Gas Industry

Gas is an essential fuel for thousands of industrial processes and the preferred fuel for many commercial purposes. Gas supplies 43 percent of the energy for industry in the U.S., including fuel requirements for 28 percent of the nation's electric utilities. The consumption of natural gas by the four sectors of the U.S. market are shown in Table 5.5 for the period 1947–1971. The use of natural gas for the generation of electric power accounted for only 12 percent of the total use of natural gas in 1950, but that figure had grown to 19 percent by 1970. Natural gas became a particularly attractive fuel for electric utilities and industry in the late 1960's when a heightened concern for pollution resulting from the burning of other fuels impelled users of coal and oil to turn to relatively nonpolluting natural gas, which has a low sulfur content.

The U.S. gas industry served 45 million customers in 1975. Of this total, 4 million were residential customers (meter connections to homes and individual dwellings), and 315 million were commercial customers such as stores and office buildings. There were also about 250,000 industrial customers in 1975.

Table 5.5

CONSUMPTION OF NATURAL GAS IN THE FOUR SECTORS
FOR THE PERIOD 1947 TO 1971 IN BILLION CUBIC FEET

Year	Household and commercial	Industrial	Transportation	Electricity generation
1947	1,087	2,906	(Negligible)	373
1950	1,586	3,602	126	629
1955	2,753	4,769	245	1,153
1960	4,123	6,074	347	1,724
1965	5,346	7,433	501	2,318
1970	6,894	9,856	722	3,894
1971	7,346	10,438	825	4,125

The total revenues for the gas utility industry were 12.5 billion dollars in 1972, with approximately half of that income coming from residential customers.

The predominant supply of natural gas comes from the states of Texas, Oklahoma, Louisiana, and Arkansas. In this area the supply far exceeds the local demand, and gas is exported via the pipeline system. The per capita gas consumption within the entire Gulf Coast area is estimated to be 350 thousand cubic feet, while an average of only 50 thousand cubic feet per person per year is consumed by residents of the East Coast of the U.S.

5.2 WORLD CONSUMPTION OF NATURAL GAS

The world consumption of natural gas was 52.6×10^{12} cubic feet (54.2×10^{15} Btu) in 1974. The U.S. used approximately 60 percent of the world's consumption of gas in that year. It can be noted, by examining Table 5.6, that several nations or regions are heavily dependent upon natural gas as a fuel. The U.S.S.R. uses natural gas to supply 20 percent of its total energy needs and presently has abundant supplies of natural gas within its borders.

Table 5.6

WORLD TOTAL OF NATURAL GAS BY REGIONS
AND NATIONS IN 1974

Nation	Natural gas (10^{15} Btu)	Total energy (10^{15} Btu)	Natural gas as a percent of total energy
U.S.A.	21.7	73.9	29
Canada	2.4	7.0	34
U.S.S.R.	9.0	44.0	20
Japan	0.13	12.4	1
Western Europe	3.7	45.7	8
Eastern Europe	1.3	13.7	10
World total	54.2	250.0	20.8

In Europe about 10 percent of the total energy needs are satisfied by natural gas. Areas that may be expected, over the next five years, to increase their consumption of natural gas are Europe and the Middle and Far East.

5.3 GAS RESERVES IN THE U.S.

The consumption of natural gas in the U.S. decreased from 1973 to 1975 by approximately 11 percent. This decrease in consumption was not due to decreased demand, but rather to a reduced supply or availability of gas. An actual gas shortage existed in the period 1974–1977, and would-be customers had to be turned away. The present shortage did not happen suddenly, but was the result of a trend over the past two decades. While the consumption of gas was increasing during the past 20 years, the exploration and drilling activities of the gas industry declined markedly. Wildcat, or exploratory, drilling declined by 40 percent over this period; and many well-drilling rigs were removed from the U.S. to more profitable sites in other parts of the world. During this period, there was a 53 percent decrease in wells completed as gas producers declined, from 909 in 1959 to 430 in 1970.

An important measure of the availability of natural gas for future consumption are the known reserves of gas. Thus, if new reserves of gas are discovered each year equal to the amount of gas consumed, we maintain a constant ratio of reserves to production. However, because the demand for gas increased over the past 20 years, while the discovery of new reserves for production has not increased, the reserve-to-production ratio has decreased from 21 in 1956 to less than 15 in 1970.

In 1969, for the first time, gross additions to proven reserves were 40 percent less than the gas produced. Production in 1968 was 19.3 trillion cubic feet, and only 13.7 trillion cubic feet of new reserves were added, a deficit of 5.6 trillion cubic feet. Production in 1969 was 20.7 trillion cubic feet, whereas new reserve additions were only 8.4 trillion cubic feet. This was the second consecutive deficit, 12.3 trillion cubic feet, or 60 percent. The nation used two and one-half times as much natural gas as was discovered. In 1970, production was 22 trillion cubic feet; and, excluding the large estimated Alaskan reserves (which are not now available for delivery), new reserve additions in the lower 48 states were only 11.6 trillion cubic feet. This resulted in a third consecutive annual deficit. As a result, severe limitations were imposed on new industrial loads, and deliveries to existing industrial customers were curtailed. [2, 3] For the period from April 1974 to March 1975, interstate pipeline companies curtailed deliveries by 1.8 trillion cubic feet.

Table 5.7

KNOWN RESERVES OF NATURAL GAS IN THE U.S.

Year	Known reserves ($\times 10^{12}$ cubic feet)	Reserve-to-production ratio
1960	260	20.2
1965	285	17.5
1967	289	15.5
1970	284	12.5
1972	238	11.5
1973	218	10.5
1974	237	11.1

The known reserves of natural gas and the reserve-to-production ratio have decreased in the U.S., as shown in Table 5.7. As of 1974, the reserves of gas, including the new discoveries in Alaska, total 247 trillion cubic feet, an 11-year supply based on present rates of consumption. If past discovery patterns prevail, projected discoverable resources of 1150 trillion cubic feet, added to the known reserves, might yield a lifetime of 40 more years for gas use.

Natural gas is a finite resource since it was produced millions of years ago and stored in the earth as a resource. The total United States resource of natural gas is difficult to estimate, but current estimates range from 1000 trillion cubic feet to 2000 trillion cubic feet. There is some general agreement that the total resources are between 1500 and 2000 trillion cubic feet, and future production rates can then be estimated. In a recent study, Duane and Karnitz conclude that, from a practical viewpoint, the maximum U.S. production rates attainable will be little, if any, higher then at present. [4, 5] The current accepted estimate of total gas resources in the U.S. is 1845 trillion cubic feet, including offshore sources and Alaska.

Duane and Karnitz examined the potential development of these resources and provided a scenario (or portrayal) of the possible consumption of gas resources. This portrayal of the possible eventual use of gas in the U.S. is shown in Fig. 5.3. The portrayal is shown for two cases of assumed total resources, Q_T, where Q_T = cumulative (produced + proved) reserves and potential reserves. Case A is shown for Q_T = 1500 trillion cubic feet, and Case B is shown for Q_T = 2000 trillion cubic feet. For each case, two possible consumption patterns are shown. The first pattern, shown with dotted lines, assumes a rapid, probably unattainable consumption of the resource, dropping off after reaching a peak in the 1980's. A more realistic pattern of consumption is represented by the dashed lines for the two possible values of Q_T.

Case A, a realistic projection of ultimate resources, shows that, when the maximum practical development of the gas resources is pursued (dashed line), the

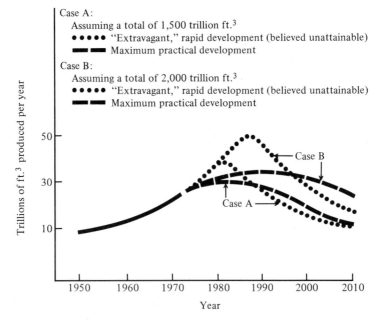

Fig. 5.3 *A portrayal of four possible consumption patterns for the use of natural gas resources of the U.S. The dotted lines show a rapid, but probably unattainable, use of the gas. The dashed lines show a realistic possible use for each of the two ultimate potentials for gas.*

maximum production is reached in 1985 with an annual production of 30 trillion cubic feet. Following this consumption pattern to the year 2000, the consumption decreases to 20 trillion cubic feet.

In terms of the practical development of gas resources in the U.S., it may be estimated that a production rate of 25 trillion cubic feet per year may be the reasonable maximum for sustained production. [4] Therefore, natural gas cannot be expected to satisfy any increase in demand for energy in the U.S. In fact, after the end of the century we will be less able to depend upon natural gas as a fuel, since the share of the total energy provided by natural gas may drop significantly below its present share (33 percent) to perhaps 20 percent of the total.

5.4 WORLD GAS RESOURCES

The U.S. has about 12 percent of the world's natural gas energy resource and currently accounts for about 60 percent of the world's annual consumption. In Table 5.8, we note that a significant resource exists in the Middle East and the U.S.S.R. However, the transport of natural gas is not readily accomplished from one continent to another, even if these nations would be willing to supply large quantities of gas to the U.S. Thus, gas is often flared at Middle Eastern oil-well sites* since it is not readily transportable to potential customers in Europe or the U.S.

Table 5.8
ESTIMATES OF WORLD RESOURCES
OF NATURAL GAS (IN TRILLIONS
OF CUBIC FEET)

Nation or region	Q_T
U.S.	1500
Canada	600
Middle East	3500
Far East	1300
South America	1400
Western Europe	300
Africa	1600
U.S.S.R. and Eastern Europe	2800
World total	13,000

Importing Gas in Liquid Form

Since natural gas is available in other parts of the world, it can be seen to be initially desirable to import some share of this gas to the U.S. to meet the demand for this fuel. Because natural gas converted to a liquified form occupies only about 1/600th the volume of normal gas, moving it by special insulated tank-ship is

---

* See footnote on Page 67.

possible. Liquid natural gas, LNG, is obtained by cooling the natural gas to 147 K where it assumes the liquid state. Currently, more than 14 LNG ships, with a capacity of 480,000 cubic meters (17 million cubic feet), are in service. LNG tanker service has proved itself since it was first used in 1964 between Algeria and the United Kingdom. Subsequently, this service has expanded to join Algeria and France, and Alaska and Japan. The east coast of the U.S. now plans to import LNG from Algeria, Libya, and Venezuela.

The transport of LNG requires special, expensive ships; and although this technology is quite mature, some experts have misgivings about the possibility of accidents, which could release and ignite the LNG, especially in harbors.

The U.S. and Russia have been negotiating for several years to send LNG to the U.S. from the port of Murmansk. The projected shipping volume could require up to 20 LNG tankers for just the U.S.S.R. route. [6] Nevertheless, it still remains problematical whether this proposed sale of LNG will actually occur. [7] Algeria also plans to sell up to 989 billion cubic feet of gas to Europe, Japan, and the U.S., beginning in 1976. [8] Again it remains to be proved whether these large amounts can be sold and shipped in the near future.

LNG is being supplied in small but increasing amounts to the U.S. in tanker ships from Libya, Algeria, and Venezuela. The economics of using LNG in any significant amount deter any rapid increase in its general use. However, the importation of LNG may supply up to 5 percent of U.S. gas needs by 1985. [9]

Gas from Alaska

With the discovery of oil on the northern slope of Alaska, natural gas was discovered also. Up to 325 trillion cubic feet of natural gas has been estimated to exist in Alaska. The transportation of the gas to the lower 48 states can be accomplished by means of a pipeline, or via a tanker to handle the LNG form.

A pipeline under consideration would progress from Prudhoe Bay, Alaska, through Northern Canada's Mackenzie River Valley and Alberta to the U.S. This pipeline project could cost up to $8 billion to construct. The project is undergoing environmental impact studies at the present time. If constructed, the pipeline could deliver up to 4 billion cubic feet a day, and ultimately supply up to 5 percent of the entire U.S. demand, perhaps by 1990.

Synthetic Gas from Coal

A synthetic gas similar to natural gas (CH_4) can be manufactured from coal. The coal gasification process encompasses a complex chemical transition of solid coal into a gaseous fuel. The process uses boiler-produced steam, which is reacted with the carbon in coal to form a hydrogen-enriched gas similar to methane (CH_4). But in the reaction, ammonia (NH_3), carbon dioxide (CO_2), and hydrogen sulfide (H_2S) are also produced. In the following sequential steps, the gaseous products are treated, cleansed, and purified to remove the NH_3, CO_2, and NH_3, and leave a gas consisting of CH_4, hydrogen (H_2), and carbon monoxide (CO). This gas, however, is lower in energy density per volume than is natural gas. In order to raise the energy density, the gas mixture is reacted with hydrogen, H_2, to raise the methane (CH_4) content. This process is accomplished at temperatures over 1000 K and very high pressures (greater than 60 atmospheres).

There are five major processes for coal gasification, but only one has been fully commercialized. The Lurgi process was first developed in the 1930's in Germany and has an overall efficiency of 70 percent. The four U.S. processes are under development, and several are in the pilot-plant stage. [10] The process being developed by the M. W. Kellogg Company feeds coal, catalyst sodium carbonate, preheated steam, and oxygen in a special gasifier vessel operating at 1700°F and 1200 psi. The process gives a raw-synthesis gas of hydrogen, carbon monoxide, methane, and impurities. Further processing takes out the salts and hydrogen sulfide, and then converts the product gas to 90 percent methane.

Coal gasification requires the use of large quantities of water and coal in the process. Commercial coal gasification will, therefore, take place near sources of easily exploited coal and sources of water. Some have advocated the placement of large coal-gasification processors in Wyoming and the Dakotas, where large amounts of coal could be strip-mined and used in the conversion process. This project will be discussed further in the next chapter, which is concerned with coal as a fuel source.

It is estimated that the cost of a coal gasification plant that will produce 250×10^6 cubic feet per day of synthetic gas might cost a total of $800 million to construct. The cost of the gas would be about $3.00 per 10^6 Btu, which is about 50 percent more than the unregulated cost of delivered natural gas in the U.S. in 1977.

Nuclear Stimulation of Natural Gas

It has been proposed that nuclear explosions be used to stimulate the production of natural gas in remote areas of the Western U.S. An estimated 300 trillion cubic feet may be available through such an approach.

Project Rio Blanco was a pilot demonstration of such an approach in Western Colorado, involving the use of three simultaneous 30-kiloton nuclear explosions.* The explosions, which occurred on May 17, 1973, were used to fracture tight gas-bearing sandstone rock formations. The explosions were detonated at depths of 5000 to 7000 feet. The Rio Blanco explosions cost about $8 million and were used along with earlier underground nuclear blasts to demonstrate fracturing of rock, which will permit natural gas trapped in rock to flow. [12, 13]

In order to stimulate a natural gas field, more than 100 nuclear detonations would be required. The prospect of such a series of nuclear explosions is viewed with consternation by the residents of Colorado, and it is doubtful that such a

---

* 30 kilotons of TNT equivalent energy.

project will be permitted to continue. Environmentalists are concerned with the possible radioactive contamination of the water aquifers which eventually flow into the state's drinking water. In any case, it will take a lot of convincing evidence that the use of high pressure hydraulic fracturing would not achieve the same desired end.

Economics of Natural Gas

The uses and prices of natural gas transported from one producing state to another consuming state in the U.S. are regulated by the Federal Power Commission (FPC). A pipeline company may not transport gas for direct sale to an industrial plant, or construct and operate facilities to make a direct sale, without the permission of the Commission. If an interstate pipeline sells to a distributor or to another pipeline for resale to an industrial customer, the terms of the sale (rate and conditions of service), as well as the facilities and transportation, are subject to FPC jurisdiction. Thus, the cases that have come before the Commission have been in the context of an applicant seeking authority to transport gas, to build and operate facilities, to make a sale for resale, or to obtain new or additional service. The statutory criterion that is applied in such cases is whether the public convenience and necessity require the issuance of a certificate. Since 1944, the Commission has examined the proposed end use to which the applicant's gas would be put, and has considered an inferior use to be a minus factor in deciding whether to permit a new sale.

The FPC, at the behest of Congress, through the National Gas Act of 1938 instituted natural-gas price controls when gas was in surplus and energy prices low. Over the past two decades, the price controls have succeeded in keeping the price of natural gas well below those of other fuels on an equivalent energy basis. In August 1976, the FPC raised the price limits on natural gas. The cost of natural gas, drilled after Jan 1, 1973, and prior to Jan 1, 1975, to be shipped interstate is now limited to $1.01 per million Btu's at the wellhead, while the equivalent cost of newly drilled oil has risen to $2.00 per million Btu's and the price for coal was more than $0.85 per million Btu's in 1976. The price of gas that was in production before Jan 1, 1973 remains limited to $0.52. The price of gas produced in wells drilled after Jan 1, 1975 is limited to $1.42. Since 87 percent of the gas delivered in 1976 was drilled prior to 1973, the impact of this price rise will be gradual.

The relatively low price of natural gas has encouraged its use to the point where demand exceeded supply by up to 10 percent during the past several years. The low price has led to its use by electric utilities and has discouraged drillers from exploring for gas.

The cost of natural gas purchases on the intrastate market for local use has risen to over $1.70 per 10^6 Btu's in 1977. These higher prices, obtainable in the Texas market, for example, have stimulated new gas-well exploration in that state for local use.

The interstate pipeline industry, the executive branch of the U.S. federal government, and the FPC have all called for deregulation of natural gas prices. [14, 15] All three groups call for an economic market, with unregulated prices, which would stimulate new wells and exploration.

On the other hand, consumer groups and many congressional representatives are opposing deregulation on the basis that it will simply raise prices to the consumers and profits to the suppliers without increasing the supply. Many senators advocate the added regulation of intrastate gas prices in order to close the gap between intrastate and interstate prices. [14, 16]

It is difficult to ascertain the level to which prices to the customer would rise if deregulation should occur. Some believe that gas prices on a customer's bill would rise 6 percent annually for several years. The issue comes down to whether the nation is willing to sustain the inflationary economic shock of paying the price for gas, after two decades of artificially restrained prices, and whether deregulation would encourage the wise exploration and utilization of domestic sources. Also, the nation must consider the effect of higher prices on making new energy sources more economically attractive to the developers.

The price of synthetic gas can range from an estimate of $1.50 to $2.00 per 1000 cubic feet, while imported LNG sells for between $1.25 and $1.50 per 1000 cubic feet. One possible course for the FPC and Congress would be to raise the maximum price level for interstate gas in steps over the next several years—in effect deregulating the price in stages. A first step might be the establishment of a maximum price of $0.75 per 1000 cubic feet for production from wells drilled prior to Jan. 1, 1973.

5.5 CONSERVATION OF NATURAL GAS

While new sources of natural gas are being explored and exploited, it is equally important to conserve the use of natural gas. [17] There are several steps that can be readily taken in order to conserve gas. One conservation move would be to discourage the use of natural gas to generate electricity for transmission to homes and commercial establishments for space heating. Consider the efficiency of space heating via a gas furnace and electric heaters, as shown in Fig. 5.4. The overall efficiency of electric heating is

$$\eta_{\text{electric}} = \eta_e \eta_d \eta_h,$$

where $\eta_d = 0.90$, $\eta_e = 0.35$, and $\eta_h = 0.98$. Therefore,

$$\eta_{\text{electric}} = 0.31.$$

The efficiency of gas heating is

$$\eta_{\text{gas}} = \eta_p \eta_f,$$

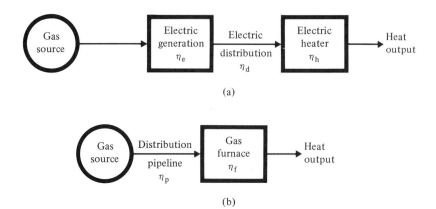

Fig. 5.4 *The use of gas for heating is compared in (a) for the generation of electricity and its use as electric heating, and in (b) as direct use in a gas furnace.*

where $\eta_p = 0.95$ and $\eta_f = 0.75$. Therefore,

$$\eta_{gas} = 0.71,$$

which is more than twice the efficiency of electric heating. Therefore, whenever possible, gas should be used for the space heating of buildings and also for cooking in homes and restaurants.

The continuous use of energy to keep a supply of hot water available in residences accounts for 15 percent of residential energy consumption and 3 percent of energy for all uses in the U.S. Thus, it would be a wise policy to encourage the use of more efficient water heaters with reduced thermal losses.

Water heaters and space heaters that operate on natural gas normally use a continuously burning pilot light to ignite the gas flame when the thermostat commands the opening of the gas input valve. It may be argued that the heat of the pilot light on the water heater is utilized effectively to maintain the temperature of the water stored in the tank. However, it is worth considering the elimination of the gas pilot light on other appliances to achieve some significant savings of natural gas. It has been estimated that a pilot light of a space heater consumes approximately 625,000 Btu's per month (584 cubic feet). Similarly, it has been determined that the pilot light of a gas range consumes approximately 150 cubic feet per month, or about 18 percent of the total gas consumption for the range. The amount of energy consumed by the gas pilot light for a space heater is also calculated to be an average of 16 percent of the total energy consumed. [17]

It is possible to use an intermittent ignition device for lighting a range or space heater, instead of a continuously burning light. An intermittent ignition device is an ignition device that is actuated only when the gas appliance is in operation; such devices now are used on many gas appliances. Over half of all gas dryers that are sold today are equipped with intermittent ignition devices. Most intermittent ignition devices on dryers use a silicon-carbide igniter—there are currently over 2 million in operation. One ignition device design uses a silicon-carbide resistance element, which heats to 1600° Fahrenheit when excited; and at that temperature the gas valve is opened, resulting in ignition. This device is safe, reliable, and has a long life.

The mandatory use of intermittent ignition devices on gas space heaters and gas ranges should be a national requirement. A savings of 317 billion cubic feet may be achieved in the U.S. with the eventual total use of intermittent ignition devices. This is a savings of 4 percent of the total consumption by residential and commercial customers, as shown in Table 5.9.

Table 5.9

ESTIMATED RESULTING SAVINGS OF NATURAL GAS BY
RESIDENTIAL AND COMMERCIAL CUSTOMERS IN THE U.S.

	By means of the eventual total use of intermittent ignition devices	An organized mandatory program for extinguishing pilot lights in the summer
Estimated U.S. savings (in billions of cubic feet)	317	27
Percent of total U.S. consumption by residential and commercial customers	4%	0.4%

An immediate and continuing saving of gas can be accomplished by an effective program that assists in the extinguishing of space-heater pilot lights during the summer months. This approach would be particularly effective in the southern and southwestern U.S. In these areas, not only is energy lost by use of a continuous pilot light, but many residences and commercial buildings use air conditioning and thus must effectively use energy to counteract the heat that is added by the pilot lights. In a limited survey in Davis, California, it was determined that one-half of the residences extinguish their space-heater pilot light during the summer months. It can be inferred that this figure is greater than or equal to the average for the state of California. During the four summer months (June through September), a savings of 2340 cubic feet per residence could be achieved. If the half of the California gas customers who did not extinguish their pilot lights were induced to do so, the savings over a four-month summer would be approximately 9 billion cubic feet of natural gas. The local gas companies could provide a regularly established and well-advertised visit to a neighborhood by a gas company employee to turn off and relight the pilot light for customers at the beginning and end of the summer period. Another incentive to the customer would be the cash savings, which could be adequately pointed out to him. A total annual conserving of natural gas can be estimated to be 27 billion cubic feet, as shown in Table 5.9 [17]

Therefore, a policy for the conservation of natural gas might include the following three approaches. The first would be the required use of natural gas for space heating of buildings, in preference to electric heating, except in remote areas where gas is not distributed. The second is the required introduction of efficient, safe, and low-cost intermittent-ignition devices for gas appliances. The third approach is the introduction of a system of incentives and customer assistance so that gas customers would extinguish their pilot lights during the summer months.

REFERENCES

1. *Energy Facts*, Science Policy Research Division, Library of Congress, U.S. Government Printing Office, Wash., D.C., 1973.

2. G. D. FRIEDLANDER, "Energy: Crisis and challenge," *IEEE Spectrum*, May 1973, pp. 18–27.

3. "Industry braces for a natural gas crisis," *Business Week*, Oct. 19, 1974, pp. 114–117.

4. J. W. DUANE and M. A. KARNITZ, "Domestic gas resources and future production rates," *Power Engineering*, Jan. 1975, pp. 36–39.

5. J. D. MOODY and R. E. GEIGER, "Petroleum resources," *Technology Review*, March/April 1975, pp. 42–45.

6. "Russian gas venture cost put at $4.5 billion," *Wall Street Journal*, May 23, 1973, p. 8.

7. "Breakdown in gas supply looms," *Power Engineering*, March 1975, p. 16.

8. "The Algerians stun their gas customers," *Business Week*, Feb. 23, 1974, p. 26.

9. W. L. LOM, *Liquefied Natural Gas*, John Wiley and Sons, Inc., New York, 1974.

10. H. PERRY, "The gasification of coal," *Scientific American*, March 1974, pp. 19–25.

11. P. GWYNNE, "Nuclear relief for natural gas?" *Technology Review*, August 1973, pp. 6–8.

12. L. J. CARTER, "Rio Blanco: Stimulating gas and conflict in Colorado," *Science*, May 25, 1973, pp. 844–847.

13. "A furious push to deregulate gas," *Business Week*, May 19, 1975, pp. 91–92.

14. "FPC publishes recommendation in national gas survey," *FPC News*, March 7, 1975, pp. 1–2.

15. "The fight to free natural gas rates," *Business Week*, May 18, 1974, p. 26.

16. R. C. DORF, "A three-step policy for gas conservation," *Public Utilities Fortnightly*, May 8, 1975, pp. 41–43.

17. R. WILSON, "Natural gas is a beautiful thing?" *Bulletin of Atomic Scientists*, Sept. 1973, pp. 35–40.

18. M. H. SONSTEGAARD, "Transporting gas by airship," *Mechanical Engineering*, June 1973, pp. 19–25.

19. "A flood of foreign LNG for U.S. factories," *Business Week*, Oct. 13, 1975, pp. 101–104.

20. R. L. GORDON, *U.S. Coal and the Electric Power Industry*, John Hopkins University Press, Baltimore, Md., 1975.

21. W. M. BROWN, "A huge new reserve of natural gas comes within reach," *Fortune*, Oct. 1976, pp. 219–222.

22. Edison Electric Institute, *Economic Growth in the Future*, McGraw-Hill Book Co., New York, 1976.

23. "The natural-gas shortage," *Business Week*, Sept. 27, 1976, pp. 66–72.

EXERCISES

5.1 Using semilog graph paper, prepare a chart showing the annual consumption of natural gas in the U.S. and the consumption per capita for the period 1920 to 1974. Estimate the rate of growth of consumption during the period 1940 to 1960.

5.2 One environmental effect of the use of natural gas is the production of nitrogen oxides as a result of combustion. [18] The environmental problems of nitrogen oxides are twofold. First, the nitric oxide forms nitrogen dioxide by a photo-chemical reaction. This process, first identified by Hagen–Schmidt, is responsible for the aesthetically unpleasant city smog. Second, nitrogen dioxide is an irritant to the human respiratory system. In several cities on the East Coast of the U.S., the burning of natural gas is the main contributor of nitric oxide (NO). Determine the levels of nitric oxide in your city or a nearby city from the air-pollution control district, and ascertain with their help what contribution comes from the burning of natural gas.

5.3 The storage of LNG in large containers has some inherent risk of explosion or fire. [18] Examine several accidents occurring over the past 5 years and determine whether adequate safety standards are established. Determine what the safety standards are for your state.

5.4 The airship has been proposed as a means of transporting commercial gas, such as natural gas from the Northern Slope of Alaska. According to one proposal, a permanent lifting-gas chamber at the top of the envelope would contain sufficient hydrogen to float the airship in the unloaded condition, the remaining space being filled with gaseous cargo during loaded flight and with air during the empty return trip. Natural gas, the most likely near-term candidate for large-scale movement by cargo-gas airship, has positive buoyancy that would have to be counterbalanced during loaded flight; this could be done by carrying liquid ballast (e.g., petroleum) in the loaded direction and returning with the ballast tanks empty. [18]

Examine some possible configurations and uses of such airships. Also, examine the economics of such an approach and compare it with the cost of a pipeline from Alaska to the U.S.

*5.5 The charges for the use of natural gas for domestic and commercial cooling and heating vary from state to state, but most utilities charge on a sliding scale basis per gas use. In Davis the following charges prevail:

	Per meter per month
First 2 therms or less	$1.57932
Next 23 therms, per therm	9.206¢
Next 175 therms, per therm	8.536¢
Next 800 therms, per therm	7.866¢
Next 49,000 therms, per therm	7.676¢
Over 50,000 therms, per therm	7.406¢

(One therm, a measure of energy, is equal to 100,000 Btu's.) Determine the rate system from your local gas utility and compare the rates. Consider the use of a decreasing price as usage increases, and indicate whether you believe this approach to be fair. Is the decreasing price scale an incentive to use more gas, perhaps wastefully?

5.6 Drilling for gas to be sold within a state has gone up 88% in the period 1971–74 in response to unregulated prices, while drilling for offshore gas declined by 20% over the same period because offshore wells are required to supply the interstate market. One option for the FPC is to regulate the price of interstate and intrastate gas at the same price. Is this approach preferable to deregulation of natural gas prices?

*5.7 The Federal Power Commission expected the nation's major interstate pipeline companies to fall short of their contracted deliveries by an average of 15% during the winter months of 1976–77. As supplies fall short, the gas supply is interrupted to consumers (based on a priority allocation system), which starts with the interruption of large industrial users and protects small commercial and residential users. The shortfall expected in California was 9% for 1976–77. As curtailments impact industrial users, they are required to shift to alternate fuels such as more expensive oil, or else shut down. The effect on the nation may be increased unemployment, and increased prices to customers of the industry. Considering the effects of unemployment and industrial shutdown, what policies for gas allocations would you propose in order to avoid disruptions to local industry?

5.8 A terminal for LNG at Cove Point, Maryland, is under construction and scheduled for operation in 1977. Gas produced in Algeria and transported as liquid in cryogenic tankers will be delivered to Cove Point and a sister facility in Savannah, Ga., at a rate of 1 billion cubic feet per day. [20] This and other projects could boost imports of LNG to 4 billion cubic feet per day. Determine what percentage of natural gas used in the U.S. would then be imported in the late

---

* An asterisk indicates a relatively difficult or advanced exercise.

1970's. Should the nation allow itself to become dependent on foreign gas, as it already is on foreign oil?

*5.9 A barrel of crude oil costs about $13 to $14 and is equivalent in energy value to approximately 6000 cubic feet of natural gas. The average price of gas sold in the intrastate market is about $1.50 per thousand ft^3. The price incentive for using gas to fuel electric generating plants is clear. Propose a scheme for limiting the use of natural gas in electric power plants using economic or regulatory approaches.

*5.10 A debate is currently being pursued regarding the means of delivery of natural gas from the Alaskan North Slope to the lower 48 states. A gas-well outlet (often called a Christmas Tree) is shown in Fig. E5.10. One proposal calls

Fig. E5.10 *A gas well outlet at Prudhoe Bay. (Courtesy of American Natural Gas System.)*

for a pipeline from Prudhoe Bay through Canada's Mackenzie Delta, that would then connect to U.S. and Canadian distribution pipeline networks. The other scheme proposes to build a pipeline from Prudhoe Bay to Point Gravina, Alaska, where a gas liquefication plant would be constructed. Eleven tankers would then carry the LNG to California, where it would be delivered to existing networks. Examine the economics and policy questions associated with each approach. Which would you recommend?

5.11 In September, 1974, the California legislature passed a bill, SB 1521, subsequently signed into law, which would result in the eventual prohibition of pilot lights on gas appliances. The bill was authored by Senator Alfred Alquist, chairman of the Senate Committee on Public Utilities and Corporations. The bill would prohibit the selling or installation in California of new residential-type gas appliances that are equipped with a pilot light. This prohibition will become effective twenty-four months after an intermittent ignition has been demonstrated and certified by the state, or by January 1, 1977, whichever is later. The bill states that the intermittent-ignition device "shall not significantly affect the price of gas appliances in competition with similar electrical appliances." However, it should be recognized that the average cost per residential customer of San Diego Gas & Electric Company to burn gas pilot lights was $32 per year in 1973. Therefore, a low-cost ignition device can pay for itself in a year or less. Discuss the possible introduction of such a law in your state and the impact of such a law.

***5.12** Several policies have been advanced with the objective of increasing the supply of natural gas in the U.S. Consider the list given below and rank them in order of desirability. List positive and negative results from each proposed action.

a) Increased leasing of government lands for the development of natural gas, particularly along the Atlantic Coast and the Gulf of Alaska.
b) Tax incentives for the exploration and exploitation of natural gas.
c) Expedition of all necessary government actions (United States and Canada) required for construction of a pipeline to bring natural gas from Alaska to the lower 48 states.
d) Intensification of efforts to establish and maintain a substantial coordinated program to promote conservation practices and to develop new techniques for energy utilization.

***5.13** Most of the energy lost in moving a gas or liquid through a pipeline is due to friction. If the flow is laminar, then the volume of flow per second, Q, is proportional to R^4 where R = radius of the pipe. We may write

$$Q = \pi R^4 \frac{(P_1 - P_2)}{8\eta L},$$

where L = length of pipe, η = viscosity, and $P_1 - P_2$ = pressure difference. Determine and compare the viscosity of gas and oil. (a) Assuming a level pipe, determine the power needed to pump a fluid or gas at a given rate through a pipeline. (b) If one wishes to increase the transportation of fluid, is it better to increase the flow rate Q or the radius of the pipe, with regard to the energy expended in moving the fluid?

5.14 About one-half of the gas used in a gas cooking stove is consumed by the pilot light. Investigate in your area the availability of gas stoves using electronic or other intermittent means of igniting the flame in such stoves. Determine whether there is an increase in price for such stoves. Are they well received by customers in your city?

6

6.1 INTRODUCTION

After decades of declining production and increasing disfavor, coal, the most abundant fossil-energy resource in the United States, has been making a strong comeback. It is one of the ironies and dilemmas of our environmentally aware age that we may use more, not less, of this relatively heavily polluting fuel.

Coal, the fossil fuel extensively used by man, was initially used by the Chinese at the time of Marco Polo or earlier. The use of coal as a major source of energy began in England in the twelfth century when pieces of black rock called "sea coles" were discovered to be combustible. Eventually, it was deduced that these rocks could be dug from strata of rock along the cliffs in England and then from holes sunk to the strata. In 1234 King Henry III granted Newcastle-upon-Tyne the right to mine coal. By the late 13th century, coal smoke was already a source of pollution in London. Coal was used as a domestic fuel, as a fuel for lime burning, and by blacksmiths and for other metallurgical processes. By 1658 the production of coal at Newcastle had reached 529,000 tons per year. By 1750 the annual production in England reached 7 million tons.

Coal in many ways was the fuel of the Industrial Revolution. Coal was used during the Industrial Revolution for metallurgical processes, glassmaking, fuel for railroads and, in general, for the steam engine. By 1860, world production of coal reached 150 million metric tons. From the period 1860 to 1910, world annual production of coal grew from 150 to 1100 million metric tons, at an annual growth rate of 4.4 percent.

During the period 1910 to 1940, world coal production grew at the relatively low rate of 0.75 percent per year. However, after 1940, the growth rate of world coal production has again risen to 3.6 percent. The world coal production for several periods is given in Table 6.1.

Table 6.1
WORLD COAL PRODUCTION

Year	Production (Millions of metric tons)
1860	150
1880	320
1900	780
1920	1200
1940	1700
1960	2100
1970	2500

Coal is actually a family name for a variety of solid organic fuels. The origins of coal were plants, which were accumulated in a bog and became a soggy mass of plant debris we call peat. When peat was compressed and burned over 300 million years ago, it became lignite. Successive invasions of the sea and piling on of layer upon layer of material resulted in the deep burial of the lignite. Deep burial often results in a rise in temperature and an expelling of the moisture, and thus lignite became bituminous coal. In some areas the layers of coal were subjected to large compressive forces, thus resulting in "hard coal" or anthracite. The main constituent of coal is carbon and hydrogen, with small added quantities of sulfur, oxygen, and nitrogen.

6.2 COAL PRODUCTION IN THE U.S.

The production of coal in the U.S. started about 1820, when 14 tons are reported to have been mined. From 1820 to 1900, the mining of coal increased rapidly to 280 million tons annual production, at an annual growth rate of 6.6 percent per year. By 1910, the annual production rate was 500 million tons per year, as shown in Table 6.2. During World War II, the annual production of coal rose to over 600 million tons. However, after 1947, the development of gas pipelines resulted in a shift of fuel use from coal to gas and petroleum. Therefore, by 1960, the annual production of coal had dropped to 440 million tons. After 1973, with the institution of higher oil prices and the growing shortages of natural gas, the production of coal again increased and reached an all time high of 635 million tons in 1975.

Because of the development of other energy sources in the U.S. over the past three decades, coal has become a continually smaller part of the set of total energy sources. In 1950 coal provided 38 percent of the total energy in the U.S. and that figure declined to 17 percent by 1972. In the future, coal may again supply over 25 percent of the nation's energy.

The U.S. is the largest user of coal in the world, as can be seen in Table 6.3. Russia, Great Britain, Poland and West Germany also use considerable amounts

Table 6.2

U.S. ANNUAL COAL PRODUCTION

Year	Production (Millions of tons*)
1850	10
1890	160
1910	500
1930	500
1935	350
1940	510
1945	600
1947	630
1950	516
1960	440
1965	513
1970	590
1975	635
1977	670
1980 (Projected)	740
1985 (Projected)	1000

* We continue to use ton = 2000 pounds and tonne = metric ton = 1000 kg. 1 ton = 0.907 metric ton.

Table 6.3

BITUMINOUS COAL PRODUCTION IN 1971

Country	Production (In millions of metric tons)
U.S.A.	495
U.S.S.R.	480
Great Britain	147
Poland	145
West Germany	111
India	70
Canada	15
Total world production	2500

of coal annually. The U.S. and the U.S.S.R. each consume about 20 percent of the world's total coal consumption annually.

The uses of coal for various purposes in the U.S. are shown in Table 6.4. Over one-half of the coal consumed in the U.S. is used by electric utilities to fuel the boilers to generate electricity. Coal is also important for making industrial steam and in the manufacture of steel. While coal was used for railroad fuel during the period 1850 to 1950, it has been essentially replaced by diesel and electric locomotives.

Table 6.4

USES OF COAL FOR VARIOUS PURPOSES
IN THE U.S. IN 1970

	Percentage of total use
Electric utility	52.0
Industrial steam	15.9
Coking coal for steel	18.1
Railroad fuel	0.2
Residential and commercial heating	1.8
Export	12.0
Total	100.0

6.3 COAL RESOURCES

The reserves of coal available are variously estimated as from 8 to 16 trillion tons. Table 6.5 gives an estimate of the world's reserves of coal with a total proven reserve of 7.6 trillion tons (189×10^{18} Btu) and an estimated total of 16 trillion tons (399×10^{18} Btu). Table 6.6 provides useful conversion factors for coal.

However, within current mining techniques and under current economic conditions, only a fraction of the coal reserves of the world are recoverable. Iden-

Table 6.5

WORLD COAL RESERVES (IN UNITS OF 10^{18} Btu)

	Proved	*Estimated total*
U.S.A.	40.0	83.4
Canada	16.0	37.0
South America	0.4	0.8
Western Europe	10.0	21.6
Eastern Europe and U.S.S.R.	120.0	250.0
Africa	3.0	6.2
Total	189.4	399.0

Note. World consumption of total energy is about 175×10^{15} Btu per year.

Table 6.6

CONVERSION FACTORS FOR COAL

One ton of coal (2000 lbs.) yields 25 million Btu.

One ton of coal = 25,000 cubic feet of natural gas
= 189 gallons of gasoline.

One ton (2000 lbs.) = 907 kilograms
= 0.907 metric ton.

tifiable and estimated resources of coal in the U.S. amount to a total of 3.2 trillion tons. However, based on current technology, economics and environmental regulations, only some 150 to 200 billion tons are recoverable. U.S. reserves, as shown in Table 6.7, lie at various depths below the earth's surface. Current mining methods limit mining to a depth of 1000 feet and, therefore, limit the total recoverable amount of coal.

Table 6.7
U.S. ESTIMATED COAL RESOURCES

Depth of overburden (Feet)	Type	Resources (Billions of tons)	Energy reserve (× 10¹⁸ Btu)
100	Strip coal	140	3.6
100 to 3000	Bituminous	959	
	Lignite	448	37.0
	Anthracite	13	
3000 to 6000	All types	337	8.8
6000 to 9000	All types	1313	34.1
Total		3210	83.5

Note. Current mining methods are not economical below depths of 1000 feet.

Life of Coal Reserves

All fossil fuels required millions of years to be formed. At the rate that we consume them, they will not replace themselves. Thus fossil fuels are a finite (or limited) source. We may define Q_R as the proven reserve of a fossil fuel and Q_P as the total accumulated quantity of the fuel removed from the ground up to any given time. The cumulative discoveries, Q_D, represents the quantities removed from the ground plus the recoverable quantities remaining in the ground. Therefore,

$$Q_D = Q_P + Q_R. \tag{6.1}$$

If Q_D is eventually finite or limited, we have the effect portrayed in Fig. 6.1. As time progresses the cumulative discoveries, Q_D, reach the finite limit called Q_F. As production Q_P increases, eventually the proven reserves, Q_R, reach a maximum and begin to reduce.

The relation between these quantities as time changes may be obtained by taking the derivative of Eq. (6.1) as follows:

$$\frac{dQ_D}{dt} = \frac{dQ_P}{dt} + \frac{dQ_R}{dt}, \tag{6.2}$$

where dQ_D/dt, for example, is the rate of discovery. When reserves reach their maximum value, their derivative is

$$\frac{dQ_R}{dt} = 0. \tag{6.3}$$

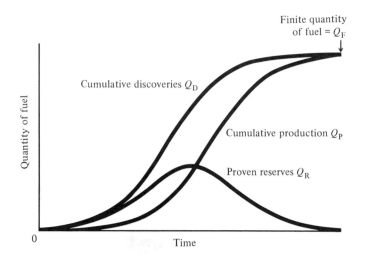

Fig. 6.1   *The consumption of a limited fuel over the lifetime of the fuel.*

Thus when reserves reach their maximum value, the curves of discovery rate and production rate will cross, with production going up and discovery going down.

The lifetime curve for coal production for the world is shown in Fig. 6.2 for two estimates of the total coal resource. In the case of the lower estimate, the peak production occurs in 2110, while for the larger estimate the peak production occurs about 2150.

A similar curve can be drawn for production of coal in the U.S. If we assume a total recoverable resource in the U.S. of 200 billion tons of coal, then a peak production rate could be attained by 2100 with essential exhaustion of all fuel by the year 2500. However, coal would remain as a significant source, supplying the current

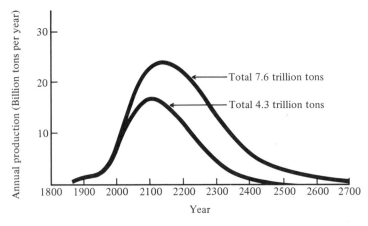

Fig. 6.2   *The rate of production of coal in the world for two estimates of total resources, Q_F.*

production of 600 million tons until the year 2300. Thus, coal could supply us with an ample fuel for several hundred years to come. In the case of coal, it is not a problem of availability so much as it is a question of environmental protection, safety in mining, and economic issues.

6.4 DISTRIBUTION OF COAL IN THE U.S.

The U.S. has, by various estimates, up to one-half of the world's coal reserves. These reserves are concentrated in seventeen states. In the past, coal has been mined in the Appalachian region of the eastern United States, particularly in the states of Kentucky, West Virginia, Pennsylvania, Ohio, and Illinois. There are large deposits to be exploited in the western states of Alaska, New Mexico, Arizona, Montana, Wyoming, Colorado, and North Dakota.

Coals occurring east of the Mississippi generally have higher heating value and higher sulfur content than those of the West. Partly for this reason, attention in recent years has been directed toward the relatively low-sulfur (less than 1 percent) coals of the West in order to satisfy requirements for reduced air pollution. Approximately 97 billion tons of recoverable low-sulfur coal lie in the western states. About one-third of this coal could be mined by stripping techniques. These Western coals are far from the industrial and residential markets of the East, and attendant transportation costs must be considered. Low-sulfur coal deposits in the East are estimated at 8.5 billion tons, with more than half of these deposits in West Virginia.

From the standpoint of coal resource availability, the East is in a dilemma: If local high-sulfur coal is used, potentially hazardous air pollution will result, unless the coal is desulfurized—an elusive technology. If low-sulfur coal is used, essential industries that depend on metallurgical coal could be starved of vital supplies if the coal were to be diverted to steam-coal applications. Pressures for relaxation of air-pollution standards are intensifying. Yet, without some stipulation that such relaxed requirements would not be continued indefinitely, there is a strong possibility that no relaxation would be granted at all. It remains to be determined whether Americans will choose clean air in exchange for energy; actually, each choice could have profound economic and social consequences.

The importance of the distinction between high- and low-sulfur coal arises because sulfur-oxide pollution-control regulations prohibit the emission of more than 1.2 pounds of SO_2 per million Btu's of heat generated by the burning of coal in new plants. To meet this standard, a coal containing 24 million Btu/ton cannot contain more than 0.7 percent sulfur by weight. Coals with a lower heat value must contain correspondingly less sulfur if they are to meet the standard.

6.5 THE MINING AND TRANSPORTATION OF COAL

Many factors influence coal-mine productivity and recovery by deep- and surface-mining methods. Coal is recovered through deep mining by three techniques: (1) conventional mining, (2) continuous mining, and (3) longwall mining. The conventional method uses a set of specialized equipment which performs specific tasks in an established sequence. Although capable of production of significant tonnages of coal, the sequential nature of conventional mining results in a lower productivity than that achievable by other methods.

The continuous mining method differs from the conventional method in that a single machine is used to mechanically remove coal from the working face. Separate steps of drilling, undercutting, blasting, and loading are not required. However, the continuous mining method usually results in greater impurities being introduced into the coal, since the mining machine is not selective. Continuous mining by underground methods is capable of high productivity, although the operation creates significant amounts of dust and fine coal particles, which represent a health hazard to unprotected workers. Efforts to control such dust have resulted in reduced mining rates and lower productivity in the deep mines where this method is used.

The longwall mining method taps a long panel of coal, often from 200 to 500 feet across and up to 1,500 feet long. Coal is removed from this panel by a shearer (or plow) which is moved across the coal face. Under optimum conditions, the longwall method is capable of the highest productivity known in deep mining, but it is limited to coal reserves where physical conditions are favorable. Also, longwall mining generates large volumes of dust and, thus, represents a health hazard.

In order of priority, new technologies are needed to (1) increase the percentage yield for a given mine or volume of coal seam worked; (2) enable the economic mining of their coal seams; and (3) mine economically at greater depths. [1] Progress in each of these areas would result in a higher utilization of the total coal resource.

Present day underground mining suffers from low productivity and low resource utilization. The modern longwall mining techniques hold great promise in meeting these challenges. In longwall mining, a flexible armored face conveyor capable of being moved forward without disassembly is used with efficient plows and shearers riding on rails mounted on the conveyor. Also, self-advancing hydraulic roof supports are utilized. Coal is mined from panels up to 5000 feet long and 300 to 800 feet wide. The shearer travels back and forth along the face, biting several feet into the coal, which then falls into the conveyor. The longwall system of mining offers important advantages that make it highly attractive. These include high productivity, high resource recovery, and improved mine safety conditions. The face equipment can be operated by a small crew of from eight to ten men, and production of up to several thousand tons of coal per day can be achieved.

New technologies are needed to increase the productivity of mines. One approach under experimentation is the use of controlled mining machines equipped with instruments to record the value of various tools and methods. One such machine is equipped with instruments that permit control and monitoring of the machine's speed, power, and other factors that can influence the formation of respirable dust. By using various combinations of operating characteristics, and experimenting with different bits on the cutting wheel, the Bureau of Mines hopes to show how standard-size continuous miners can be operated in ways that will generate less fine-size, breathable coal particles, without reducing coal production. [2]

Methods of continuous transportation of coal away from the coal face are needed to keep up with continuous mining machines and to facilitate overall automation of the mining operation. Many approaches are being investigated, including hydraulic, pneumatic, and flexible conveyors.

An intriguing possibility for the future is oxygen-free mining. Such a system would have many advantages. Without oxygen, pyritic material in coal does not oxidize to form acid and sulfate; there could be no fires or explosions and hence there would be no need for explosion-proof equipment; miners would not come in contact with dust, so black-lung disease would be eliminated and there would be no need for rock-dusting; and mine ventilation could be reduced. Under such a system, the miners would wear space-type life-support suits. This technology would be particularly attractive for deep mines where potentially dangerous gas might be present.

In spite of many recent developments in deep mining technology, a number of problems remain. Largely because of economic factors, many mines still do not use the technology that would contribute to highest productivity or resource recovery.

Coal Slurry Pipelines

The most common method of transporting coal is by railroad. An efficient method of transporting coal from mine to power plant is the use of a pipeline. A patent was granted in 1891 for transporting coal mixed with water, as a slurry, through a pipe. An 8-inch-diameter pipe of this type became operational in London in 1914. There are several versions of pipelines that carry pulverized coal mixed in water over distances of 10 to 300 miles. The coal and water are mixed in equal amounts, by weight. At the destination, centrifuges whip the coal particles out of the slurry, creating a pulverized fuel that is ideal for big boilers.

The Black Mesa Coal-Slurry pipeline, owned by the Southern Pacific Railroad, is a 273-mile system capable of transporting over 5 million tons of coal annually through the 18-inch pipeline. The coal transported by this pipeline would require about 150 rail cars each day to accomplish the task.

Construction of pipelines of 1000 miles in length would enable coal mined in North Dakota or Montana to be transported to the power plants of the Midwest. One limiting factor in the use of such ambitious pipelines is the necessity for large amounts of water for the slurry. In addition, the pipeline companies would need the rights to lay their pipes, usually underground, through the farmlands of the Midwest and across existing railroad rights-of-way.

Several companies are pursuing the construction of long-distance pipelines. [3] Slurry pipelines are estimated to consume 20 percent less energy per ton of coal moved than for a railroad system. Reliability would no doubt improve as well. The Black Mesa pipeline functions more than 99 percent of the time.

Nevertheless, railroads oppose the establishment of new, competing, slurry pipelines, on the basis that they will undercut the railroad's base business. Coal is currently about 25 percent of railroad freight business in the U.S. Railroads move about 30 million tons annually in some 390,000 cars.

The question of availability of water is also of paramount importance to the future of coal slurry lines.

The economics of pipelines depend upon the amount of coal transported per year (throughput) and the distance transported. For a large pipeline moving 6 million tons per year over a distance of 1000 miles, the cost can be estimated to be one-half cent per ton-mile for a total cost significantly less than the equivalent cost of railroad transport of coal. The total yearly transportation cost for such a pipeline would be $15 million, including the cost of pulverizing the coal. Also, the energy consumed in the transportation of the coal is considerably less for a pipeline, compared to railroad transport.

For the transportation of relatively small amounts of coal from small mines, railroads and barges on lakes and rivers remain the economical and efficient means of transporting coal. To move coal over a distance of 100 miles in small quantities (less than 300 tons per day), it might cost $4 per ton by railcar and $1 per ton by barge on a river.

Health and Safety Considerations

Mining coal in deep mines avoids much of the landscape disruption that accompanies surface mining. For this reason, many have advocated decreasing the number of surface mines and shifting the emphasis to underground mines. Unfortunately, deep mining has a serious set of associated hazards. Under the best of circumstances, the underground coal mine represents an extremely hostile and hazardous environment for the miner; this fact is reflected in the much higher injury and death rates experienced in underground mining. It is difficult, but important, to protect the workers from the health and safety hazards that are inherently present in a very confined environment. This requires better prediction of mine conditions and control of the mining environment. That means improved provision of ventilation, control of methane gas, prevention of fires and explosions, dust suppression and its removal, and control of other mine-worker hazards.

Over the seven-year period 1965 to 1972, 1412 lives were lost in the underground mines. [1] The fatality rate for underground mining, per ton of coal mined, was five times as great as that recorded for the surface-mining (strip mines) industry. In recent years, collapsing mine roofs accounted for 40 percent of the deaths in underground mines, and coal haulage accidents for about 20 percent.

A similar safety disparity between surface and deep mines holds for nonfatal injuries as well. The range of accidents among the top ten coal producers went from 2.72 to 72.13 injuries per million man-hours for the period 1968 to 1971.

Since the passage of the Coal Mine Health and Safety Act of 1969, most companies have greatly strengthened their safety programs, and this increased emphasis on safety may bring significant improvements in fatality and injury rates. Indeed, the wide range of safety performances cited above makes it clear that—even without a technological breakthrough—much can be done to narrow the safety gap existing between deep and surface mining.

Workers in deep mines are exposed to fine particles of the dust from rock and coal. These workers contract respiratory diseases due to the dust lodged in their lungs. These diseases are called pneumoconiosis (black-lung disease) and silicosis. Ventilating systems and protective masks help to prevent the dust from reaching the worker's lungs.

Surface Mining

During the past 50 years, the proportion of coal mining occurring at the surface has increased significantly. In 1925, surface mining accounted for only 4 percent of the total, while surface mining accounted for approximately one-half of the total coal produced. In the eastern U.S., strip mining disturbs about 0.6 acres per 1000 tons of coal removed.

The coal found in the Appalachian Mountains was originally deposited in horizontal layers. Over millions of years, erosion and weathering have cut deep valleys into the highland plateau and shaped the landscape we observe today. The coal outcrops thus exposed roughly follow the contour lines, as on a topographical map. Therefore, strip mining proceeds along the coal strata, with the overburden of earth removed from above the coal bed during each cut typically being cast down the hillside or stacked along the outer edge of the bench. The exposed coal is then removed and a second cut is made to uncover more coal. Finally, when the overburden becomes too thick for economical stripping, augers—up to seven feet in diameter—are used to drill horizontally several hundred feet into the coal seam to bring out additional coal.

The mining proceeds along the mountainside using this combination of stripping and augering. Behind is left a steep, nearly vertical, highwall on the upslope side, and piles of overburden or spoil material stacked along the outer bench, or cast downslope.

In other parts of the U.S., where coal seams lie near the surface under level terrain, strip mining is simpler. An area is cleared of overburden and then the coal is removed. The overburden from the next area is moved onto the previously stripped area.

Surface mining has increased in production over the past decade, primarily as a result of the introduction of giant excavation and haulage equipment.

Although surface-mining methods are inherently less hazardous to workers than deep-mining methods, they result in disturbances of the surface environment that are matters of intense public concern. These disturbances include erosion, scarred landscapes, and surface and subsurface water degradation.

The nature of environmental damage from surface mining depends significantly upon the climate and terrain of the mining region. Large-scale operations in the water-poor regions of the southwestern U.S. may have significant and unforeseen environmental effects.

A method, called the block-cut system, useful in mountain strip mining, may reduce the area of disrupted land and improve the potential for economic land rehabilitation. In the block-cut method the spoil from the first section ruined is stored on the downslope side of its bench, or off-site if possible. Then this first mined area is available to hold the spoil from sections immediately adjacent on both right and left, and in turn these mined areas are used for storing the spoil from the next sections. Finally, the spoil from the first section is distributed to the last mined sections as mining of the property is completed.

Reclamation of surface areas disturbed by strip mining is a complex technical and institutional problem. Improved reclamation techniques may be required, especially to deal with difficult terrain and disturbed lands—as well as stringent rehabilitation criteria. These new techniques will undoubtedly increase mining costs, and the institutional question to be decided is how such costs should be shared by those who enjoy the benefits of surface-mined coal and the energy it produces. The total cost of reclamation of stripped land ranges from $1000 to $4000 per acre, depending upon the extent of the reclamation, the type of surface conditions, water availability, and climatic conditions.

The production of surface-mined coal has already resulted in more than 3000 square miles of disturbed land. At present rates in the U.S., over 100 square miles are being stripped annually. As a result, extensive tracts of unrestored land have accumulated and whole areas, particularly in the Appalachian Mountains, have suffered environmental damage.

With over 100 billion tons of strippable coal in the U.S., a system of environmental protection must be perfected and implemented. Efficient reclamation methods remain to be proved and fully utilized. The coal recovery factor for surface mining is much higher than for underground mining, approaching 80 to 90 percent. In addition, during the process of stripping away the overburden to reach thick coal seams, thinner seams are often encountered and profitably extracted. Indeed, this increased productivity should lead to renewed efforts towards the achievement of acceptable surface-mining methods.

Over the past several years the U.S. Congress has passed legislation regarding strip mining. However, these bills have been vetoed by the President on the basis that they would decrease the amount of coal mined in the U.S. at a time when it is sorely needed in order to avoid even greater dependence upon imported oil. Many states have enacted strip-mining laws, however. Legislation to stop the worst abuses of strip mining and to support reclamation is continually being discussed and pursued. Strip mining is being more and more brought under regulatory standards. As long as it is less expensive to strip mine, even including the cost of reclamation, surface mining will continue to increase in extent, while deep mining remains relatively constant or decreasing.

Coal in the Western United States

A modern-day coal rush is underway in the Western United States. For example, the deposits in the area of Wyoming, Montana, and the Dakotas incorporate the Fort Union coal formation. This basin may contain up to 40 percent of the U.S. reserves of coal. About 20 million tons of coal are strip-mined each year from Montana alone and the figure may reach 80 million tons if developments are allowed to proceed. This coal is highly desirable because of its low content of sulphur, sodium, and ash.

One study of the potential of this region suggested that 42 electric generating stations be constructed at the mouths of strip mines to convert the coal to electricity and send it via high-voltage transmission lines to the Midwest. Some critics of these proposed projects state that the water requirements for cooling these plants would be too extensive for the area. Western coal, most of it strip-mined, now accounts for more than 10 percent of the 600 million tons of coal the U.S. produces annually, up from 5 percent in 1966. In Wyoming, for instance, coal production has jumped from three million tons in 1968 to 14 million tons in 1975, and may reach 40 million tons by 1980 and 100 million tons by 2000. Meanwhile, production in West Virginia has fallen from 145 million tons in 1968 to 115 million tons last year, and may drop further. Western strip-mined coal may account for one-half of the U.S. coal production by 1985. [4, 16]

The costs of transporting either the electric power generated from the coal, or the coal itself, to the markets of the Midwest are as high as the costs of mining the coal. The question of how to transport the energy in large amounts will have to be solved before the production of Western coal reaches its predicted levels.

Nevertheless, Western strip mines offer the advantages of lower cost since the coal seams are often more than 100 feet thick and near the surface for easy stripping. As a result, some mines can yield 100 tons per man-day of labor, compared with 12 tons in an Eastern deep mine and about 35 tons in an Eastern strip mine.

Many Westerners are concerned about the potential for effective restoration of the land after stripping. Unlike Appalachia, where the hills and hollows make restoration difficult and costly, the plains are flat and would be easy to regrade after mining. But the soil is fragile, and rainfall is scant, making revegetation a problem. If the vegetation does not take root, the wind can scatter the topsoil, making grazing impossible.

The North Central Power Study carried out by the U.S. Bureau of Reclamation examined the potential for mine-mouth coal-fired steam electric power plants on the Tongue and Powder River Basins in Montana and Wyoming. [6] The deposits were estimated to be capable of producing 156,000 megawatts of electric power for 35 years. The study considered an estimated water requirement to be 855,000 acre feet (1.06×10^{12} liters) per year. The power was to be transmitted to the Midwestern U.S.

While less than 10 percent of the coal mined in the U.S. was extracted from Western mines in 1975, this proportion will probably increase significantly over the next decade. It is estimated that only 10 percent of the coal reserves east of the Mississippi can meet the standards of the Clean Air Act, which was scheduled to go into effect on July 1, 1975, but has been delayed temporarily.

Nevertheless, reclamation of the stripped lands remains an issue of contention. Many claim that the land has been restored adequately, while others dispute this claim. Estimates of reclamation costs range from $1000 to $4000 per acre, but this may amount to only 20 cents per ton maximum. It isn't easy for outsiders to understand the concern of local ranchers who require about 40 acres of land for each head of cattle. The National Academy of Sciences study says that projections indicate a disturbance due to coal mining of only about 300 square miles of Western lands by the year 2000, a figure less than 1 percent of the total area of eight Western coal states, and only 15 percent as much as the nearly 2,000 square miles of land already disturbed by surface mining in the East. Moreover, only a fraction of the total acreage to be mined will ever be disturbed at any one time. So the question remains unsettled, but hotly debated.

Mohave and Navajo Power Projects

Black Mesa is a barren plateau on the Navajo and Hopi Indian reservations in northeastern Arizona. The largest strip mine in the U.S. is located where the Peabody Coal Company has leased 65,000 acres of land from the tribes. The Black Mesa mine feeds coal into two electric generating stations called the Mohave and Navajo generating stations. The Mohave station consists of two 755,000-kW units and utilizes a slurry pipeline to receive the coal from the Black Mesa mine. The Navajo station consists of three 750,000-kW units. The coal at Black Mesa is soft-bituminous, low-sulfur coal, and is relatively close to the surface.

The station at Four Corners produces 2085 megawatts of power alone. The Navajo and Four Corners plants supply power to the metropolitan areas of Los Angeles, Las Vegas, Phoenix, and Tucson via transmission lines. These plants use 30,000 tons of stripped coal daily. Regrettably, the Navajo and Four Corners plants also produce significant emissions of sulfur and nitrogen oxides and limited quantities of particulate matter.

The Hopi and Navajo tribes will receive about 25 cents per ton for the coal, some $75 million over the life of the project. The plants currently generate about 6000 megawatts and transmit this power to the user regions. In some sense, this is a case of the metropolitan areas simply moving the waste products and environmental damage attendant upon a generating plant to another area, the Four Corners, while consuming the power themselves. The first plant of the Four Corners project began operation in 1963 and the full project may be operational by the early 1980's. However, opposition to one of the proposed plants, the Kaiparowits plant in Utah, may result in that plant never being completed. In addition, limited success has been achieved in the attempt to reclaim the more than 1500 acres of land strip-mined on the reservations.

A 78-mile railroad has been constructed solely to carry coal from Black Mesa to the Navajo Generating Station near Page in Northern Arizona (near the Glen Canyon Dam). The Navajo Plant has a total capacity of 2250 megawatts and uses water from Lake Powell for the plant's cooling water. The Navajo Plant was completed in 1976 and incorporates an extensive environmental quality-control system. About $200 million was expended on environmental protection equip-ment, one-fourth of the total cost of the plant.

Coal from Black Mesa is also supplied to the Mohave plant in Southern Nevada by means of a 430-kilometer pipeline. The coal is mixed with water to form a slurry and transported via the pipeline. A slurry pipeline is a relatively efficient means of transporting coal when sufficient water is available at a low price.

The Navajo and Four Corners projects are two of the largest of their kind in the world; and if coal is to remain a primary fuel over the next 25 years, we can expect to find several similar developments in the Western United States.

Environmental Effects of Coal Production

One of the accomplishments of the environmental movement of the late 1960's was the fostering of the passage of the Clean Air Act of 1970. Provisions of the Act set standards for sulfur emissions that would rule out, in effect, a major portion of the coal from the fields that now supply power plants in the East.

The primary strategies aim at controlling sulfur-oxide emissions and limiting the sulfur content of the fuel or limiting the SO_2 emitted at the stack. Most states have a sulfur standard for coal calling for sulfur content less than 0.67 lbs. of sulfur per million Btu.

The most promising flue-gas desulfurization processes utilize scrubbing of flue gases with lime or limestone slurries, forming a sludge waste product. Most desulfurization processes are plagued with problems of reliability, as well as investment and operating costs.

Electrostatic precipitations work well to control fly-ash problems. However, a 1000-megawatt coal-fired plant will require 2.4 hectares (6 acres) of land annually for fly-ash disposal if the ash is piled to a depth of 7.5 meters, so the solid-waste disposal problem is not trivial.

Air-pollution problems from fossil-fuel plants can be reduced somewhat by the use of tall stacks or the construction of mine-mouth generating plants. Tall stacks 150 meters high or more can help disperse the pollutants and thereby lower pollution concentrations at ground level; but they obviously do not decrease the amount of air pollution and may simply transfer the problem to a new locale; in addition they are likely to be rather unaesthetic.

The environmental effects of coal mining include effects on the air, the water, and the land. Air pollution occurs from burning of refuse banks and in under-ground mines. The effects of underground mining include the introduction of methane and dust into the air.

Effects on the water in a region of mining include acid mine drainage and surface-water pollution. The effects on land include the refuse and spoil banks, surface subsidence, erosion, and strip pits. The use of water for coal also competes with water needs for agriculture. Nevertheless, effective environmental controls for mining can be instituted and incorporated into the economics of coal production. This will undoubtedly take place over the next decade in the U.S. and in Europe as coal is exploited as a fuel, but environmental safeguards are demanded.

6.6 ECONOMICS

The cost of coal to the purchaser, available at the mine, has risen over the past several years. In addition, the consumer must account for the cost of coal delivered to his electric generating plant or steel mill. Contracts for coal negotiated in 1975 called for a price of about $15 per ton. Of course, if a long-term contract was not held, but rather the consumer bought coal on the open market, it might cost upwards of $25 per ton.

Today it costs about $50 million to open a new deep mine with an annual capacity of 2 million tons. [8] It is estimated that to double the capacity of our coal mines by 1985, over $10 billion of investment capital will be required.

In the U.S., bituminous coal, at the mine mouth, costs about 60 cents to 90 cents per million Btu in 1976. Lignite coal costs about 15 cents per million Btu, but is less useful than bituminous coal because of the difficulty of pulverizing it for furnaces. In 1970, coal cost about 20 cents per million Btu, so it has experienced a threefold increase in price over the past six years. By contrast, petroleum, which cost 50 cents per million Btu in 1970, now costs from 1.50 to 2.50 dollars per million Btu in 1976. It is interesting to note that natural gas still costs about 60 cents per million Btu and is relatively easy to transport by means of pipelines. Is it any surprise, therefore, that the use of natural gas grew so rapidly over the past several decades?

The ability of the coal industry to double production by 1985 will depend upon (1) the opening of about 400 new coal mines; (2) an increase in coal-mining employment from 158,000 men in 1973 to about 250,000 men in 1985; (3) a cumulative investment in coal mines of up to $20 billion; and (4) a cumulative investment in railroad equipment of at least $7 billion, or an equivalent transportation system, for the coal produced.

6.7 COAL GASIFICATION

In the 1920's, nearly every major city in the Eastern U.S. had its gashouse, where gas was manufactured from coal for lighting and cooking. Now interest is reawakening in the idea of the gasification of coal as natural gas is declining in availability.

Gasification of coal involves not only heating the coal, as in distillation, but also the subsequent reaction of the solid residue with air, oxygen, steam, or various mixtures of them. The distillation step releases a certain amount of gas that has a fairly high Btu content because methane (CH_4) and other higher hydrocarbons contained in the coal are among the first components to emerge as the coal decomposes. The gasification step produces a gas that is essentially a mixture of hydrogen and carbon monoxide.

The basic steps to produce a synthetic natural gas from coal are as shown in Fig. 6.3.

First the coal is distilled or devolatilized by heating it in the absence of air. [9] The next step is gasification involving steam and oxygen as the feed materials. The

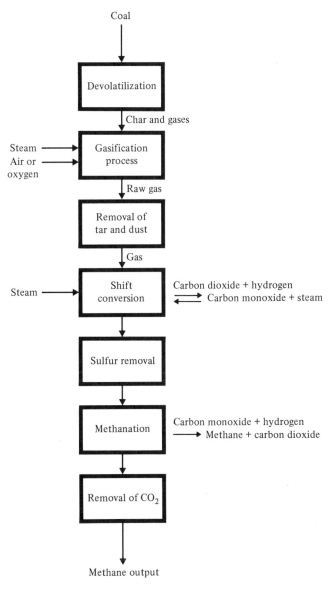

Fig. 6.3 *A schematic of the coal gasification process.*

Table 6.8

COAL GASIFICATION PROCESSES

Name of process	Heat input	Reactor type	Status
Lurgi	Oxygen	Downward–moving bed	Near demonstrated
Hygas	Oxygen	Fluidized bed	Pilot plant
Bi-gas	Oxygen	Entrained flow	Will build pilot plant
CO^2 acceptor	Air	Fluidized bed and dolomite	Pilot plant
Cogas	Air		Bench-scale data
Synthane	Oxygen	Entrained/fluidized	Will build pilot plant

oxygen reacts with part of the carbon to provide heat and raise the temperature high enough for the balance of the carbon to react rapidly with steam to produce a mixture of carbon monoxide and oxygen. Then the raw gas is treated to remove tar and dust. This gas is passed over a shift catalyst which serves to speed up the reaction, yielding a desired mixture of carbon dioxide, hydrogen, carbon monoxide, and steam. The sulfur is removed and then the methane is produced in a catalytic reaction. Finally, the carbon dioxide is removed, yielding the high-Btu methane output.

Several specific processes are being thoroughly investigated to determine the characteristics of each process. The overall reaction that one wishes to achieve is:

$$\text{Coal} + \text{water} \rightarrow CH_4 + CO_2.$$

Present processes cannot achieve this in a single step, so several steps are employed. Several processes under study and experimentation are summarized in Table 6.8.

The economics of coal gasification are not fully known since the processes under study are in the pilot-plant phase. It has been estimated that a large-sized coal gasification plant would cost more than $800 million and be capable of providing 250 million cubic feet of gas per day. It is estimated that the first output from a Lurgi gasification plant would cost about $3.00 per million Btu (1000 cubic feet). [18] The HYGAS pilot plant is shown in Fig. 6.4.

Underground Gasification

The technology of underground gasification of coal is about a century old. Activity in this process reached a peak after World War II when the process was competitive with other energy sources, but it could not compete with low-cost petroleum. The oxidizing agent, air, was cheap, but it included N_2, which diluted the products. These were useful when burned locally in large installations. One of the problems of underground gasification is to obtain suitable communication between the input and output holes. In most of the earlier installations, underground mining preparation was involved. Now, however, a process might be achieved that would

Fig. 6.4 *The HYGAS pilot plant is operated by the Institute of Gas Technology. The pilot plant tests the HYGAS process for making synthetic gas from coal. The plant is designed to process 75 tons of coal a day to make 1.5 million cubic feet of synthetic gas. A commercial-sized plant would use 17,000 tons of coal a day to produce 250×10^6 cubic feet of gas. (Courtesy of the Institute of Gas Technology.)*

avoid the necessity for underground labor; and furthermore, oxygen could be used rather than air. Called *in situ* or in-place gasification, the technique could be applied where deep-lying deposits exist. One idea is to ignite the coal underground and siphon off the gas caused by the heat. It is estimated that gas produced *in situ* could cost about 50 to 70 cents per thousand cubic feet.

Magnetohydrodynamic Power

Magnetohydrodynamic (MHD) generators that convert heat from combustion gases directly into electricity constitute one possible effective use of coal. MHD generators operate at elevated temperatures, typically, 2400°C; and technical problems associated with the endurance of the equipment remain to be resolved. Also, the hot residues from coal combustion are extremely corrosive.

The MHD generator is basically an expansion engine in which hot, partially ionized gases flow down a duct lined with electrodes and surrounded by coils that produce a magnetic field across the duct. Unlike the gas in a turbine, the expanding gas propels only itself, and the movement of the electrically conducting gas through the magnetic field generates a current in the gas that is collected at the electrodes. Thus MHD generators are compact, have no moving parts, and can potentially accommodate temperatures and corrosive gases that would destroy conventional turbines. Very high temperatures would be necessary to ionize combustion gases; but with the addition of small amounts of potassium or other alkali metals, temperatures in the range 2000° to 2500°C would provide sufficient ionization to allow the process to work. The overall efficiency of such an electric power plant could be 50 percent compared with 35 percent normally achieved in steam power plants. Several projects studying pilot MHD plants are being pursued in the U.S. and the U.S.S.R. [18]

Perhaps the best advantage of MHD power generation is the potential environmental advantages of these plants over traditional coal-burning plants. The sulfur of the coal can be removed in the MHD process and nitrogen oxide emissions can be reduced.

6.8 COAL AS A FUEL FOR THE NEXT 100 YEARS

Oil and gas currently provide about 75 percent of U.S. energy. But domestic production of these two fuels has been declining since 1970 and the U.S. will turn to coal and solar and nuclear power over the next several decades. The question remains, however, whether the production of coal can meet the demand during the next 25 years. The issues that will have a significant impact on whether coal can be used as the primary fuel of the next 25 years are whether:

1. SO_2 emissions can be controlled at electric power plants;
2. strip mining can be permitted where reclamation can be adequate;
3. the mines can attract enough laborers to do the work;
4. research and development of conversion processes for producing synthetic fuels from coal can proceed without delay;
5. water is available and price competition can be met; and
6. capital is available for investment.

A recent study, using a computer model, showed that energy demand can be met through the production of coal over the next 30 years by reducing delays in synthetic oil and gas development, temporarily relaxing SO_2 standards for power-plant emissions, and banning construction of oil and gasfired power plants. Price guarantees, direct subsidies, or accelerated exports would also be used to encourage the smooth expansion of coal production after 1976. [15] Whether the U.S. government will adopt this policy remains to be determined.

REFERENCES

1. E. A. NEPHEW, "The challenge and promise of coal," *Technology Review*, Dec. 1973, pp. 21–29.

2. J. PAPAMARCOS, "King Coal's last chance," *Power Engineering*, March 1974, pp. 39–45.

3. "The fight over moving coal by pipeline," *Business Week*, July 27, 1974, pp. 36–37.

4. "The coal industry's controversial move west," *Business Week*, May 11, 1974, pp. 134–138.

5. "Slurry power," *Newsweek*, Oct. 13, 1975, p. 84.

6. *North Central Power Study*, U.S. Bureau of Reclamation, Washington, D.C., 1971.

7. J. T. DUNHAM, *et al.*, "High-sulfur coal for generating electricity," *Science*, April 19, 1974, pp. 346–351.

8. "The hog in coal expansion," *Business Week*, Jan 27, 1975, pp. 127–130.

9. H. PERRY, "The gasification of coal," *Scientific American*, March 1974, pp. 19–25.

10. O. HAMMOND and M. B. ZIMMERMAN, "The economics of coal-based synthetic gas," *Technology Review*, August 1975, pp. 43–51.

11. H. C. HOTTEL and J. B. HOWARD, *New Energy Technology*, MIT Press, Cambridge, Mass., 1971.

12. E. FALTERMAYER, "The clean synthetic fuel that's already here," *Fortune*, Sept. 1975, pp. 147–154.

13. N. PRECODA, "Soviet mine wastes," *Environment*, Nov. 1975, pp. 15–19.

14. K. DYBECZYNSKA, "Coal: Safe or dangerous fuel?" *Electronics and Power*, Oct. 18, 1973, pp. 453–454.

15. R. F. NAILL *et al.*, "The transition to coal," *Technology Review*, Nov. 1975, pp. 19–29.

16. G. ATWOOD, "The strip mining of Western coal," *Scientific American*, Dec. 1975, pp. 23–29.

17. P. M. BOFFEY, "Energy: Plan to use peat as fuel stirs concern in Minnesota," *Science*, Dec. 12, 1975, pp. 1066–1070.

18. A. L. HAMMOND, "Coal research: Gasification faces an uncertain future," *Science*, Aug. 27, 1976, pp. 750–753.

EXERCISES

6.1 A large percentage of the low-sulfur coal in the U.S. lies in deep strata underground. Strip mining, if allowed to increase in use, could exhaust strippable reserves over the next 100 years, while deep-mine reserves will last for over 300 years. Therefore, strip mining will be used over the next 25 to 50 years under regulated circumstances while new technologies are developed for the production and safe mining of deep mines. Investigate the current and newly developing technologies for safe productive deep mining.

6.2 Of the following fuels, the biggest U.S. reserves are those of:

a) natural gas
b) petroleum
c) coal
d) tar sands

Choose one and rank in order the others.

6.3 Coal may be transported from mine to power plant by coal-slurry pipeline or by railroad. If an existing railroad is available over the planned transportation route, then railroad may be less costly overall. Determine the coal freight charges from your local or regional railroad and compare with an estimate of slurry pipeline costs.

6.4 The United States produces about 640 million tons of coal today. It is estimated that 750 million tons will be produced in 1980 and 1000 million tons will be produced in 1985. What are the effects of this rapid growth of coal production? Can we achieve this and still maintain environmental standards?

6.5 In Eastern Kentucky, the Tennessee Valley Authority (TVA) has constructed electric power plants dependent upon coal. The Paradise power plant, for example, uses coal obtained from local strip mines. This plant and other TVA plants provided an incentive for strip mining. Explore the TVA use of coal and compare the electricity rates for TVA with your city's rates. Obtain information from TVA regarding their policies covering land reclamation.

*6.6 Mining of coal is a large operation in the Soviet Union. In the U.S.S.R., abandoned and operating surface mines cover a total area of nearly 5.5 million acres; this will increase by 100,000 acres every year. [13] The two largest sources of coal in Russia are the Donbass basin and the Kuznetz Coal Basin, which yielded 241 and 141 million tons, respectively, in 1974. In the Kuzbass basin 35 million tons were strip-mined in 1972 and this number is growing rapidly. The Kansk–Achinsk coal basin in Central Siberia holds an estimated 500 billion tons of coal suitable for strip mining. Determine the safety and health safeguards used for the mining of coal in the U.S.S.R. Locate several large coal fields in China and determine what methods are used for coal mining. Compare the methods in the U.S.A., the U.S.S.R., and China.

6.7 Poland has extensive coal deposits and very limited resources of natural gas and petroleum. Therefore, it is heavily dependent upon the use of its coal resources, which are estimated to be 100 billion tons. Poland is estimated to use coal to produce 90% of its electricity. [14] An alternative is to develop an extensive series of nuclear power stations. It is known that the sulfur dioxide emitted by the coal plants (2.7 megatons) is the greatest danger. The coal used contains an average of

---

* An asterisk indicates a relatively difficult or advanced exercise.

1.7% of sulphur. It is estimated that the annual emission of dust amounts to 5 megatons. Examine the various current methods for controlling the release of sulphur dioxide and flue gases at electric power plants. What policy would you recommend for Poland: nuclear power or controlled use of coal?

6.8 A slurry pipeline is planned for the proposed mine at Gillette, Wyoming, by Bechtel, Inc., of San Francisco and the Middle South Utilities Co. [5] The officials say that the pipeline would deliver coal at an average cost of $8 a ton over a 25-year period, versus $25 per ton which they estimate the railroads would charge for the 1000-mile haul. The pipeline would use 12 gallons of water to deliver 1 million Btu's of energy; a coal-gasification plant would consume as many as 30 gallons for each million Btu's, and a mine-mouth power plant would use 100 gallons. Investigate the costs and benefits of a slurry line versus a railroad transportation plan for Wyoming.

6.9 A new power plant was recently constructed in South Dakota called the Big Stone Power Plant. This plant has a capacity of 150 megawatts of electricity and will consume 2.5 million tons of coal annually. The plant has a precipitator to trap fly ash and a 500-foot-tall stack. Contact the Montana–Dakota Utilities Company and Bechtel Corp. of San Francisco to determine the operating experience of this plant in regard to environmental impacts.

6.10 Sulfur dioxide may be removed, in part, by using several processes including wet limestone or lime-scrubbing systems. In these systems (1) dry lime or limestone is injected into the boiler and the partly reacted material is removed in a wet scrubber; or (2) slurries of lime or limestone are reacted with sulfur dioxide in scrubbing towers to form calcium sulfates and sulfites, which are collected and impounded. [7] The average increase in electricity cost due to the addition of these devices may be about 5%. Investigate the method of sulfur removal used by your local power company if they use coal, or contact Consolidated Edison of New York City for further information.

6.11 One of the oldest fuels in service to man is peat, which covers about 1% of the earth's surface and represents the equivalent of 100 billion tons of oil. Though low in heat value, peat burns cleanly. Peat is vegetable matter in the early stages of becoming coal. Much of the peat supply is found in the U.S.S.R., Northern Europe, and Ireland, as well as North America. In an undrained bog, peat is about 93% water. No effective and economical way of drying peat has been found other than natural exposure to summer air and sun. Traditionally, peat is cut into brick-sized chunks, a process that can now be done by machines like huge reapers. But increasingly peat is milled for more rapid drying, into a dust that can be fed into power plants or compressed into briquettes for domestic heating. When peat is thoroughly dried, it has a heat value of up to 8,000 Btu/lb, or roughly two-thirds that of coal. But in power plants it may also be burned with a moisture content of above 50% and a caloric value as low as 3,300 Btu/lb—perhaps the lowest of all commercially used fuels. Nevertheless, at its current price in Ireland of less than $6 per ton, milled peat is still highly competitive.

Explore the history of peat use and list the advantages and disadvantages of its use as a fuel. What locations of peat bogs can you find in North America?

*6.12 The process of coal utilization for electric power generation is shown in Fig. E6.12. Determine the energy expended as a percent of the total energy input for each of the steps in the process. How much of the energy originally stored in the coal is delivered to the final use?

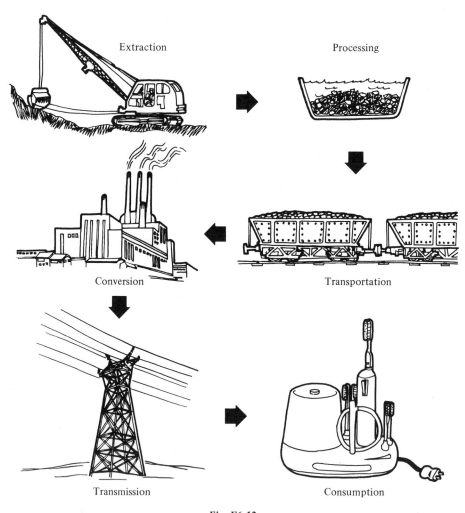

Extraction

Processing

Conversion

Transportation

Transmission

Consumption

Fig. E6.12

6.13 In the Northern reaches of Minnesota are some of the largest and most desolate peat bogs. The potential for using this peat as a fuel in a gasification scheme is being pursued by the Minnesota Gas Company and the Minnesota

Department of Natural Resources. [16] It is planned to use the HYGAS process to generate methane. Explore the environmental problems of harvesting (mining) the peat and converting it to gas. Examine the economics of the situation by obtaining data from the company and the state department involved in the project.

*6.14 In the long run, in order to double the production of coal from the level of 1975, the problem of capital may be the overriding issue. It has been estimated to cost $20 billion over the decade 1975 to 1985 in order to double production. Explore the economics of the coal industry and determine whether it is feasible to double production by 1985.

6.15 The one resource that may limit the use of Western coal is water. Explore the need for water for coal gasification and determine the water availability in Montana and Wyoming.

*6.16 Methanol (CH_3OH) can be made from coal as well as wood. To produce methanol, coal is converted under high temperature into a synthetic gas and then passed over a catalyst to yield methanol. [12] Investigate the production of methanol from coal and its use for an industrial, residential, and automotive fuel.

6.17 The world's largest dropline for moving overburden covering coal seams near the surface is the Big Muskie owned by Central Ohio Coal Co. The bucket capacity is 220 cubic yards (168 cubic meters) and measures 21 feet by 25 cubic feet by 13 feet. Investigate the use of large droplines in strip mining and the speed of clearing overburden using such machines.

6.18 In 1924, Germany faced an energy crisis. World petroleum supplies were declining and Germany did not want to become more dependent upon foreign sources. Germany's largest chemical concern, I. G. Farben, with a strong background in the development of processes and production plants for synthetic ammonia and methanol, decided to make a heavy investment into development of processes for making gasoline from coal. It took the company five years to develop a suitable process, only to be confronted with the world depression of 1929. Explore the history of the Farben process and its use during World War II by Nazi Germany.

*6.19 The environmental damage caused by strip mining can be viewed in at least four ways. One is to accept the damage, on the basis that the cost to society of controlling it is excessive. The second is to insist that enough remedial work be done to reclaim the area after mining. The third is to forestall the problems by requiring preventive measures during the mining cycle. The last is to avoid the problems by not mining the area at all. In recent years the trend has been toward preventive measures such as segregating spoil, burying toxic material, and incorporating grading operations into the mining cycle. As a result, revegetation has been more successful and certain of the hydrologic problems have been avoided. Prepare a position paper on the reclamation requirements and preventive measures you advocate for strip mining in the U.S.

Petroleum

Petroleum has assumed an important and growing role as an energy source in the world. Its superior qualities and the ease with which it can be transported made petroleum the preferred fuel after 1920. The result is a U.S., European, and Japanese economy heavily dependent upon the availability of oil. The world economy is rapidly approaching the same dependence that industrialized nations now experience.

Nevertheless, we understand that oil, like natural gas and coal, is finite in supply. The increased use of petroleum around the world is causing an accelerating problem of supply and demand and resulting in price increases everywhere. The U.S. has become increasingly reliant on oil as a primary fuel, as well as a supply for the production of chemicals and other materials.

7.1 CONSUMPTION AND PRODUCTION OF PETROLEUM

The annual production and consumption of petroleum in the U.S. since 1920 is given in Table 7.1. The growth in consumption of oil is also shown in Fig. 7.1—an average annual growth of 5 percent per year. U.S. consumption grew by a factor of twelve over the 50-year period from 1920 to 1970.

As domestic production of oil in the U.S. fell short of meeting the demand, imports have grown to 45 percent of our current use of oil. By 1985, imports of oil may account for one-half of our total U.S. consumption, thus causing us to be even more heavily dependent upon other nations.

The annual consumption of petroleum in the world is shown in Fig. 7.1 to be growing at a rate of 5 percent per year. Oil and oil products are heavily used in all industrialized nations, and other less developed nations are also becoming dependent upon oil. Table 7.2 compares the growth of oil imports to Western Europe and Japan with those to North America. It is clear that Europe and Japan are

Table 7.1

ANNUAL CONSUMPTION AND PRODUCTION OF PETROLEUM IN THE U.S.

Year	Consumption (Millions of barrels*)	Domestic production (Millions of barrels)	Imports as percent of consumption
1920	434	443	2%
1930	862	898	4%
1940	1,285	1,353	5%
1950	2,375	1,974	17%
1960	3,611	2,575	29%
1970	5,365	3,517	34%
1974	5,900	3,500	40%
1980 (Projected)	7,200	4,000	44%

* One barrel of petroleum equals about 6.0×10^6 Btu or 6.3×10^9 joules.

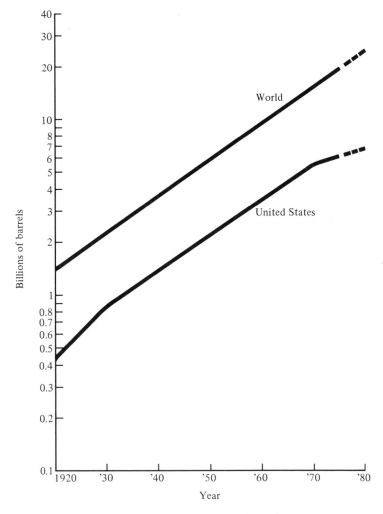

Fig. 7.1 *Annual consumption of petroleum*

Table 7.2
GROWTH OF OIL IMPORTS
(IN BILLIONS OF BARRELS PER YEAR)

Year		North America	Western Europe	Japan
1950		0.40	0.51	0.07
1960		0.80	1.31	0.22
1970		1.20	4.67	1.53
1975		2.15	4.79	1.84
1980	(Projected)	4.0	5.50	3.29

Table 7.3
WORLD PRODUCTION OF
PETROLEUM IN 1975

Nation or region	Annual production (Billions of barrels)	
United States	3,720	
Canada	694	
Latin America	1,752	
Subtotal		6,166
West Europe	183	
Africa	2,190	
Middle East	7,994	
Far East and Oceania	1,022	
Subtotal		11,389
Communist nations		4,344
World total		21,899

heavily dependent upon imported oil for their source of a primary energy supply. Currently, these three regions of the world consume two-thirds of the world's production of petroleum.

The production of petroleum in 1975 is given in Table 7.3. The primary exporter of oil is the Middle East, which produced 36 percent of the world's petroleum in 1975. The Middle East produced 46 percent of the world's petroleum, exclusive of that produced in Communist nations (which is not exported to Western Europe or Japan in substantial amounts).

7.2 USE OF PETROLEUM IN THE U.S.

Petroleum is the primary fuel for our transportation system; the automobile is totally dependent upon the availability of gasoline, a refinery product of petroleum. Table 7.4 shows that transportation uses over 53 percent of the petroleum consumed in the U.S. Residential and commercial use is primarily for fuel oil for

Table 7.4
USES FOR PETROLEUM IN THE U.S. IN 1971

Type of use	Consumption $(10^{12}\ Btu)$*	Percent of total
Residential and commercial	6,545	21.6%
Industrial	5,091	16.6%
Transportation	16,267	53.3%
Electricity generation	2,417	7.9%
Other	172	0.7%
Total	30,492	100.0%

* One barrel of oil equals about 6×10^6 Btu.

Table 7.5
DEMAND FOR PETROLEUM IN THE INDUSTRIAL SECTOR

Year	Fuel use (Million barrels)	Nonfuel use (Million barrels)	Total (Million barrels)	Industrial use as percent of total use of oil
1950	362	79	441	19%
1960	444	200	644	18%
1970	547	415	961	17%

heating. Industry uses petroleum for heating, for diesel engines, and in chemical processes.

Industrial use of petroleum is shown in Table 7.5. Nonfuel use of petroleum has grown from 18 percent in 1950 to 43 percent in 1970. Petroleum is a most important factor in the chemical industry, and the supply should be maintained to keep our economy strong.

An oil refinery turns crude oil into gasoline, diesel oil, lubricating oil, kerosene, and jet fuel. Gasoline is the chief product of a refinery (45 percent average) and fuel oil of various grades ranks second in importance. Fuel oil is used to heat homes and buildings and to supply power for factories and railroads. Jet fuel contains a mixture of gasoline, kerosene, and oils with low freezing points. Asphalt for roads is another product of crude oil.

Petrochemicals are chemicals made from petroleum; examples are plastics, nylon, fertilizers, and medical drugs.

7.3 PETROLEUM RESOURCES

The availability of petroleum as a world resource and its annual production depend upon several parameters, some of which relate to the geology of the earth and to the techniques for oil production, but many of which are dependent upon economics, governmental regulations, material and equipment supply, and similar factors. A prediction of future oil supplies will be found in a composite assessment of several questions: (a) How much oil is there to be found? (b) How effectively and rapidly can new oil deposits be located? (c) How much of the oil that has been found or will be found will be recovered? and (d) How fast can known oil be produced? The answer to every one of these questions can be at best an estimate, and the uncertainties in the estimates arise from both nonphysical and physical factors.

Since the beginning of commercial use of crude oil in the U.S. in 1859, the U.S. oil industry has produced about 100 billion barrels (1.6×10^{13} liters) and the total proven reserves of oil discovered thus far amounts to 34.2 billion barrels, as shown in Table 7.6. Our current oil consumption rate in the U.S. (including imports) is about 6 billion barrels per year. If we depended upon our domestic

Table 7.6
ESTIMATED PROVED WORLD RESERVES OF PETROLEUM (IN 1974)

Region (with selected countries)	Proved reserves (Billions of barrels)	Annual production (Billions of barrels)
I. North America	44.5	4.01
Canada	7.2	0.60
U.S.	34.2	3.20
II. South America	21.1	1.48
Argentina	2.6	0.15
Venezuela	14.6	1.09
III. Western Europe	20.1	0.11
Norway	5.5	0.01
Great Britain	12.0	Negligible
IV. Eastern Europe	50.5	3.49
U.S.S.R.	47.5	3.01
V. North Africa	35.3	1.00
Libya	23.2	0.55
Algeria	9.3	0.36
VI. Africa, Central and Southern	22.6	0.98
Nigeria	18.3	0.82
VII. Middle East	326.0	7.95
Iran	68.1	2.20
Kuwait	70.9	0.83
Saudi Arabia	96.9	3.00
Iraq	35.7	0.74
VIII. Far East Asia	16.0	0.51
China (mainland)	14.8	0.26
India	0.9	0.06
IX. Oceania	19.4	0.74
Indonesia	12.0	0.50
Australia	2.5	0.14
Total world	556.2	20.34

supply alone, we would exhaust our proven reserves in about six years. At our current domestic production rate of 3.2 billion barrels per year, our proven reserves would last only ten years. Clearly, the U.S. will remain heavily dependent upon imported oil, if our consumption rate is to be maintained around 6 billion barrels per year.

Examining Table 7.6, one notes that the Middle East accounts for 39 percent of the total world production of crude oil and 59 percent of the world's proven reserves. [1]

Russia is the world's leading oil-producing nation, with a production in 1975 of 3.22 billion barrels, an increase of 22 percent in production over 1970. Russia exported about 803 million barrels of oil and oil products in 1975, about 9 percent of its total production, mostly to Eastern Europe. [2]

The Middle East is the primary region holding proven reserves of crude oil, and the region exports the major portion of its production to Western Europe, Japan, and the U.S.

Oil is discovered by small and large companies by means of geological exploration and the drilling of exploratory wells. In 1974, 33,470 new wells were drilled in the U.S. Exploration is dependent upon economic factors, tax incentives, and governmental regulations. The right to drill is obtained from the land owner, usually in the form of a lease, which conveys the right to explore and produce oil in exchange for a share of the financial return. A breakthrough in drilling technology could accelerate the rate of discovery of petroleum. Drilling technology has steadily improved, but there have been no major innovations in the past 50 years. Although many exotic drilling techniques have been proposed and are being investigated, the rotary drilling technique still remains the most efficient and economic method for oil-well drilling in the earth's crust.

While exploitation of proven reserves continues, exploration and discovery of new reserves must continue if petroleum is to be produced at a constant rate. Table 7.7 shows the annual discoveries, averaged over a five-year interval. In

Table 7.7

TOTAL DISCOVERIES OF OIL FOR 5-YEAR INTERVALS
IN THE WORLD (EXCLUDING U.S.S.R. AND CHINA)

	Average annual discovery over 5-year period (Billions of barrels)	Average annual consumption over 5-year interval (Billions of barrels)
1935–40	24	3.0
1940–45	8	4.0
1945–50	22	4.8
1950–55	18	5.5
1955–60	17	7.5
1960–65	18	10.0
1965–70	18	14.0
1970–75	18	18.0

recent years, the additional discoveries added in a given year have barely kept pace with the total consumption in that year. Thus, we are not adding any net amount of petroleum reserve to our total of proven reserves.

In the U.S., it is estimated that the total resource of petroleum is 200 billion barrels. We have already extracted 100 billion barrels (by 1975), and *proven* reserves amount to 34 billion barrels. Therefore, 66 billion barrels remain to be discovered. An estimated curve indicating the history of oil production in the U.S. is shown in Fig. 7.2. This curve shows a declining domestic production, with an annual production of less than 1 billion barrels by 2010.

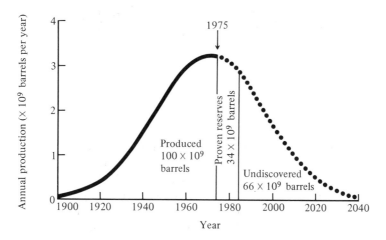

Fig. 7.2 *Petroleum production in the U.S., assuming a total resource of 200 billion barrels.*

Of course, the *actual* ultimate recoverable petroleum in the U.S. may differ from the estimate. The U.S. Geological Survey and the National Academy of Sciences differ on the estimate of the ultimate resource of crude oil. [3, 4] The Academy estimate is 213 billion barrels as a total resource, while the Survey estimates an ultimate resource of less than 200 billion barrels.

Estimates of the total world resource of crude oil range from 1350 to 2000 billion barrels. [5, 6] Of this, it is estimated that the Middle East has 630 billion barrels and the U.S.S.R. and China combined are estimated to have 500 billion barrels. Figure 7.3 projects the course of future petroleum production in the world for two values of the total resource. For the larger value, a peak in annual production of 36 billion barrels is reached in the year 2000. Current world annual production is 20 billion barrels and increasing.

The ratio of proven world reserves to total production fell by a factor of 4 during the period 1950 to 1975 because world production significantly increased, from an annual production of 5 billion barrels in 1950, to the current level of 20 billion barrels per year.

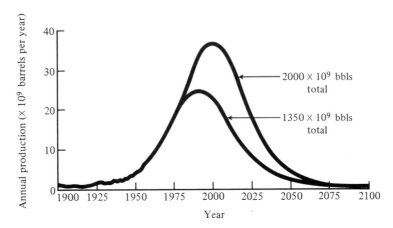

Fig. 7.3 *Petroleum production in the world for two estimated values of total oil resource.*

Because oil is formed from marine life, oil deposits are widely distributed, especially in coastal areas and beneath the continental shelves. Oil is also found in inland regions that once were submerged, such as the Gulf Coast region of the U.S. and the Middle East. Locating deposits of oil is not as straightforward as finding mineable coal. Even in a proven field, not every well that is drilled will yield oil. The search for oil is, therefore, not certain to be everywhere successful, and the estimates of crude oil resources remain inexact.

7.4 THE PROCESSING OF PETROLEUM

Coal, which is essentially pure carbon, can be directly used as a fuel, but oil must be processed prior to use. Crude oil consists of a variety of hydrocarbon compounds, which can be separated by a fractional distillation process based on the fact that the different hydrocarbon compounds have different boiling points. About one-half of a typical crude oil can be separated into gasoline. Compounds with higher boiling points yield kerosene, fuel oils, and lubricating oils.

The fractional-distillation process is carried out in refinery plants usually located near sources of crude oil. Large capital investments and long lead times are required for the construction of new refineries. As demand for petroleum products increases throughout the world, an increased number of refineries is required. If the oil is delivered in tankers, preferred sites for refineries are on the coast.

There are many environmental concerns regarding refineries and it is difficult to obtain a permit to build in the U.S. at present. Currently, refineries are operating near capacity in the U.S., and new refineries are needed, as the demand for petroleum products continues to grow.

Imported oil is brought to the U.S. by oil tankers and delivered to refineries on the coasts. The oil brought from the Alaskan North Slope will be delivered via tanker to the Pacific Coast of the U.S. The U.S. is currently considering building large port facilities especially to accommodate large (super) tankers of imported oil. Tankers weighing 500,000 tons when fully loaded require large port complexes to off-load oil rapidly. On Canada's east coast, one superport already is operating. It was built by Gulf Oil Canada, Ltd., at Port Hawkesbury, Nova Scotia, to serve its 100,000 barrels-a-day refinery there. The port has unloaded crude from the 326,000-ton *Universe Japan*, with no problems.

Secondary and Tertiary Recovery

The proportion of oil recovered from a deposit is usually about 30 percent. If the heretofore unrealized resource could be economically recovered, then the oil fortune of the U.S. and other nations would increase substantially. Secondary recovery relates to oil obtained by the augmentation of reservoir energy, often by the injection of air, gas, or water. Tertiary recovery means the use of heat and methods other than fluid injection to augment oil recovery, and takes place after secondary recovery. Many believe that secondary and tertiary recovery could double the proportion of resource recovered from a given deposit. Although most oil is still being recovered without such methods, in the United States fluid injection has already been widely applied and was responsible for about 35 percent of total oil production in 1974.

In the 1980's at least 40 percent of the known oil in place should be recoverable, as compared to 30 percent today; if this goal is achieved, it will increase recoverable reserves by huge quantities. This ratio could be increased if the methods of secondary and tertiary recovery can be made more economic and energy-efficient. For example, at present the steam-injection process uses energy equivalent to about one-half of a barrel of oil to obtain one barrel of oil for sale. Nevertheless, as the U.S. oil industry is freed from price regulation over the next several years, we can expect new and improved secondary and tertiary methods.

In California, for example, higher oil prices have stimulated steam injection. In more than half the state's deposits the oil is extremely thick and only 10 percent of this crude can be produced by natural well pressure. By injecting steam to lower the viscosity and then flush out the oil, producers have lifted the average recovery of the thick oil fields to about 25 percent in the past decade.

7.5 THE OIL INDUSTRY

The price of crude oil and its products gasoline and fuel oil have risen sharply since 1972. In late 1973, under pressure of an embargo of oil exports from the Middle East, the prices of crude oil exported from these Middle Eastern nations quadrupled. With this sharp rise, the shape and makeup of the oil industry in the U.S. and the world came under severe scrutiny.

The U.S. oil industry, at the end of the 19th century, essentially consisted of a monopoly held by John D. Rockefeller and the Standard Oil Company. In 1911 this trust was broken up into several smaller companies, which have grown into the dominant oil companies of today. The current eight largest oil companies in the U.S., in order of sales volume, are: (1) Exxon, (2) Texaco, (3) Mobil, (4) Gulf, (5) Standard of California, (6) Standard of Indiana, (7) Shell, and (8) Atlantic Richfield. Combined, these eight companies account for 55 percent of all gasoline sold in the U.S. and 58 percent of the refining capacity in the U.S. The largest company, Exxon, holds assets exceeding 22 billion dollars and its sales amount to more than $20 billion per year.

The prime charge against these oil giants by the Federal Trade Commission was put in these words:

> American consumers have been forced to pay substantially higher prices for petroleum and petroleum products than they would have had to pay in a competitively structured market.

Yet the oil companies point out that there are 239 refineries in the U.S. operated by 127 companies. The largest refinery accounts for less than 9 percent of the total U.S. refining capacity.

The return on equity held by the stockholders of these U.S. oil corporations rose to 21 percent in 1974. [7] These companies earned record profits in 1974 and came under the full scrutiny of consumers and the Congress. Profits have steadily risen from 10 percent of operating revenue to currently over 16 percent.

Nevertheless, the oil industry requires large accummulated earnings, in many experts' opinions, in order to be able to invest in continuing exploration and drilling of new wells. The U.S. oil industry has been averaging about $5 billion per year of capital investment for exploration and development over the past decade. [5] In order to increase domestic oil exploitation, investment per year may have to double for several years. The industry estimates that $800 billion will be required for capital spending over the period 1970 to 1985.

The big oil companies are beset by demands from the public for regulation, tax reform, and even splitting up the large companies. Some critics have alleged that these oil companies have engaged in price-fixing in order to increase profits. [8]

Under this wave of consumer unease regarding the oil industry, Congress repealed the 22 percent oil-depletion tax allowance in March 1975 and limited foreign tax credits. The depletion allowance extended to the owner of the oil well a tax deduction based on the reduction in value caused by the removal and sale of the oil. These tax breaks were provided to the oil industry in earlier years when profits were relatively low, in order to encourage the reinvestment of earnings in new exploration and drilling programs.

Large oil companies are under scrutiny regarding their large energy-related holdings in coal and uranium. There is considerable discussion, also, regarding the possibility of breaking up the companies. Presently, large oil companies are vertically integrated; that is, the company holds the process of (1) production, (2) refining, (3) marketing, and (4) transportation and pipelines. Critics are demanding that each company be broken up into four companies, for the four stages of the petroleum process.

The earnings on foreign operations of the seven largest multinational oil companies nearly doubled, from $3.2 billion in 1972 to about $6.2 billion in 1973. This is, in part, due to the jump in Middle East crude-oil prices, which increased by a factor of four over a two-year period, as can be seen in Table 7.8. [9] The seven multinational companies, often called the "Seven Sisters," are Exxon, Royal Dutch/Shell, Texaco, Mobil, British Petroleum, Socal, and Gulf. These firms have been involved in the Middle East oil industry since after World War I. A recent book, *The Seven Sisters*, attempts to show how this group acted as a cartel in maintaining world control of oil, eventually to be overtaken by the Middle Eastern oil-producing nations themselves. [10] Now, the oil companies, of necessity, must cooperate with the oil-producing nations; and we have what might be described as a cartel of companies in league with a cartel of oil-producing nations. How dependent each of these two groups is upon the other remains to be seen as the drama of oil-production economics and diplomacy is played out over the next decade.

Table 7.8
MIDDLE EAST CRUDE-OIL PRICES

Period	Posted price	Arab tax
1968–69	$1.80	$0.88
Feb. to May 1971	2.18	1.26
Jan. to Mar. 1973	2.59	1.51
June 1973	2.90	1.70
Oct. 1, 1973	3.01	1.77
Oct. 16–Dec. 31, 1973	5.12	3.05
Jan. 1, 1975	10.46	7.00
Jan. 1, 1976	11.51	Not available
Jan. 1, 1977	12.66	Not available

7.6 THE OIL-PRODUCING NATIONS AND OPEC

The annual cost of importing oil to the U.S. in 1976 was $35 billion. The U.S. is importing over 40 percent of its oil, principally from the 13-nation Organization of Petroleum Exporting Countries (OPEC). OPEC includes not only the Middle Eastern nations of Saudi Arabia, Iran, Kuwait, Iraq, Qatar, and the United Arab Emirate, but also Venezuela, Ecuador, Indonesia, Nigeria, Gabon, Algeria, and Libya. These countries hold four-fifths of the proven crude reserves in the non-Communist world; and for some, oil provides over 90 percent of government revenues.

OPEC states its principal aims as "the coordination and unification of the petroleum policies of member countries and the determination of the best means for safeguarding their interests, individually and collectively." The Organization emerged in the late 1950's when, due to an excess supply of oil, the multinational oil companies cut posted prices upon which they paid royalties and taxes to the exporting nations. Angered at this action, the original OPEC nations met and formed the new group on September 14, 1960. The oil companies quickly rescinded their price cuts, but it was too late and OPEC emerged slowly during the 1960's to challenge the multinational companies. By mid-1972 Iraq, Libya, and Algeria had nationalized all or part of their oil industries.

Also, in mid-1967 came the closing of the Suez Canal in the Six-Day War, and an oil shortage was created. This shortage again permitted several OPEC nations, particularly Libya, to demand and obtain a large increase in the tax payments on its exported oil.

Toward the end of 1970, the producers consolidated new tax demands through OPEC, and began to act as a single group—and more stridently. Every OPEC member, with the exception of Indonesia, increased its demands for higher oil prices and taxes. A settlement was reached, with the overt aid of the U.S. State Department, in Tehran in February 1971, which yielded price increases of 45 to 80 cents a barrel. With this success, OPEC became a strong cartel. [19]

The purpose of any cartel is to maximize the earnings and power of its members. By mid-1972 OPEC turned its attention to the question of ownership of the international oil companies and their operations and assets on OPEC soil. The agreements reached in Riyadh by the end of 1972 provided for the producing governments to acquire percentage shares starting at 25 percent and working up gradually to 51 percent.

Then in 1973, as an aftermath of the 18-day Arab-Israeli War in October, the Arab oil-producing nations instituted an oil embargo on specific nations including the U.S. As a first step, the OPEC nations instituted a large price increase for their oil and began to reduce exports to embargoed nations at a rate of 5 percent a month. By the time the embargo ended in 1974, oil had doubled in price and supply had been constricted to match demand. The rise in the oil prices of the Middle Eastern nations is shown in Table 7.8. In the three-year period 1973 to 1976, the price of oil had quadrupled!

Professor Adelman of M.I.T. has argued on several occasions that the U.S. needlessly allowed OPEC to emerge as a powerful cartel. [11, 21] This is due, in his view, to the willingness of the multinational companies to effectively fix the price worldwide for the cartel, and to serve as the collecting agency for the new increased prices. Also, Adelman believes that many oil producers outside the cartel have used taxing and regulatory systems, in effect, to safeguard high cartel prices. Professor Adelman particularly scores the Tehran agreement as a capitulation to the oil monopoly. He suggests that the multinational oil companies give up their producing concessions in OPEC countries and become open-market buyers of crude. (Perhaps they might still operate the wells under management contracts of some kind.) Instead of receiving royalties and taxes on oil production, as they do under present arrangements, the members of OPEC would have to sell their oil in order to get any money from it. Presumably, open-market competition would break up the cartel.

The Economic Impact of Oil Price Increases

The increase in the OPEC oil price from $10.46 a barrel to $11.51 a barrel in January 1976 yielded another transfer of 10 billion dollars to the OPEC nations. This increase cost Italy $700 million more in 1976, for example.

The OPEC price increases of late 1973 and early 1974 knocked $20 billion of real growth off 1974's United States Gross National Product. [12] The price hikes boosted the 1974 inflation rate, as measured by the Consumer Price Index, from 8.1 percent to 11 percent. The January 1976 price hike boosted the U.S. import bill by $2 billion. [22]

The current price elasticity of oil is estimated to be only -0.15 in the U.S. Therefore, a 100-percent price increase leads to only a 15-percent reduction in consumption of oil. However, over the next five years, consumers have a good chance of substituting other forms of energy and reducing their dependence upon oil by means of conservation. Therefore, it is entirely possible that if OPEC holds to its high prices, alternative sources of energy will be developed because of the economic price incentive. This could lead to an excess supply of exportable oil from OPEC nations and an eventual drop in the price. The demand for crude-oil exports from OPEC nations lessened in 1975, and production curtailments had to be effected in Saudi Arabia. As importing nations carry out conservation programs and switch to alternative energy sources, the OPEC will experience severe strains on the cartel. Nevertheless, OPEC in 1976 is strong and cash-rich. By 1980 or 1985, the cartel may be enjoying a new international marketplace. The multinational oil companies and OPEC may both find themselves characters in a revised drama.

7.7 OIL-PRODUCING NATIONS OUTSIDE THE MIDDLE EAST

The Middle Eastern nations are not the only oil-producing nations with significant resources. About 70 percent of the U.S. crude-oil imports come from non-Arab countries. For example, Nigeria is a large supplier to the U.S., currently sending 800,000 barrels a day to the U.S.

The United States hopes to import oil from Russia and establish an initial agreement to purchase oil from Russia while selling $1 billion worth of U.S. wheat and corn to that country. Russia has committed itself to purchase about six million tons of American grain over the period 1976 to 1981. In the possible oil deal, crude oil of up to 300,000 barrels a day may come to the U.S.

Another large producer of oil is mainland China, heretofore not present in the international market for oil exports. Many predict that China may export 25 percent of its production by 1980, or about 347 million barrels annually.

7.8   U.S. OIL POLICY

The effect of importing high-priced oil into the U.S. is to distort the trade balance, unless the U.S. can continue to export large and increasing amounts of agricultural products. However, as world oil prices increase, domestic oil supplies increase and the need for oil imports decreases. The increasing prices also boost demand for natural gas, natural gas prices drop, and gas reserves associated with oil wells increase because of increased domestic oil exploration.

With imported oil costing over $11 a barrel, coal may become the predominant fuel for electric-power generation by 1980. Also, oil imports may drop so that by 1985 they would be less than imports in 1975.

However, if the U.S. is to shift from importing foreign oil, it will have to engage in extensive exploration and capital investment. Should the U.S. drain its own reserves over the next decade while developing alternative resources, or should it continue to use significant amounts of imported oil, thus maintaining its own reserves? Some experts believe the U.S. should build up at least a six-month reserve of imported oil on the East coast.

The U.S. does hold a large reserve under the U.S. Navy, established between 1912 and 1923 in California, Wyoming, and Alaska. These reserves, set aside for emergency needs of the Navy, contain an estimated 11 to 38 billion barrels. Most of the Navy oil reservoir is in North Slope Reserve No. 4 in Alaska (an area bigger than Indiana), which has been largely unexplored and would take several years to bring into production. One to five billion barrels remain in the Elk Hills Reserve No. 1, which has been partially developed and could be brought up to a production rate of 160,000 barrels a day in a few months. The maximum production of Elk Hills in California could rise, over a three-year period, to a rate of 300,000 barrels a day (about 3 percent of the total U.S. production).

The U.S. Congress acted in late 1975 to remove price controls on domestic oil. The legislation extends government control over oil prices for 40 months and rolls back the average price of crude oil produced in new wells in the U.S. from $8.75 a barrel in late 1975 to $7.66. But, starting in February 1977, crude prices could start rising again—at an annual rate of about 10 percent—until they reached a peak of no more than $11.21 a barrel. The bill's conservation provisions include mandated fuel economy for automobiles and more use of U.S. coal, thus working against an increase of oil imports.

7.9   OFFSHORE OIL

The U.S. and Europe are entering a period when increased demand for petroleum products may have to be satisfied by increasing imports from OPEC countries. As an alternative, many coastal nations are seriously considering the full exploration and exploitation of offshore oil resources.

The lands extending from the coast into surrounding areas form a continental margin that holds substantial amounts of oil and gas. The U.S. has been actively extracting oil and gas from the continental shelf for more than 25 years, mostly in the Gulf of Mexico and off the coast of California. Exploration is also under way off the coast of Alaska, and exploration off the East Coast is under study.

Thus, oil companies hope to expand oil production from the ocean floor over the next decade. Estimates of the U.S. offshore resources of oil and gas are given in Table 7.9. Despite these great potentials, only 3 percent of the U.S. continental margin has been leased for development. From this 3 percent, the United States obtains about 18 percent of its oil production and 15 percent of the domestic natural gas production. The blowout in the Santa Barbara Channel, California, in 1969 focused attention on adverse environmental effects, halted or slowed the Federal leasing of offshore tracts, and ushered in a wave of concerted effort in opposition to offshore drilling and production.

Nevertheless, exploration is proceeding. The locations in the U.S. where the promise is greatest include Georges Bank on the continental shelf off Cape Cod, the Baltimore Canyon Trough across the shelf off Delaware, the Southeast Georgia Embayment, and Blake Plateau off the southern coast of the Atlantic seaboard. In addition, there is great interest in the 6.5 million acres stretching from the seaward side of the Santa Barbara Channel islands south to the Mexican border. This area lies as far as 40 miles off the California coast.

Of the 100 billion barrels of estimated U.S. offshore resources, only about 40 billion barrels can be economically exploited within the framework of today's oil prices and technology. However, as the technology of deep-sea drilling and production improves, we might expect to exploit a greater percentage of this resource.

In an address to the nation on January 23, 1974, President Nixon directed the Secretary of the Interior to increase the acreage leased on the Outer Continental Shelf (OCS) to 10 million acres, beginning in 1975. This more than tripled the acreage the Department of the Interior had originally planned to lease. The basic objective of the proposed acceleration in OCS development was to increase domestic production as rapidly as possible and reduce dependence on expensive and unstable foreign supplies of oil. However, a number of questions about the feasibility and desirability of the proposal have since been raised by the

Table 7.9
U.S. OFFSHORE OIL AND GAS RESOURCE ESTIMATES

	Offshore oil (Billion barrels)	Offshore gas ($\times 10^{12}$ cu. ft.)
Gulf of Mexico	30	220
Pacific	20	20
Atlantic	20	60
Alaska	30	190
Total	100	490
	(Billions of barrels)	(Trillions of cubic feet)

Congress and representatives of nearly every coastal state. Residents of these coastal states remember the breakup of the tanker *Torrey Canyon* in 1967 off Land's End in England and the Santa Barbara oil well blowout in 1969, two of the worst oil-spill disasters in history. The potential environmental impact of offshore oil drilling is significant. In addition, the states are disputing ownership of these offshore sites, claiming that the states own the mineral rights of their offshore lands. The U.S. government has maintained that offshore lands are federal lands, and the U.S. Supreme Court agreed in 1975. [14] In addition, the U.S. government has supported an internationally recognized 12-mile territorial sea limit and a coastal economic zone extending 200 miles offshore. [15]

There is a serious question, if an additional ten million acres of offshore area were to be leased for oil exploration, whether oil companies could mobilize sufficient drilling rigs and other equipment to successfully exploit these leases. It has been proposed that leasing 5 million acres would provide sufficient exploration opportunities for another five years. Nevertheless, it is estimated that while the OCS share of domestic oil produced today is only 17 percent of the total, this might be increased to 30 percent by 1985. This effort would require extensive capital investment in equipment offshore and at the coast. Also, the cost of producing offshore oil is significantly higher than producing onshore oil.

Environmental Impacts of Offshore Oil Production

The potential environmental impacts of offshore oil production are serious and difficult to evaluate. Escaping oil may contaminate the fishing shoals and seabeds, eventually destroying commercial fishing in the area of the oil leak from an oil well. In addition, oil leaking from an oil well could eventually pollute vast stretches of beaches on the coasts. It is feared that marine algae and sea grasses could be severely damaged while fish, seabirds, and other wildlife would have significant mortality rates as a result of oil spills. One indirect effect of spilled oil is the increased concentration of chemical detergents used to disperse the oil. In addition, an oil spill could cause the recreational and resort sector to experience irreparable economic damage.

An example of the conflict over potential environmental impact on commercial fishing occurred in Alaska in 1975. Oil and gas leases over a 98,000-acre area in the lower Cook Inlet have been sold for $25 million. The focus of the controversy is Kachemak Bay, totaling less than 5000 acres, and acknowledged to be one of the most biologically productive bodies of water in the nation, and perhaps the world. Although relatively small, the bay is among the most important breeding grounds and supports the most productive fisheries in Alaska. The annual first wholesale value of the bay catch exceeds $7 million. The catch includes all five species of salmon, three species of crab, and at least two species of shrimp, as well as herring and halibut. [15]

As noted recently, the environmental costs of offshore oil development are localized in the immediate vicinity of the development in a highly visible manner, while the economic benefits are spread across the entire country. [17] This conflict is clearly seen in the issue of leasing the area off the shore of Southern California, which may hold up to 14 billion barrels of oil and 30 trillion cubic feet of natural gas. The residents oppose the rapid exploitation of these coastal areas because of the impact of the necessary transportation, refining, and distribution facilities that would be built on the coast, as well as the potential for oil spills on the beaches. (*Case in point*: The severe winter of 1976–1977 saw several tanker groundings—and a few near misses. The breaking up of the tanker *Argo Merchant* off Nantucket released about seven million gallons of heavy oil which posed a threat to the George's Bank fishing grounds. Another tanker was lost completely and no one knows what became of her cargo. Chesapeake Bay was the site of another grounding, but that ship was later refloated with little loss of oil into the waters. Cape Cod Canal was closed temporarily after a barge was shattered by ice blocks; residents on the Cape feared damage to the beaches when spring would allow the oil-clogged ice to break up and melt. Public indignation resulted in Congressional studies aimed at producing legislation to impose new and stricter regulation of foreign shipping and Coast Guard inspection to ensure that safety measures were being complied with.)

Most of the nation's experience with offshore oil drilling has been good. For almost a decade, companies have been producing oil from California tidelands extending from north of Santa Barbara to the Oxnard–Ventura area some 30 miles south of Santa Barbara. Production from 610 wells drilled along that narrow strip now totals 56,000 barrels per day. The City of Long Beach, California, has been operating offshore oil wells since 1939, and it has never had a spill or leak of any consequence. This series of wells is considered to have the most advanced environmental controls of all tidelands operations. The measures taken extend not only to the prevention of water pollution but to the prevention of odor, noise, and visual pollution as well.

The reef effect of offshore oil platforms provides one environmental benefit. Another potential benefit may be the development of a platform as a feeding place for marine life.

North Sea Oil

The discovery and exploitation of oil under the North Sea is being pursued by Norway and the United Kingdom. The total resource of oil under the North Sea is estimated at 40 billion barrels. The North Sea holdings of the United Kingdom is estimated to be 20 billion barrels. However, at best, the production of the British offshore oil wells might yield 2.5 million barrels per day by 1980. The exploitation of these resources is particularly important since Britain imported 7 billion dollars worth of crude oil in 1974. However, recovery of North Sea oil requires overcoming severe climatic problems, drilling in deep water, and abating environmental concerns. A blow-out during drilling in the spring of 1977, which posed a threat to the fishing industry and the beaches on the shoreline of the Netherlands, was finally brought under control when a team of experts was brought in from the U.S. to cap the new well.

7.10 OTHER SOURCES

Shale Oil

Oil shale is a sedimentary rock that contains the solid hydrocarbon wax kerogen in tightly packed limey mud and clay. The kerogen may be decomposed at elevated temperatures (450°C), resulting in an oil suitable for refinery processing. The U.S. is estimated to have over 1000 billion barrels of shale oil, and the world resource is over 5000 billion barrels. The primary obstacle to use of this oil shale is finding an economical, environmentally acceptable process for extracting the oil from the shale. Also, the question remains whether a favorable net energy will be obtained on a commercial basis.

Most U.S. shale-oil deposits are within Colorado, Utah, and Wyoming. Several companies are developing pilot projects to explore the oil-shale recovery process. It is projected that oil produced from shale could reach 100,000 barrels per day by the early 1980's. Optimistic estimates of production of oil from shale call for 300,000 barrels per day by 1985. This level might yield two or three percent of the total domestic U.S. production of oil. A successful exploitation of oil shale might yield 10 percent of the U.S. domestic production of oil by the year 2000.

Nevertheless, in spite of thirty years of research and development by the oil industry, it remains marginally economical to produce shale oil on a commercial basis. Two different approaches to getting the oil out of the rocks have been tested on a pilot scale. One approach involves mining the oil-bearing rocks, crushing them, and heating them in a surface retort; and the other involves mining or blasting out an underground cavern and heating the shale in place by burning gases. The former method has been more extensively tested, and is reckoned to be closer to commercial application, but the second method, the *in situ* approach, could have significant advantages if it proves feasible on a large scale.

There are severe environmental problems associated with oil-shale exploitation. With the existing technology, about 3 barrels of water is required in order to yield one barrel of oil. Also, after processing, most companies plan to dispose of waste rock above ground by depositing it in nearby canyons. How readily native grass and shrubs can grow on saline-rich spent shale is an open question. Also, there is a fear of increasing the salinity of the nearby Colorado River. Besides the potential for polluting the Colorado River, an oil-shale industry could pollute the air with dust from mining, crushing, and disposal operations, and with sulfur-dioxide emission from the retorting process. Some local impact on plants would occur, and there is considerable doubt whether shale operations could meet the recent court ruling that air quality not be degraded when it is initially purer than environmental standards. [18]

The *in situ* process of extracting oil from shale is particularly attractive since it eliminates the necessity for disposing of large quantities of spent shale, and may also require substantially less water per barrel of oil produced. However, this process is not fully developed; it is still only in the research stage and has not been proved economically feasible. One proposed *in situ* process consists of two steps: (1) a continuous mining process to convert the shale to rubble, and (2) *in situ* retorting (heating) of the rubble to free the oil. One method for breaking the rock into rubble is to use underground nuclear explosives.

If the above-ground mining and retorting process is used, a significant amount of rock waste is produced, since we might expect to obtain a yield, at best, of 25 gallons of oil per ton of shale. Whether oil can be extracted commercially from shale depends upon overcoming the potential environmental effects in the form of surface land despoilage, pollution of the Colorado River drainage system with alkaline salts and waste-derived muds, and derangement of the regional water supply. It has been estimated that it may cost 80 million dollars to construct a 50,000-barrels-a-day facility incorporating the necessary environmental protection methods. Whether oil shale will be a significant source of oil over the next 30 years is still relatively doubtful.

Tar Sands

The presence of the Athabasca tar sands in Canada has been known for many years. These tar-sand deposits cover an area of 9000 square miles in the province of Alberta. The tar sands are beds, or layers, of a mixture of sand, water, and bitumen. The water and bitumen form a film around each tiny grain of sand. When a handful of sand is compressed, it leaves a discernible oily stain and smell. The deposits lie under the earth's surface and are estimated to contain approximately 600 billion barrels of oil. Perhaps half of this could be eventually recovered. Only 10 percent of the deposits, however, have a thin enough overburden—up to 300 feet—to permit recovery by open-pit mining operations. These near-surface deposits might yield up to 20 billion barrels of oil. To obtain oil from the deeper deposits will require *in situ* processes. Among methods proposed to liquefy the oil are controlled underground fire, steam injection, emulsion injection, and underground atomic explosions. In all these plans, the idea is to heat the reservoir and apply pressure sufficient to cause the heavy oil to migrate to drilled recovery wells.

One commercial plant at the Athabasca region is scheduled for operation in 1978 with a production of 125,000 barrels a day. The plant will use the hot-water separation method. Hot water and steam will be used to separate the sticky oil, called bitumen, from the sand and other solids. The bitumen will be upgraded by a fluidized bed-coking process, and then treated with hydrogen to remove impurities, producing high-grade synthetic crude oil. The plant uses two barrels of tar sand to yield one barrel of oil.

This commercial plant follows a pilot plant built in 1967 that produced 45,000 barrels per day. Other plants are being prepared for the Athabasca area. However, the economics of large-scale plants are difficult to estimate. The plant currently under construction, which will produce 125,000 barrels a day with a 25-year life, may cost more than 1.8 billion dollars to construct, along with the required open-pit mine and associated equipment.

The magnitude of the economic and environmental problems associated with producing oil from tar sands may not yet be fully understood. The complexity of the process, the size of the plant, and the equipment and capital expense have all been underestimated initially. Environmental problems may include the siltation of streams and lakes from the destruction of natural drainage patterns, destruction of salmon-spawning areas, contamination of ground water and the Athabasca River, and air pollution.

U.S. tar-sand deposits are located principally in Utah, and are estimated to contain 20 billion barrels. However, it is difficult to estimate the percentage of these reserves that are economically recoverable while maintaining compatibility with environmental constraints. The characteristics of the Utah deposits are significantly different from the Athabascan deposits, so new technology would need to be developed. The Utah tar sands could not be expected to ultimately produce more than 500,000 barrels per day, and we cannot expect any production before 1985. The value of the Utah deposits remains to be proved.

REFERENCES

1. "Productive capacity grows as world demand falters," *World Oil*, Aug. 15, 1975, pp. 41–44.

2. D. A. RIGASSI, "U.S.S.R. becomes the world's leading oil producer," *World Oil*, Aug. 15, 1975, pp. 121–123.

3. R. GILLETTE, "Oil and gas resources: Academy calls U.S.G.S. math misleading," *Science*, Feb. 28, 1975, pp. 723–727.

4. *Mineral Resources and the Environment*, National Academy of Sciences, Washington, D.C., Feb. 1975.

5. J. D. MOODY and R. E. GEIGER, "Petroleum resources: How much oil and where?," *Technology Review*, March 1975, pp. 38–45.

6. "World crude resource may exceed 1,500 billion barrels," *World Oil*, Sept. 1975, pp. 47–50.

7. "The oil companies did spectacularly well last year," *Forbes*, Jan. 1, 1975, pp. 216–217.

8. "Big oil's barrel of problems," *Newsweek*, August 11, 1975, pp. 63–64.

9. "The new shape of the U.S. oil industry," *Business Week*, Feb. 2, 1974, pp. 50–56.

10. A. SAMPSON, *The Seven Sisters*, Viking Press, New York, 1975.

11. M. A. ADELMAN, *The World Petroleum Market*, John Hopkins Press, Baltimore, Maryland, 1973.

12. "Where oil's new price hurts," *Business Week*, Oct. 13, 1975, pp. 34–35.

13. L. J. CARTER, "Energy: A strategic oil reserve as a hedge against embargos," *Science*, Aug. 1, 1975, pp. 364–366.

14. D. SHAPLEY, "Offshore oil: Supreme Court ruling intensifies debate," *Science*, April 11, 1975, pp. 135–136.

15. D. Shapley, "Law of the sea: Energy, economy spur secret review of U.S. stance," *Science*, Jan. 25, 1974, pp. 290–292.

16. M. Panitch, "Offshore drilling: Fishermen and oilmen clash in Alaska," *Science*, July 18, 1975, pp. 204–206.

17. J. W. Devanney, "Key issues in offshore oil," *Technology Review*, Jan. 1974, pp. 21–25.

18. W. D. Metz, "Oil shale: A huge resource of low-grade fuel," *Science*, June 21, 1974, pp. 1271–1275.

19. "Don't blame the oil companies: Blame the State Department," *Forbes*, April 15, 1976, pp. 69–81.

20. D. A. Rustow and J. F. Mugno, *OPEC: Success and Prospects*, New York University Press, New York, 1976.

21. M. A. Adelman, "How to cope with OPEC," *Fortune*, December, 1976, pp. 189–190.

22. "How OPEC's high prices strangle world growth," *Business Week*, Dec. 20, 1976, pp. 44–50.

EXERCISES

7.1 Examine the potential for increased imports of oil from the U.S.S.R. to the United States. What percentage of the U.S. imports of oil in 1980 will be Russian? Is this dependence upon Russian oil a wise foreign-policy move?

7.2 The average annual discovery of new reserves of oil is about equal to the world's consumption of oil. As oil consumption increases in the world over the next decade, what economic effects will occur as the ratio of proven reserves to consumption diminishes? What price for oil in the export market do you predict for 1980?

7.3 Petroleum production in the world reached 22 billion barrels in 1975. Petroleum production in the world, estimated for the next 100 years, is shown in Fig. 7.3. Draw an estimated curve for a total world resource of 1650 billion barrels.

7.4 Contact the State agency that reviews applications for new oil refineries in your state or a nearby state. Identify the location and the firm that proposes to build or expand a refinery and contact the firm for further information regarding its size, cost, and environmental impact. Determine whether there is any local opposition to this refinery.

*7.5 For some time, the United States Congress has been considering proposals to break up the oil companies. Some of these proposals would limit petroleum companies to one functional stage of the business. Companies would be directed to reorganize so that exploration and production, transportation, refining, and marketing operations were each separately owned and managed. This is called *vertical divestiture*. A different set of proposals would exclude oil companies from

---

* An asterisk indicates a relatively difficult or advanced exercise.

participating in the development of non-oil energy sources and require disposal of investments already made in coal, nuclear, synthetic fuels, and solar energy. This is sometimes called *horizontal divestiture*.

The advocates of divestiture claim that the breakup of big oil companies would increase energy-industry competition, lower prices to consumers, and encourage the development of more domestic supplies. Explore the pros and cons of legislation to break up the large oil companies.

7.6 The construction of a new 250,000-barrels-a-day oil refinery in the U.S. costs from $500 to $750 million. Could small, independent refining companies afford to build such a refinery? How many new refineries are needed in the U.S. over the next decade?

*7.7 Most scientists who study offshore oil-well technology point to downhold safety devices (or storm chokes, as they are commonly called) as being a weak point in production systems. These devices are intended to shut down a well automatically when an accident occurs. Two alternatives for dealing with this problem are to make the storm choke more reliable, or to replace it with another valve.

Examine the literature on the reliability of these valves, and provide your recommendations for improved technology for storm chokes.

7.8 The City of Long Beach, California, built four artificial islands to camouflage the oil drills. Some islands have 45-foot waterfalls and plants. Obtain further information on the extent of the effort to control the environmental effects of the offshore oil wells near Long Beach. (Contact the Department of Oil Properties, Long Beach, California 90801.)

*7.9 The capacity of U.S. crude-oil refineries increased steadily from 6.5 million barrels per day in 1950 to 14 million barrels per day in 1973. U.S. refineries run at about an average of 90% of total capacity. Analyze the need for additional capacity, if any, and the cost of adding an oil refinery (in dollars per thousand barrels of capacity).

7.10 The drilling and equipping of exploratory oil wells consumes about 43% of the capital expenditures of oil companies in the U.S. In light of this high-risk capital expenditure, do you find the oil-depletion allowance or some alternative, to be necessary and useful as an incentive for oil companies to develop new wells?

8

Electricity serves as a carrier of energy to the user. Energy present in a fossil fuel or a nuclear fuel is converted to energy in the form of electricity in order to transport and readily distribute it to customers. By means of transmission lines, electric power is transmitted and distributed to essentially all the residences, industries, and commercial buildings in the U.S.

Electricity is a relatively useful energy carrier since it is readily transported with low attendant losses, and improved methods for safe handling of electricity have been developed over the past 80 years. Furthermore, methods of converting fossil fuels to electric power are well developed, economical, and reliably safe. Means of converting solar and nuclear energy to electric energy are currently in various stages of development or of proven safety. Geothermal energy, tidal energy, and wind energy may also be converted to electric energy. The kinetic energy of falling water may also be readily used to generate hydroelectric power.

8.1 HISTORY OF ELECTRIC POWER

The first generator for producing electric current was built in 1831 by a young Englishman named Michael Faraday. Faraday mounted a 12-inch metal disk on an axle between the poles of a magnet. When, by means of a crank, the disk was rotated through the magnetic field, an electric current was generated and flowed through the connecting wires. A sketch of his dynamo is shown in Fig. 8.1.

A modern generator has two basic parts: the rotor and the stator. The rotor is an electromagnet that spins inside the stator, which is a stationary part that contain many coils of copper wire. When the rotor spins, an electric current is set up in the stator coils. The rotor may be turned by harnessing the power of

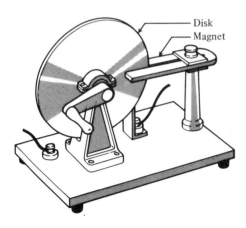

Disk
Magnet

Fig. 8.1 *A sketch of Faraday's dynamo.*

steam or by falling water in a turbine. Of course, the rotor can also be turned by a windmill, or other rotating force.

An electric generator can produce either direct current or alternating current. Modern electric power is usually alternating current (ac); many generators rotate at speeds of as high as 3600 revolutions per minute.

Thomas A. Edison helped perfect the electric generator and put electric power to common use in the late 1800's and early 1900's. Edison invented the stock-ticker in 1869 and the phonograph in 1877, both of which depended solely on the availability of electric power. Other engineers had developed versions of the electric light, but Edison's work was directed toward the construction of small lamps useful in homes and offices. In 1879 Edison developed the incandescent lamp with a filament. Arc lamps had been used in Europe for municipal lighting, and the first municipal lighting system in the U.S. was built in San Francisco in 1879.

Edison saw the implications of parallel wiring of lamps and electric devices, which was essential to the development of the new industry. Edison was the first to have the idea to sell electricity as a carrier of energy. This required distribution networks connected to a central power plant. The first such installation was the Pearl Street Power Station in New York (in 1882), which had a power output of 792 kilowatts.

In addition to its use in lighting, electricity was used for motive power. It was used to power elevators developed by Otis in the 1880's and to drive electric streetcars in the 1890's.

With the invention of the triode vacuum tube in 1907 by Lee Deforest, the U. S. entered the age of radio and eventually of television. In addition, the vacuum tube and its later counterpart, the transistor, became components in the computer, an important element in today's society. All these devices are dependent upon electricity for the delivery of energy for their operation. Thus, one might call the last half of the 20th century the electric half-century.

8.2   CONSUMPTION OF ELECTRICITY IN THE U.S.

The United States consumed about 3.7 billion kilowatt-hours of electric energy in 1900, for a per capita consumption of 49 kilowatt-hours per person per year. By 1974 total consumption had grown to 1862 billion kilowatt-hours, with a per capita consumption of 8612 kilowatt-hours per year. The growth in the consumption of electric energy is shown in Table 8.1. The average annual growth rate averaged 7.5 percent per year over the period 1900 to 1970. This growth rate is expected to decrease over the next 25 years as population levels off, conservation practices are developed, and more efficient use of electricity is achieved. Nevertheless, the consumption of electric energy is expected to grow to 2750 billion kilowatt-hours and 4000 billion kilowatt-hours in 1980 and 1990, respectively. [14] In order to supply this demand, the capacity of electric-power plants will have to be commensurately increased from 475 million kilowatts in 1974 to 600 million kilowatts in 1980 and then to 900 million kilowatts in 1990. This increase in plant capacity can be achieved over the next decade only if nuclear-power plants that are now planned are built and permitted to operate at near full power, or new large coal-fired plants are constructed.

Table 8.1
ELECTRIC ENERGY USE IN THE U.S.

Year	Population (Millions)	Generating capacity (Million kW)	Consumption (Billion kWh)	Consumption per person per year (kWh)*
1900	76		3.7	49
1920	106		57.5	540
1930	123		115.0	935
1940	132		141.8	1,074
1947	144	52.3	255.7	1,774
1950	152	68.9	329.1	2,161
1955	166	114.5	547.0	3,297
1960	180.7	168.0	753.4	4,169
1965	184.2	236.1	1,055.0	5,433
1970	204.8	340.4	1,529.8	7,469
1971	207.0	367.5	1,614	7,800
1974	213.0	475	1,862	8,741
1975	216.0	500	2,037	9,430
1980 (Projected)	229	600	2,750	12,009
1985 (Projected)	236	750	3,500	14,830
1990 (Projected)	242	900	4,000	16,529

* Note that 1 kWh = 3413 Btu = 3.6×10^6 joules.

In order to add 440 million kilowatts of new capacity at an average cost of $500 per kilowatt, approximately 220 billion dollars of investment capital would be required over a period of less than 15 years. Whether this can be achieved will depend on the availability of capital, interest rates, and other economic factors.

8.3 CONSUMPTION OF ELECTRICITY IN THE WORLD

World electric-power consumption has grown at a rate of 7 percent a year over the past 50 years, as shown in Fig. 8.2. Per capita consumption in other indus-trialized nations in the world (as shown in Table 8.2) is about one-half of the per capita consumption in the United States. However, per capita consumption in European nations and Japan has grown more rapidly than in the U.S. in recent

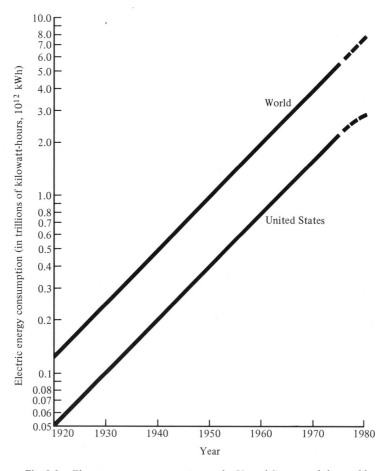

Fig. 8.2   *Electric energy consumption in the United States and the world.*

Table 8.2

ELECTRICAL ENERGY CONSUMPTION PER CAPITA
(IN KILOWATT-HOURS PER ANNUM PER CAPITA)

	U.S.A.	U.K.	Japan	Hungary	Netherlands	Yugoslavia	Federal Republic of Germany
1930	935	280	230	100	210	60	
1940	1074	620	470	200	420	50	
1950	2160	1120	550	320	730	150	860
1960	4169	2630	1270	760	1440	490	2220
1970	7500	4440	3100	1460	3300	1220	4000

years, and these nations may approach the per capita consumption of the U.S. by the end of this century. As the less developed nations build electric-power plants, the total energy used for electricity will grow rapidly. Many of these nations are planning to use hydroelectric power plants and nuclear power plants to produce their electricity.

In Table 8.3, the consumption of electricity in several nations of the world is shown for 1971. It is clear that many heavily populated, less developed nations, such as Brazil or India, are currently using small amounts of electricity; their consumption could grow rapidly if an inexpensive fuel source became available.

Table 8.3

ELECTRICITY CONSUMPTION IN
SEVERAL NATIONS IN 1971 IN
BILLIONS OF KILOWATT-HOURS

	Consumption (10^9 kilowatt-hours)
U.S.A.	1614
U.S.S.R.	800
Japan	379
West Germany	260
Great Britain	250
Canada	215
France	150
Yugoslavia	29
India	59
Brazil	48
Mexico	31

The electrical energy consumed per capita is related to the affluence of a nation. In Fig. 8.3, the electric energy consumption per capita is shown versus the gross national product per capita for several nations for 1968. As a nation

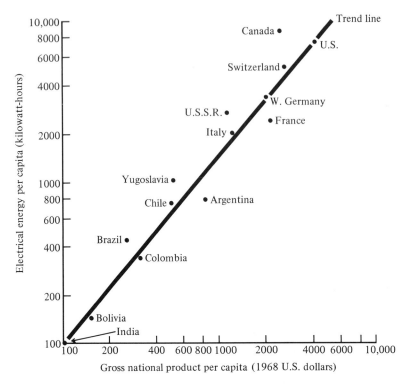

Fig. 8.3 *The electrical energy consumption per capita versus the Gross National Product per capita for several nations.* [*Note that both scales are logarithmic.*]

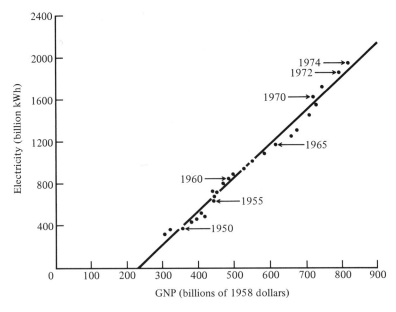

Fig. 8.4 *The growth of electricity consumption and the Gross National Product in the United States.*

increases its GNP per capita, it also increases its per capita use of electricity, following a trend shown in Fig. 8.3. This growth in the use of electric energy in the U.S. versus the growth of Gross National Product is illustrated in Fig. 8.4 for the period 1948 to 1974. As affluence and wealth increases, nations tend to shift to the use of electricity, which delivers energy to the home and office in a convenient, safe, and clean form.

Table 8.4
INDEX OF ENERGY USE AND ELECTRICITY
USE FOR SELECTED NATIONS IN 1975

Country	Index of energy use [Energy/(GNP in dollars)]	The percent of total energy used to generate electricity
U.S.A.	2.7	26
U.S.S.R.	3.2	24
Canada	2.8	28
Japan	1.5	38
France	1.2	24
Sweden	1.6	42
West Germany	1.6	28
United Kingdom	2.1	28
Mexico	1.8	16
Israel	1.0	34
World	2.4	22

Another informative measure of electric energy consumption is the percent of total energy used in a nation to generate electricity, as shown in Table 8.4. Approximately 26 percent of the total energy is used in the U.S. to generate electricity. Note that Sweden uses 42 percent of its total energy for electricity, while the world's average is 22 percent. The growth in percent of total energy converted to electricity in the U.S. is shown in Fig. 8.5. The installed electric capacity in eleven nations of the world in 1968 is also provided in Table 8.5.

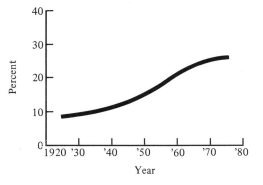

Fig. 8.5 *Percent of total energy converted to electricity in the U.S.*

Table 8.5

INSTALLED ELECTRIC CAPACITY
(IN MEGAWATTS) IN 1968

1.	United States	310,125
2.	U.S.S.R.	142,504
3.	United Kingdom	59,628
4.	Japan	52,650
5.	West Germany	47,054
6.	Canada	35,993
7.	France	34,133
8.	Italy	30,264
9.	Spain	14,910
10.	India	14,314
11.	Sweden	13,731

8.4 STRUCTURE OF THE ELECTRIC POWER INDUSTRY IN THE U.S.

There are hundreds of privately and publicly owned electric power companies, often called utility companies, in the U.S. The public companies may operate as Federal, municipal, or regional utilities. All utilities are subject to regulation by the Federal Power Commission, which oversees performance to customers and regulates prices to consumers. Private companies are regulated to permit a reasonable rate of return on capital investment, usually about 8 percent. The structure of the electric utility industry is summarized in Table 8.6. While investor-owned companies dominate the industry, several large public utilities contribute heavily to the statistical total for power generation. [2]

For example, the Tennessee Valley Authority (TVA) is a Federally operated utility, which had a gross production of over 100 billion kilowatt-hours in 1972. This production amounted to about six percent of the total production of electricity in the U.S. TVA has 48 hydroelectric plants, 16 coal-fired plants, and a total installed capacity of 17,700 megawatts. By contrast, Pacific Gas and Electric Company of Northern California produced 70 billion kilowatt-hours of electric energy in 1972.

The classification of customers in 1973 in the U.S. is shown in Table 8.7. The great majority of customers are residential users (83 percent of the total), but the

Table 8.6

STRUCTURE OF THE ELECTRIC UTILITY INDUSTRY
IN THE U.S. IN 1970

Type of system	Number of systems	Number of plants
Investor owned	250	1923
Federal	2	⎫
Public nonfederal	700	⎬ 800
Cooperative	65	⎭
Total	1017	2723

Table 8.7
CLASSIFICATION OF CUSTOMERS IN 1973 IN THE U.S.

Type	Number of customers (Millions)	Energy consumption (Billions of kWh)	Percent of total
Residential	67.84	608.4	35.5%
Commercial and industrial	8.62	1,041.4	60.6%
Street and highway lighting	0.11	14.3	0.8%
Other public authorities	0.15	43.7	2.5%
Railroads and railways		4.7	0.3%
Interdepartmental	4.77	5.0	0.3%
		1,717.8	100.0%

largest portion (60.6 percent) of electric energy is consumed by the commercial and industrial sector. Electricity is heavily used to heat, cool, and light commercial buildings. It is also consumed in large amounts for industrial processes such as for the production of aluminum and other metallurgical processes.

The end uses for electricity are summarized in Table 8.8. The greatest use of electricity is for industrial drives (such as motors in steel mills and on assembly lines). Refrigeration is the next largest use, and includes home refrigeration as well

Table 8.8
END USES FOR ELECTRICITY

End use	Percent of total electricity use
Industrial drives	39.7
Refrigeration	11.6
Lighting	10.8
Water heating	6.2
Electrolytic processes	5.8
Air conditioning	5.6
Space heating	3.3
Direct heat	2.6
Television	2.6
Cooking	2.1
Clothes drying	1.0
Other (includes transportation, computers, etc.)	8.7
Total	100.0

as refrigeration in food-processing and distributing plants. Lighting in commercial and residential locations is the third largest user of electricity. It is clear that the development of more efficient motors, refrigerators, lights, and heaters could conserve a significant amount of electric energy.

The residential use of electricity in the U.S. for several years is recorded in Table 8.9. The number of customers has grown significantly over the past ten years, and the use per customer has also grown. A customer is defined as a single meter connection to the distribution system; several persons will normally live in each household.

Table 8.9
RESIDENTIAL USE OF ELECTRICITY IN THE U.S.

Year	Customers (Millions)	Use per customer (kWh)	Revenue per kWh (cents)	Average annual bill (Dollars)*	Residential revenue (Million dollars)
1964	56.3	4,703	2.31	$108.64	6,041
1970	64.0	7,066	2.10	$148.39	9,416
1974	71.0	7,907	2.83	$223.77	15,703
1980 (Estimated)	80.3	9,682	4.14	$401.32	31,917

* All dollar figures in constant 1975 dollars.

The use per customer may be approaching a level of saturation, since over 95 percent of homes in the U.S. now have a refrigerator and a television set, and electric lighting is used in essentially all homes. Potential growth in the use of electricity in the home could occur with increased use of air conditioning, which now is used by 30 percent of the households, and with the use of dishwashers,

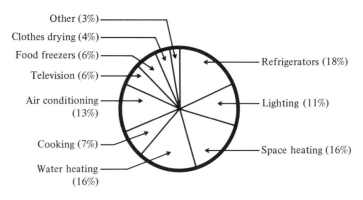

Fig. 8.6 *The uses of electricity in an average household in 1970 in the U.S., as a percentage of the total energy use. In 1970, 7000 kilowatt-hours were consumed by the average household.*

currently in use in 25 percent of the households. Figure 8.6 shows the use of electricity in an average home.

During the period 1950 to 1970, the price of electricity, expressed in constant dollars, declined steadily from an average of 2.8¢ per kWhr in 1950 to 2.10¢ per kWhr in 1970. As a result of the 1973 increase in oil prices, the cost of electricity has started to rise and the increased price of electricity will begin to cause a decline in the growth of the use of electricity in residences.

Electric Power Plants

Power plants for producing electricity may use the energy of falling water, geothermal steam, a steam produced from nuclear reactors, or fossil-fuel boilers. Fossil fuels provide the largest part of the energy used to generate electricity, about 80 percent today. The types of power plants and the percent of electricity generated by each type is given in Table 8.10. Since fossil-fuel boilers convert only about one-third of the input energy to output electricity, the energy attributable to fossil-fuel plants is relatively high. Large turbine generators at the Moss Landing plant in California are shown in Fig. 8.7.

Table 8.11 lists the fossil-fuel inputs to produce electric power for 1971. Since then coal has increased its share to close to 58 percent, while natural gas has decreased its share to approximately 20 percent because of the limited availability of natural gas.

Geothermal electric-power plants in the U.S. today contribute a capacity of 516 megawatts, while hydroelectric power plants contribute 60,000 megawatts capacity. It is anticipated that hydroelectric and geothermal power plants will grow moderately in capacity over the next decade, but that these sources will continue to provide only about 16 to 18 percent of our total electric energy over the next 20 years.

It is planned that an increased use of coal, as shown in Fig. 8.8, will enable the use of electricity to continue to grow over the next decade. Also, planned additions of nuclear-reactor power plants will provide a significant portion of the increase in electric-power production. It is planned to add 175 megawatts of nuclear capacity over the next decade, while adding only 129 megawatts of fossil-fuel generating plants. [3]

Table 8.10
ELECTRICITY GENERATION IN 1968

Type of power plant	Electricity generated (10^9 kWh)	% of total	Attributable energy (10^9 kWh)
Fossil fuel	1067	80	3384
Nuclear	38	3	38
Hydroelectric	222	17	222
Total	1327	100	3644

Fig. 8.7 *Two of the seven turbine generators at Moss Landing on Monterey Bay, California. The big pipes carry steam from the boiler to the turbine. Each generator shown has a capacity of 120,000 kilowatts. Two newer generators produce 750,000 kilowatts each. Moss Landing is the second largest power plant in the world. (Photo courtesy of Pacific Gas and Electric Company.)*

Table 8.11
FOSSIL-FUEL INPUTS TO ELECTRICAL POWER

	1971 actual	
Type of fossil fuel	*(10^{15} Btu)*	*% of total*
Coal	7.7	55
Petroleum	2.4	22
Natural gas	4.1	23
Total	14.2	100

The installed generating capability in the U.S. is increased each year in order to meet increasing demand for electricity. Table 8.12 summarizes the net additions for the different types of power plants since 1964, and provides projections for 1978 and 1985. By 1978, it is expected that the net additions of nuclear-power plant capacity will be 36 percent of the total addition in that year, and that the

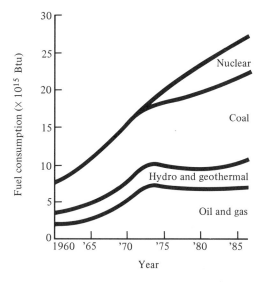

Fig. 8.8 *The actual and predicted energy sources used for the generation of electricity in the U.S.*

Table 8.12

GENERATING CAPABILITY NET ADDITIONS IN THE U.S.
(IN MEGAWATTS) FOR SEVERAL SELECTED YEARS

Year	Conventional hydroelectric	Pumped storage hydro	Fossil-fuel steam	Nuclear steam	Combustion turbine	Total
1964	1,974		9,418	45	299	11,736
1970	1,789	313	16,800	2,513	6,126	27,741
1974	720	1,087	18,874	9,196	6,236	36,113
1978 (Projected)	899	2,031	13,855	10,140	1,006	27,931
1985 (Projected)	0	2,500	12,000	16,337	3,500	34,337

percentage will rise to 48 percent by 1985. As we will note later (in Chapter 12 on nuclear power), this planned increase in nuclear-power generating capacity may not actually be achieved because environmental and safety considerations imposed on nuclear-power facilities may delay the installation and operation of these nuclear plants.

In 1930 the largest single steam unit for generating electricity was 200 megawatts, and the average size was 20 megawatts. By 1970, the largest unit in service was 1150 megawatts, and the average size of all units under construction was about 450 megawatts. Capital costs per kilowatt, and operation and maintenance costs per unit of energy generated, are less for large units than for smaller ones. The maximum size of conventional 3600-RPM, single-shaft generating units expected to be in service by 1980 is approximately 1500 megawatts. It is not

Table 8.13

ANNUAL CAPITAL EXPENDITURES IN THE USA
(IN MILLIONS OF 1975 DOLLARS)

Year	Generation	Transmission	Distribution	Other	Total
1964	1,819	1,056	1,800	267	4,940
1970	6,860	2,123	3,264	529	12,776
1974	12,504	2,450	4,576	1,024	20,555
1980	10,410	2,064	5,570	902	18,946
(Projected)					

expected that the size of high-pressure, high-temperature, single-shaft turbine generators will exceed 2000 megawatts by 1990. As we can determine from Table 8.13, over 12 billion dollars was expended in 1974 to increase the generating capacity of the U.S. electric utilities.

In 1948 there were only two steam electric-generating plants in the U.S. with capacities over 500 megawatts. In 1970, the largest plant was TVA's three-unit, 2558-megawatt Paradise plant. The largest current plant is the four-unit, 3200-megawatt Monroe plant of Detroit Edison Company. It is expected that plants having 5000 megawatts of capacity will be in service by 1980.

All of the high-pressure, high-temperature, fossil-fueled, steam electric-generating units, 500 megawatts and larger, have been designed to serve as "base load" units and built for continuous operation at or near full load. This is also true of large nuclear-reactor units. In order to provide storage of energy generated at off-peak hours, pumped hydroelectric schemes have been developed. The 8827 megawatts of pumped-storage capacity installed at the end of 1974, however, represented only 2 percent of the total electric generating capacity. Plans through 1980 call for another 8900 megawatts to be installed for storage of off-peak power. Electricity generated from fossil-fuel or nuclear plants at off-peak hours in the night is used to pump water up to a reservoir, where it is later allowed to fall through turbines to generate electricity at times of peak need. A pumped-storage system at Cornwall, New York, now in the planning stage, would yield 2000 mega-watts when releasing water from its reservoir. [4] This project involves the use of a reservoir 1000 feet above the Hudson River and a tunnel from the river to the reservoir for pumping the water up at night and releasing it down during times of peak demand.

Gas turbines are also used for times of peak power demand. It was not until the 1960's that gas-turbine units were first used for this purpose. In 1966, 2500 megawatts of capacity of gas-turbine units were added to utilities, while in 1971 over 8000 megawatts were added. The advantage of a gas-turbine unit is that it is a low-capital-cost unit. However, as the price of natural gas rises, this advantage may be lost.

In 1903, the first turbine using superheated steam at 513°F at 200-psi pressure achieved an efficiency of about 10 percent. Today steam pressures up to 3500 psi and temperatures of 1000°F are commonplace. It is possible for modern turbines for generating electricity to achieve an efficiency close to 40 percent. Figure 8.9 shows the increase in overall efficiency in converting fossil fuel to electricity for the period 1925 to 1975 in the U.S.

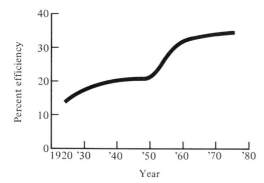

Fig. 8.9 *Overall efficiency of converting fossil fuel to electric energy for 1925 to 1975 in the U.S.*

A superconducting electric generator is under development that operates with its electromagnetic rotor at $-452°F$. This prototype machine is rated at 5 megawatts and uses superconducting niobium–titanium alloy wire wound into the electromagnet. Superconducting generators will be smaller, more efficient, and less expensive to build. [5]

Electric-Power Transmission

Electric-power plants are not often located at the point of end use; they are generally situated away from the metropolitan areas at dam sites or near sources of cooling water. The electric power generated is then transmitted to the user by means of transmission lines. There are energy losses at each stage of energy movement (which are shown in Fig. 8.10 for a coal-fired system). The average

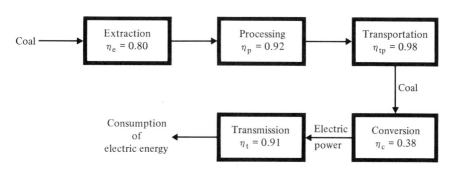

Fig. 8.10 *The movement of energy through the various stages of a coal-fired electric power system.*

efficiency of each step is shown, and the overall efficiency is then:

$$\text{Overall } \eta = \eta_e\eta_p\eta_c\eta_{tp}\eta_t = 0.25. \tag{8.1}$$

The efficiency of power transmission is normally about 90 percent; in other words, about 10 percent of the power transmitted is lost in that process.

Transmission has several functions in a power system. Conceptually, these may be classified as (1) energy transportation, (2) system integration, and (3) interconnection. Energy transportation, as a transmission function, has as its purpose the movement of electric energy from generating sources (i.e., power plants) to major load centers for distribution to the ultimate consumer. System integration as a transmission function allows a combination of generating sources and large distribution centers to operate as a synchronous whole in the most reliable and economical manner. The interconnection function of transmission allows separate systems to interchange power with each other in order to enhance reliability of service and to secure additional economies. It is through interconnection that the individual power systems of the United States have grown into vast networks.

The entire northeastern U.S. is tied together in a network. On November 9, 1965, at 5:16 P.M., in a power station in Ontario, Canada, a protective relay opened, triggering a complex sequence of overloads and line disruptions. Within twelve minutes, thirty million people in the U.S. and Canada had lost electric power. The blackout lasted from a few minutes to over 13 hours in several locations. New York City experienced great difficulties with subways, elevators, and trains stalled at the rush hour.

The large Pacific Intertie network and California's interconnected system survived a severe test of their reliability on March 21, 1975, when more than three million kilowatts of power feeding into California from the Pacific Northwest was cut off at 12:55 A.M. This represented a third of the power being used at the time; the cutoff was due to an equipment failure. However, in this case the network shifted the load to an alternate transmission line and most outages were restored within five minutes. Of course, this occurrence after midnight had small effect compared to the rush-hour failure in New York ten years earlier.

The main loss in above-ground power lines is the loss due to the resistance of the wires, which results in a heat loss. The power loss is

$$P_{\text{loss}} = I^2 R_{\text{line}}, \tag{8.2}$$

where R_{line} is the line resistance. The power transmitted is

$$P_{\text{trans}} = IV. \tag{8.3}$$

Solving for I in Eq. (8.3) and substituting in (8.2), we have:

$$P_{\text{loss}} = \frac{(P_{\text{trans}})^2 R_{\text{line}}}{V^2}. \tag{8.4}$$

It is evident that, in order for a given amount of power to be transmitted, the power loss varies inversely with the square of the voltage and directly with the line resistance. The resistance of the lines can be reduced by using low-resistivity materials and making the lines large in diameter, since

$$R_{\text{line}} = \frac{\rho l}{A},$$ (8.5)

where l is the length, A is the cross-sectional area, and ρ is the resistivity.

The common method of reducing transmission power loss is to use high-voltage lines. Voltages used range from 100,000 volts up to 750,000 volts.

The ease with which voltage can be stepped up or down by using transformers is the main reason that alternating current is in common use. The large voltages used for transmission are then stepped down, prior to distribution within a city.

There are several advantages to using direct current for very-high-voltage transmission lines, which largely compensate for the difficulty in changing the voltage in the case of dc. Only two wires are needed, instead of the three needed for three-phase ac, thus realizing a savings in wire cost. Direct-current lines are a little easier to insulate, since the peak and the rms voltages are the same. In ac lines the peak voltage is about 40 percent higher than the rms voltage; one must insulate for the peak voltage.

The disadvantage of dc transmission is that expensive terminal facilities are needed for converting ac from the generators into dc for transmission, and then converting dc to ac at the end-use location. Consequently, dc lines are commonly used only for long-distance transmission. The Pacific Intertie running from the Dalles Dam on the Columbia River to the Sylmar station in Los Angeles is about 1300 kilometers long and carries 800,000 volts dc. Line losses are about 30 percent greater for ac transmission lines than for dc lines for the same power transmitted. [6]

Large ac power lines are commonly used in the U.S. The Bonneville Power Administration has under construction a 500,000-volt, double-circuit line with an ultimate capacity of 5000 megawatts, to transport power from the Grand Coulee Dam in eastern Washington to Seattle. [7] One complete three-phase (3-wire) circuit is built on each side of the 200-foot tower. This tower system uses a 135-foot-wide right of way and strings 782 towers over the total distance of 174 miles. The total cost of the project is $72 million.

Typical ac transmission lines use 230 kV and 345 kV. However, many long-distance lines use 500 kV and 765 kV. Studies are underway to investigate the characteristics of 1100-kilovolt ac transmission. A 500-kilovolt transmission line system is shown in Fig. 8.11.

New technologies for underground power transmission are being studied. Underground transmission cables insulated with compressed gas, cables cooled to the temperatures of liquid nitrogen (called cryoresistive transmission lines), and cables cooled even more so that they become superconducting are all being developed as options to the conventional underground power lines. The conventional cables are insulated with oil and paper, and operate at ambient temperatures. Only about 30 percent of the transmission of power over 100,000 volts is currently accomplished by underground lines. Distribution lines operate at

Fig. 8.11   This tall tower supports wires carrying 500,000 volts of alternating current power. (Courtesy of Pacific Gas and Electric Company.)

lower voltages, require much simpler technologies, and are being routinely installed underground in many cities.

Cables cooled to low temperatures can carry more power with less loss, but the requisite refrigeration also uses energy, thus balancing the reduced losses in the line. Cooling aluminum cable with liquid nitrogen (at 77 K) reduces the resistivity by a factor of ten, while the use of liquid hydrogen (at 20 K) reduces the resistivity by a factor of 500 from its value at ambient temperature. Unfortunately, there is a trade-off between the increase in conductivity as temperature is reduced and the decrease in efficiency of refrigeration as lower and lower cryogenic conditions are sought. In fact, the resistivity advantage at liquid-nitrogen temperature is just about cancelled out by the refrigeration penalty incurred. However, at liquid-hydrogen temperature, the increase in conductivity is about 10 times the additional refrigeration premium.

8.5   THE ECONOMICS OF ELECTRIC POWER

The cost of electric power decreased from 1931 to 1971 while the cost of living increased, as shown in Fig. 8.12. However, the price of electricity has risen over the past several years due to increased fuel costs. Furthermore, in recent years,

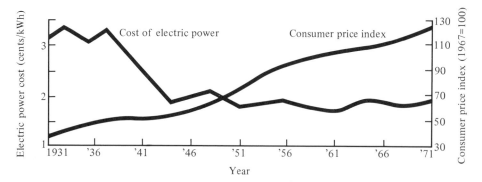

Fig. 8.12 *The average cost of electric power and the cost of living during the period 1931 to 1971.*

because of increased interest rates, inflation, and the economic recession, utilities have experienced difficulties in obtaining additional funds for financing capital construction of new plants. The eventual result of these trends may be that the demand for electricity will exceed the supply in the 1980's, because the new plants we need will not be available. Alternatively, the demand for electricity may level off, so that the supply is in fact adequate. Electric power companies must, however, initiate construction of new plants about seven years ahead of the anticipated demand.

The average investment cost of a fossil-fuel steam electric plant is about $400 per kilowatt of capacity. While fuel costs are increasing, utilities are attempting to build larger, more efficient plants. Furthermore, increased automation of plant operations is expected to permit some reduction in the number of plant employees. A typical 1000-MW coal-fired facility might cost upwards of 400 million dollars and take six to eight years to construct. [8]

All of the high-pressure, high-temperature fossil-fueled steam electric-generating units of 500 megawatts and larger have been designed to serve as base load units operating continuously. The 1625-megawatt Haynes plant of the Los Angeles Department of Water and Power is shown in Fig. 8.13.

Over $150 billion was the total investment in electric utilities by the end of 1975. With the increased cost of production, fuel, and transmission, it is expected that rates will continue to rise over the next decade. Due to increased costs of investment capital, over 170,000 out of the 360,000 megawatts that utilities are planning to build have been cancelled or delayed. [8] If utilities need to meet increased demand, they may use gas turbines, which currently take only about two years to install, at a cost of $200 per kilowatt. But the turbines require natural gas, which is difficult to obtain and is increasing in price. In effect, the utilities would be substituting higher operating costs for lower construction costs.

The comparative operating costs of electric power plants in 1973 are shown in Table 8.14. With a load factor greater than 0.55, the nuclear power plant yields

Fig. 8.13 *The Haynes steam plant uses gas and oil for fuel. The 1625 megawatt plant contains six generators and was completed in 1967. Pacific Ocean water is used for condenser cooling. (Courtesy of Los Angeles Dept. of Water and Power.)*

Table 8.14

COST COMPARISONS OF ALTERNATIVE POWER-GENERATING
PLANTS (1973 DOLLARS) (MILLS* PER KWH AT GENERATING PLANT)

Item	Type of power-generating plant			
	Nuclear	*Coal*	*Oil*	*Gas turbine*
Fuel cost	2.50	5.56	15.83	23.54
Operating and maintenance expenses*	1.50	2.75	2.65	2.50
Fixed charges Load factor = (0.7) Load factor = (0.4)	14.84 25.97	11.48 20.10	10.45 18.30	5.12 8.96
Total costs Load factor = (0.7) Load factor = (0.4)	18.84 29.97	19.79 28.41	28.93 36.78	29.88 32.76

* One mill = 0.1¢ = 10^{-3} dollar.

the lowest cost. However, several factors have changed these figures upward in the past several years.

Electric Load Variations

A key to the problem of the economics of the electric power utilities is the variation in the demand for electric power during each day, each week, and throughout the year. Demand fluctuations result in cyclic peak-load curves for utilities. For example, a peak demand occurs at 4 P.M. on a weekday in the summer and at 6 P.M. on a winter weekday, while the demand will be different on a weekend. A typical demand variation for the weekly cycle and the daily cycle is shown in Fig. 8.14. Depending upon the part of the country, the peak demand may occur in the summer in the South Central States due to air-conditioning loads, while the peak demand in the Northeastern U.S. occurs in the winter. The ratio of the high-to-low-power demand may be as high as 2 to 1. Unfortunately, plant capacity must be available for the peak demand and may, on the other hand, stand idle for much of the time. The ratio of the average kilowatt load over an interval, to

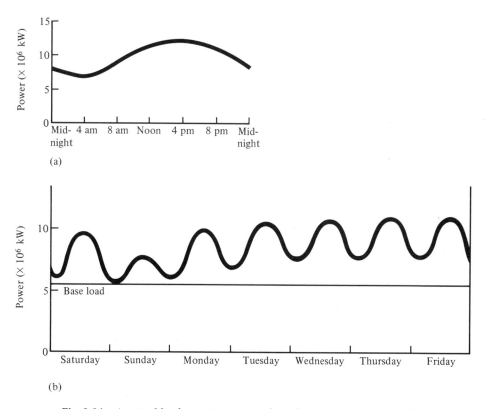

Fig. 8.14 *A typical load-variation pattern for a day (a) and for a week (b).*

the peak load over that period, is called the *load factor*. The average load factor for a utility ranges from 50 to 60 percent.

One possibility is to manage the load so as to reduce the peaks and increase the low points, so that the load factor is closer to 70 or 80 percent. One powerful method of evening out the power-demand cycle is the use of variable rate structures for the electricity purchased.

Electricity Rate Structures

The shifting of load demands to times of low use of electric generating capability would decrease the need for new power plants and increase the economic use of the existing plants by increasing their load factor.

The unit rates paid by industrial consumers have been lower than those paid by commercial and residential customers in the past because of lower distribution costs. Within each customer classification, the price structure has followed a decreasing block structure. For example, on a monthly charge, the total charge may be based as follows:

a) 10¢/kWh for the first 100 kWh,
b) 5¢/kWh for the next 100 kWh,
c) 3¢/kWh for the next 300 kWh,
d) 2¢/kWh for all use beyond (c).

This type of structure tends to encourage users to increase consumption of electrical energy. The Wisconsin Public Service Commission recently stated that the system of reducing unit charges for electricity for bulk users should be modified in favor of "flat" rates, except in cases where the declining rate can be proved to encourage the most efficient allocation of energy. It also ordered utilities to inaugurate a system of peak-load pricing, with higher rates set for summer months when air conditioning puts the greatest stress on the system. In addition to calling for a winter–summer price differential, the commission directed that different day- and nighttime rates be implemented for large industrial users.

The Vermont Public Service Board has recently authorized a rate structure based on the time of day that the electricity is consumed and on the period of the year. The 24-hour residential charge for the Central Vermont Public Service Corporation is as follows: [9]

	Peak season (Dec. 1–Apr. 30)	Off season (May 1–Nov. 30)
Monthly customer charge	$7.00	$7.00
Capacity charge:		
Peak hours	10.32¢/kWh	1.65¢/kWh
(8 A.M.–11 A.M.)		
(5 P.M.–9 P.M.)		
Off peak hours	1.65¢/kWh	1.65¢/kWh

Note that there is no block discount for using more electricity but, instead, a discount for using off-peak power. Customers can switch their washing and drying, ironing, and dishwashing to off-peak periods easily to obtain the reduced rates.

There are numerous benefits of utilizing time-of-day pricing including (1) cost minimization for the utility, (2) equity and fairness in the tariff structure, (3) load factor management, (4) conservation, and (5) consumer savings. [10] Time-of-day pricing of electricity has been practiced in France and Britain for almost 20 years and it has been estimated that each country has realized a savings in annual cost of a quarter of a billion dollars. Since the U.S. has about ten times the installed capacity, a savings of 2.5 billion dollars per year might be achieved.

A basic requirement for load management is the existence of a deferrable load such as an electric water heater. In Europe, where load-management techniques have been used for years, utilities and government agencies have encouraged the development of another load segment with a large energy storage capacity— electric storage heaters. These heaters consist of an attractively packaged and well-insulated pile of magnesite bricks, which are heated electrically to a temperature up to 700°C during off-peak hours. Electric storage units for individual rooms include a fan and an air-mixing device to keep the air emerging from the unit at a roughly constant temperature as the central core cools down during the day.

Since special meters recording the time of use are required for the time-of-day pricing scheme, this method will probably be applied at first with commercial and industrial users. However, the cost of a meter, about $75, can be quickly recovered by a residential customer who now pays $75 to $100 per month to the utility and receives no reduction in rates for off-peak use.

As rates for electricity are developed to put into effect higher charges for peak usage and encourage off-peak energy consumption, many utilities and their large industrial users will tend towards higher load factors and significant savings in capital construction and energy conservation. [17]

8.6   ENVIRONMENTAL CONSIDERATIONS

The use of fossil fuels or nuclear reactors to generate electric power results in a myriad of environmental considerations. Coal mining significantly affects the environment, as do the emissions of fossil-fuel electric power plants. The potential effects of a nuclear power plant are significant and will be thoroughly discussed in Chapter 12. The control of sulfur and particle emissions from a power plant is a costly and only partially successful venture.

The environmental electric power issue is perhaps exemplified by the proposed Kaiparowits project north of the Colorado River in Southern Utah. It is feared that the 3000-megawatt plant will foul the air and impact nearby Bryce Canyon and Zion National Park. The plant would use 12 million tons of coal per year and generate 12.2 million tons of soot, 34.3 million tons of sulfur dioxide, and 250 million tons of nitrogen oxide. The power would be delivered to Los Angeles and Phoenix, and many new jobs would become available for residents of the area. Nevertheless, it is far from settled whether this project will ever meet the needs and desires of the ultimate power consumer as well as of the environmentalists. [12] The project was temporarily suspended in 1976 due to environmental concerns.

The environmental impact of the use of large amounts of water to cool electric power plants has been studied intensively over the past decade. The effects of the use of cooling towers, cooling ponds, and spray ponds, as well as of water from a river or lake, have been investigated over the past several years. Thermal power plants require substantial amounts of water for cooling purposes. While some environmental effects result from warming the cooling water, the principal one is the evaporation of about five pounds of water for each kWh produced. This consumption is about the same whether the cooling is accom-

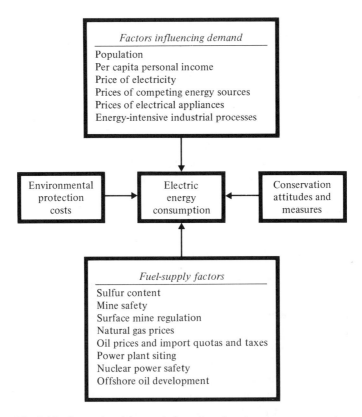

Fig. 8.15 *Interrelated factors influencing electric energy consumption.*

plished by once-through, cooling towers, ponds, or other methods. It would not, of course, apply for air cooling. The power production considered feasible in 1985 would result in the consumption of about 7 million acre-feet per year of water, or about 27 gallons per day per person.

8.7 SUMMARY AND THE FUTURE

The electric energy consumption of the U.S. and the world is continuing to grow and is influenced by a myriad of interrelated factors, as illustrated in Fig. 8.15. As all these factors change over the next decade, we can expect that electric energy consumption may level off in the developed nations. Conservation attitudes and measures will continue to influence our use of electricity and the availability of nuclear power will significantly influence whether we can continue to expand our use of electricity.

The characteristics of the three primary means of generating electric power are summarized in Table 8.15. The U.S. is currently basing its projections of future additional sources of electric power on the availability of proven, safe, nuclear power. Whether this will be achieved by 1985 is not now known.

Table 8.15
CHARACTERISTICS OF ELECTRICAL ENERGY SOURCES

   I. Conventional fossil-fuel steam electric
      1. Air and thermal pollution (60% heat rejection)
      2. Best location near fuel source and cooling water
      3. Large world reserves available
      4. Technology well developed and proven
      5. Relatively high operating cost

  II. Hydroelectric power plants
      1. Clean, nonpolluting, and no heat rejection
      2. Environmental impact is dam and reservoir
      3. Only available where suitable sites exist
      4. Largely already exploited in industrialized countries
      5. Technology well developed and proven
      6. Very small operating cost, while high capital cost per kW
      7. 50- to 100-year economic life

 III. Nuclear fission reactors
      1. Thermal pollution (65% heat rejection)
      2. High capital costs (higher than fossil plants and lower than hydro)
      3. Technology available
      4. Safety and environmental effects not proven

Other means of generating electricity from solar power, wind power, or the oceans may be available by 1990 or later. However, these means are probably long-term solutions to the problem of achieving continuous, pollution-free, inexpensive electric power.

REFERENCES

1. "KWh sales recovering," 26th Annual Electrical Industry Forecast, *Electrical World*, Sept. 15, 1975, pp. 41–50.

2. P. Sporn, *The Social Organization of Electric Power Supply in Modern Societies*, M.I.T. Press, Cambridge, Mass., 1971.

3. R. C. Rittenhouse, "Power generation growth patterns," *Power Engineering*, April 1975, pp. 42–44.

4. L. J. Carter, "Con Edison: Endless Storm King dispute adds to its troubles," *Science*, June 28, 1974, pp. 1353–1358.

5. J. Papamarcos, "Cryogenics in generation," *Power Engineering*, October 1975, pp. 30–37.

6. M. Klein, *et al.*, "HVDC to illuminate darkest Africa," *IEEE Spectrum*, Oct. 1974, pp. 57–58.

7. "Highest Capacity 500-kV line in the Western World," *Transmission and Distribution*, March 1975, pp. 24–26.

8. "Utilities: Weak point in the energy future," *Business Week*, Jan. 20, 1975, pp. 46–54.

9. W. A. Gilbert and R. V. DeGrasse, "Prospects for electric utility load management," *Public Utilities Fortnightly*, Aug. 28, 1975, pp. 15–19.

10. C. J. Cicchetti, "The design of electricity tariffs," *Public Utilities Fortnightly*, Aug. 28, 1975, pp. 25–33.

11. H. Glavitsch, "Computer control of electric power systems," *Scientific American*, Nov. 1974, pp. 33–34.

12. J. L. Dotson, Jr., "Duel in the sun," *Newsweek*, Oct. 27, 1975, pp. 10–11.

13. T. J. Nagel, "Electric power's role in the U.S. energy crisis," *IEEE Spectrum*, July 1974, pp. 69–72.

14. G. D. Friedlander, "Energy's hazy future," *IEEE Spectrum*, May 1975, pp. 32–40.

15. D. A. Haid, "Power transmission via the superconducting cable," *Mechanical Engineering*, Jan. 1976, pp. 20–25.

16. Edison Electric Institute, *Economic Growth in the Future*, McGraw-Hill Book Co., New York, 1976.

17. "Bigger electric bills ahead for big business," *Business Week*, November 29, 1976, pp. 55–56.

18. M. L. Baughman and P. L. Joskow, "The future outlook for U.S. electricity supply and demand," *Proceed. of the IEEE*, April, 1977, pp. 549–561.

EXERCISES

8.1 Consider your own use of electricity.

   a) List five electrical devices that you use directly in your usual routine, picking those that you think account for the largest share of your personal electricity consumption. (Count all lights as one device.)

b) Give the power rating of each device in watts. (This information can usually be found somewhere on the item, with watts abbreviated "W.")

c) For each device, estimate how many hours it is running each day (on the average). Combine this information with the answer to (b) to obtain the average daily electricity consumption per device (remember, lump all lights together) in kilowatt-hours.

d) Obtain your average power demand in kilowatts (for these 5 devices) by adding up the kilowatt-hours per day and dividing by 24 hours.

e) What is the greatest total wattage you ever have on *at one time* (your personal *peak* demand on the utility system)?

8.2 Suppose a toaster-oven draws 1,500 watts and operates 2 hours per day. Suppose a furnace fan draws 300 watts and operates 20 hours per day. Which device uses more electric energy in a day? Which creates the greatest peak power demand?

8.3 Contact your local power company and determine which electric transmission line they are connected to. Determine the voltage and the location of the primary source of power.

8.4 Extensive investigations into the commercial development of the Zambezi River Valley in Northwestern Mozambique are being pursued. [6] A narrow gorge of the river, called "Cabora Bassa," seem the ideal site for construction of a multipurpose dam for:

a) Generating low-cost electricity to develop the area's rich mineral resources;

b) Regulating the river flow for flood-control purposes;

c) Irrigating widespread agricultural areas.

Further, financing of the project could be amortized by the sale and transmission (over a distance of 1400 km) of large surplus blocks of electricity to the neighboring Republic of South Africa.

The first stage involves building a power plant with an output of more than 2000 megawatts. Both ac and dc transmission were considered and a dc voltage of $\pm 533,000$ volts was selected (1066 kilovolts line to line). Investigate and discuss the details of the hydroelectric power plant and the transmission of the power over this long distance.

8.5 It is estimated that one hundred thousand miles of transmission lines will be built in the U.S. during the next decade. Each mile of right of way slashes through forest or farmlands and impacts the aesthetics of the area. Explore and list the environmental impacts of electric transmission lines.

8.6 Alternating-current electricity is used and transmitted at a constant peak voltage and at a frequency of 60 Hertz ($\omega = 377$ radians/second). For a voltage of $E_{peak} \sin \omega t$, show that the root-mean-square voltage or effective voltage is $E_{rms} = E_{peak}/\sqrt{2}$.

8.7 The question of whether to use coal or oil in fossil-fueled plants is a difficult issue.

One example is Consolidated Edison's Ravenswood plant, Unit No. 3, providing 1000 MW in New York City, which formerly used 0.9-percent sulfur coal (equivalent to perhaps 1.25-percent sulfur oil) and in the winter of 1973–74 burned up to 1.5-percent sulfur residual fuel oil part of the time. The city and Federal environmental agencies refused a request to convert back to coal. This decision, aimed at a marginal reduction in SO_2 emission, resulted in an additional fuel cost of about $100 million per year and also diverted almost $1\frac{1}{2}$ million gallons per day of liquid hydrocarbons from other purposes. Explore the environment–energy interaction and state how you would have decided this case.

*8.8 Electrically-driven heat pumps provide thermal energy for space heating that is appreciably greater than the thermal equivalent of the input electrical energy. This is accomplished by extracting thermal energy from the cold outside air, boosting the temperature of this energy and delivering it to the inside of the house. The coefficient of performance (COP) is usually about 2.0 or more.

From the homeowner's economic standpoint, the energy savings achievable by using the heat pump instead of resistance heating must be balanced against higher capital and maintenance costs. These costs have tended to retard the widespread use of heat pumps; in 1970, only 11% of the electrically heated households in the U.S. had heat pumps. Investigate the availability of heat-pump systems in your community and determine the COP and price of a unit for an 1800-square-foot house.

8.9 The U.S. now has approximately 1200 miles of transmission lines that use 765 kilovolts. These lines are surrounded by intense electric fields, and persons living near the lines are exposed to electromagnetic radiation. Investigate the potential effects of exposure to these lines, and determine the regulations regarding high-voltage power lines in effect in the Soviet Union. State your position regarding the potential hazard associated with high-voltage lines.

8.10 Underground ac transmission cables operating at superconducting temperatures are under development. The resistance of a cable decreases rapidly as the cable is held at low temperatures (about 5°K). For a large, 1000-megavolt-ampere cable, the line loss due to line resistance is reduced to a negligible amount compared to the refrigeration energy required to cool the line. [15] Investigate the utility of superconducting electrical transmission cables and estimate when the first system will be installed in the U.S.

8.11 Newly available energy-saving frostless refrigerators are said to use 35% less energy than a pre-1976 frostless refrigerator. Obtain the power ratings of these new refrigerators, and compare with a comparable 1974 model. Also, obtain and compare the prices for the models.

---

* An asterisk indicates a relatively difficult or advanced exercise.

9

Transportation

The transportation of goods and people is integral to the needs of any society. Transportation technologies have evolved over the centuries, and now people may travel long distances at high rates of speed. Transportation systems use significant amounts of energy and are important to the whole social and economic fabric of any nation.

In the United States, transportation accounts for about 25 percent of the total energy use. About half of the petroleum consumed in the U.S. is used for autos, trucks, and airlines. Transportation is estimated to be responsible for about 20 percent of the gross national product (GNP).

The automobile manufacturing industry is a very important industry in the U.S. The auto industry has produced over 10 million vehicles per year during most of the past several years.

9.1 AUTOMOBILES AND RAILROADS

In 1900 there were about 8000 cars in the U.S., principally used as amusements for the rich, as horses are today. People still depended on the horse and carriage and the electric streetcar for transportation. By 1914, less than 10 years after Henry Ford introduced the Model T, more cars than carriages and buggies were made in America. By 1919, there were seven million cars registered in the U.S. In 1929, the year of the stock-market crash, there were already 23 million cars, or one for every five persons in the U.S.

Railroad passenger traffic reached its peak use in the early 1920's; the automobile became dominant after that. Table 9.1 shows the evolution of the use of railroads in the U.S. Only during World War II, when gasoline for private cars was rationed, was there a significant temporary increase in train travel.

Table 9.1

RAILROAD USE IN THE U.S. IN THE
20TH CENTURY

Year	Passengers (Millions)	Passenger-miles (Millions)
1900	577	16,038
1910	972	32,338
1920	1,270	47,370
1930	708	26,876
1940	456	23,816
1950	488	31,790
1960	327	21,284
1969	302	12,214
1975	269	9,582

The peak use of urban public transit was in 1926; since then, except for the period 1941–1945, the use of urban transit has declined in the United States, as shown in Table 9.2. Bus ridership increased for a period, but only at the expense of electric streetcars.

Table 9.2

PUBLIC-TRANSIT PASSENGERS IN MILLIONS FOR THE
PERIOD 1930 TO 1970 IN THE U.S.

Year	Electric railway	Trolley coach	Motor bus
1930	13,000	16	2,479
1940	8,325	534	4,239
1950	6,168	1,658	9,420
1960	2,313	657	6,425
1970	2,116	182	5,034

Table 9.3

MOTOR FUEL USE AND MILES OF TRAVEL BY MOTOR VEHICLES FOR
1921 TO 1972 IN THE UNITED STATES

Year	Highway motor fuel use (Billions of gallons)	Miles of travel by motor vehicles (Billions of vehicle-miles)	Motor vehicle registrations (Millions)
1921	3.9	55	8
1930	14.8	206	24
1940	22.0	302	29
1950	35.7	458	49
1960	57.9	719	65
1970	92.3	1,121	105
1972	103.2	1,250	115

Automobiles increased individual mobility and influenced the shape of our cities, the range of our vacation travel, and our commuting habits. There are more licensed drivers than registered voters in the U.S. Motor-vehicle registrations increased from 8 million in 1921 to over 115 million today, as shown in Table 9.3.

The long-term trend, from 1925 to 1970, was that motor-vehicle miles traveled approximately doubled every fifteen years. That amounts to a growth rate of about five percent a year. By 1971, a full one-sixth of U.S. employment depended directly on the manufacture, sale, servicing, and use of motor vehicles. One-quarter of the total retail sales dollars went for purchases related to trucks, buses, and cars. In 1969 the estimated area in highway right-of-way was 40,000 square miles, which is slightly larger than the large state of Virginia. In one year, 1972, we spent $22 billion on building and maintaining highways.

Passenger Traffic

The market distribution of energy use for transportation in the U.S. is shown in Fig. 9.1. Passenger travel accounted for 55.6% of the total energy use for transport in the U.S. in 1971. More than 86% of the intercity passenger traffic is by private automobile, as shown in Table 9.4. The total volume of traffic, measured in passenger-miles, doubled during the twenty years 1950 to 1970. During this period, travel on railroads and buses decreased significantly, while air travel increased by a factor of five, as shown in Table 9.4.

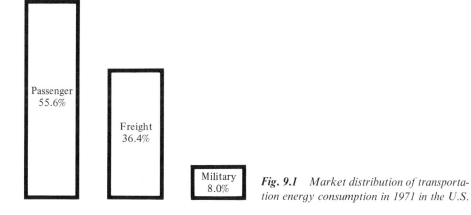

Fig. 9.1 *Market distribution of transportation energy consumption in 1971 in the U.S.*

Freight Traffic

The portion of freight moved by railroad currently accounts for 39 percent of the total freight traffic (measured in ton-miles), as shown in Table 9.5. Approximately 22 percent of the freight movement is by trucks using our vast system of

Table 9.4

INTERCITY PASSENGER TRAFFIC IN THE U.S., IN BILLIONS OF
PASSENGER-MILES, AND THE PERCENT TRAVELLED ON EACH MODE

Year	Total volume (10^9 Passenger miles)	Private auto (Percent of total)	Airlines (Percent)	Bus (Percent)	Railroad (Percent)
1950	508	86.2	2.0	5.2	6.4
1955	716	89.0	3.2	3.6	4.0
1960	784	90.1	4.3	2.5	2.8
1965	920	88.9	6.3	2.6	1.9
1970	1,185	86.6	10.0	2.1	0.9

Table 9.5

INTERCITY FREIGHT TRAFFIC IN THE U.S., IN BILLIONS OF
TON-MILES, AND THE PERCENT MOVED BY EACH MODE

Year	Total traffic ($\times 10^9$ ton-miles)	Railroads (Percent of total)	Trucks (Percent)	Inland waterways (Percent)	Oil pipelines (Percent)
1940	651	63	10	18	9
1950	1,094	57	16	15	12
1960	1,330	45	22	16	17
1970	1,921	40	22	15	22
1975	2,400	39	22	14	23

highways. The use of inland waterways remains an important and efficient means of moving many goods such as coal, ores, and other materials. Oil pipelines efficiently move petroleum to refineries and to market for later use as fuels or chemicals. In this case, we are using energy to deliver energy for later use, as we also do when we expend energy to deliver gasoline by tank truck.

While airlines transport a small percentage of the total freight moved within the U.S., they do consume a significant fraction of the energy used for freight movement. Airlines consumed approximately 4 percent of the total energy use for domestic freight transport in 1975, while railroads used 9 percent of the total energy for such use. Domestic air-cargo energy use has sustained an annual average growth rate of 16 percent since 1955. Although freight transportation by air is more expensive than truck or rail, it has gained rapidly in popularity because of speedy delivery and safe handling.

9.2 ENERGY USE IN TRANSPORTATION

Petroleum is the primary fuel in the transportation sector since it is used for trucks, autos, and buses. As shown in Table 9.6, petroleum supplied 95 percent of the energy consumed in the transportation sector in 1970. The use of coal for transportation has declined from 20 percent of the total energy in 1950 to less than 1 percent in 1970.

Table 9.6
DEMAND FOR ENERGY INPUTS IN THE
TRANSPORTATION SECTOR IN THE U.S. (in 10^{12} Btu)

Year	Coal	Petroleum	Natural gas	Electric energy	Total
1950	1,701	6,785	130	24	8,640
1960	87	10,372	360	18	10,836
1970	8	15,592	745	16	16,361

The shift to petroleum as the primary fuel has occurred because of the shift from railroad transportation to the increased use of autos, trucks, and airplanes, as shown in Table 9.7. The energy used by autos and trucks has grown by 5 percent per year over the period 1955 to 1971, while the energy consumption of airlines has grown by 15 percent per year over the same period. The demand for oil for transportation uses has grown so rapidly that approximately 70 percent of the U.S. domestic production is consumed in the transportation sector.

Table 9.7
TREND OF TOTAL ENERGY CONSUMPTION IN THE
U.S. FOR THE PERIOD 1955 TO 1971 FOR
SEVERAL MODES OF TRANSPORTATION

Mode	1971 Consumption (Trillion Btu)	1955–1971 Average annual rate of growth	1971 Percent of total
Automobiles	8652	4.6	48.2
Trucks	3445	4.7	19.2
Pipelines	1590	6.0	8.9
Military	1443	2.3	8.0
Airlines	698	15.2	7.1
Water	544	−0.9	3.9
Railroad	123	−2.7	3.0
Buses	89	1.3	0.7

Energy Intensiveness of Various Transportation Modes

Let us examine the average energy-intensiveness of the various modes of transportation, where intensiveness is measured as the energy required to move a passenger a given distance in the care of passenger transport. The energy-intensiveness of six modes of transport for moving people is shown in Table 9.8. The airplane is the most energy-intensive and requires about four times the energy required per passenger mile for a bus. [1]

Automobiles have the greatest claim on energy resources used for urban transportation. A major reduction in energy expended, and presumably in urban pollution problems, could be achieved if the use of automobiles for trips of two

Table 9.8

AVERAGE ENERGY-INTENSIVENESS* OF VARIOUS
MODES OF TRANSPORTATION FOR PEOPLE

Mode	Btu/Passenger mile	Total percent of transportation demand
Airplane	7150	9.3
Automobile	5400	88.4
Train	2620	0.7
Bus	1700	1.2
Walking (3 mph)	524	Less than 0.1
Bicycle (8 mph)	310	Less than 0.1

* Energy intensiveness of transportation modes varies with the load factor and the distance travelled.

miles or less could be significantly reduced. For small distances, bicycles or walking are extremely energy efficient. The use of buses and rapid transit rather than autos for urban commuting would also result in a significant reduction of energy use for transportation. Of course, another simple method of reducing energy expended for urban commuting is to encourage car-pooling, that is, to increase the number of passengers per vehicle.

Table 9.9

ENERGY-INTENSIVENESS OF
VARIOUS MODES OF
TRANSPORTATION FOR
FREIGHT

Mode	Btu/Ton-mile
Pipelines	450
Waterways	540
Railroads	680
Trucks	2,340
Aircraft	37,000

The energy-intensiveness of various modes of transportation for freight is shown in Table 9.9. Moving freight by railroads and waterways is preferable to the use of trucks and aircraft. Most waterways are underutilized and could be used to much greater advantage in both the U.S. and Europe. The desire for increased flexibility and speed have resulted in an increase in the use of airlines and trucks for freight transport. In the future, the increased use of trains and waterways for freight traffic could significantly moderate the demand for energy in the transportation sector.

Trucks move about half of the nonfuel, nonbulk, manufactured and general products in the U.S. [2] Heavy tractor trailers, capable of carrying 10 to 50 tons of cargo, get about 4 miles per gallon at 55 mph, for an energy efficiency of 50 ton-miles per gallon. Today's freight trains deliver up to 250 ton-miles per gallon. The use of truck-trailers moved piggyback by train would provide the efficiency of trains, while retaining the final delivery flexibility of trucks.

The use of airplanes for freight movement over distances less than 1000 miles is particularly inefficient compared to the use of railroads or trucks.

9.3 THE AUTOMOBILE AND ENERGY CONSERVATION

About 60 percent of the energy used in the transportation sector is used in the form of gasoline for approximately 110 million automobiles and small personal trucks in the U.S. [3] Virtually every aspect of American life—industrial, commercial, cultural, and recreational—is organized around the existence of motor vehicles. It is a reasonable assumption that this dominance of automobiles will continue to prevail over the next one or two decades. Therefore, it is imperative that we examine the automobile in terms of improving its energy efficiency. The average 1974 auto achieved 14 miles per gallon during tests; the President of the U.S. has announced a 40 percent increase in fuel economy as a goal for 1980. The goal for new-automobile performance in 1980 might be 20 miles per gallon. The fuel economy of new 1975 autos averaged 15.9 miles per gallon—an improvement of 13.5 percent in one year. The fuel economy of automobiles varies with the weight of the auto, as shown in Fig. 9.2 for 1975. In order to significantly increase the fuel economy of automobiles beyond 20 miles per gallon, it is important that an increasing percentage of vehicles be light in weight, specifically less than 3000 pounds.

In addition, it is important to consider the use of other engines. The present Otto-cycle engine achieves about 22 to 25 percent efficiency. However, under normal driving conditions, the energy delivered to the wheels is only about 10 percent of the energy stored in the fuel. The use of diesel engines and engines operating in a "stratified charge" would increase engine efficiencies to over 30

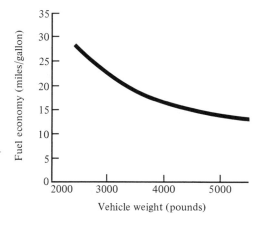

Fig. 9.2 *The fuel economy of U.S. autos as a function of the vehicle weight for 1975.*

percent. Stratified-charge engines use an air–fuel mixture that is normally lean (the air-to-fuel ratio is greater) and therefore improves the efficiency of the engine. This is accomplished by bringing together air and fuel in combustible proportions in selected locations in the cylinder.

About half of all auto trips are under five miles; therefore, autos specifically designed for efficient short-haul service would be desirable. According to the U.S. 1970 census, 60 million Americans commute by private automobile (51 million travel alone and nine million in car pools); 4.2 million use a bus or streetcar; 1.8 million use a subway or an elevated-railway line; 500,000 use railroads; 300,000 use taxis, and 5.7 million live close enough to their jobs to walk to them. At the low speeds used for most commuting travel, the primary factors militating against efficient travel are traffic congestion and the weight of the vehicle. Improved highway systems supporting lightweight commuter autos could increase fuel economy significantly.

The energy consumed in the construction of an average-sized American car of 3600 pounds includes the energy requirements of mining, materials production, manufacturing, and shipping, to name the most important. It has been estimated that it requires about 100 million Btu to produce an average automobile from primary metals. Therefore, the production of 8.2 million new autos each year requires about 0.82×10^{15} Btu, or about 1 percent of the total energy consumption of the U.S. [4]

Significant auto-fuel savings can be achieved by closer adherence to recommended maintenance practices for the engine and the vehicle. Further improvements can result from energy-conscious driving habits. Car fuel efficiency is very sensitive to rates of acceleration and speed. [5]

Automobile air conditioners reduce gasoline fuel economy about 9 percent, and the typical automatic transmission reduces fuel economy by about 6 percent. In addition, steel-belted radial tires can save 10 percent on gasoline mileage. The aerodynamic drag of an automobile is proportional to the velocity of the car raised to the second power. Therefore, reducing both the highway speed and the drag coefficient would result in significant improvement in fuel economy. The aerodynamic drag is represented by the equation

$$\text{Drag} = 0.65 C_{\text{d}} A v^2, \tag{9.1}$$

where A is the frontal drag area in square meters and the speed is in meters/second.

The application of a continuously variable transmission to a conventional engine could result in 30 percent improvement in fuel economy. A continuous transmission selects the gear automatically, thus maximizing fuel economy. Such a transmission can be built using a hydromechanical or traction drive. [6]

A price increase in gasoline can achieve a reduction in total gasoline consumption. A recent study showed that price increases of 15¢ to 45¢ per gallon can achieve annual reductions in gasoline consumption (by 1980) of 16 percent to 41 percent, respectively. [7] One method of increasing prices is through gasoline taxes. Over long periods, however, the impact of the tax declines as incomes rise and inflation progresses.

Therefore, in summary, substantial savings in fuel consumption could be achieved by reducing vehicle weight and air resistance, by using better and harder tires together with better suspensions, by cutting the waste of energy by accessories and high speeds, by designing better transmissions and, if possible, by achieving efficient operation with a shorter warm-up period. Beyond these possibilities, engine efficiencies must be considered.

The ubiquitous automobile consumes more than half the fuel used in transportation, or roughly 13 percent of the total direct energy used in the United States. Moreover, the share being consumed by automobiles has been rising over the past decade. If we can devise policies that save even modest percentages of the fuel consumed by automobiles, the absolute savings in energy will be very large. [20] Therefore, it is important to carefully examine all the potentially useful policies for increasing the fuel economy of the automobile.

The Electric Automobile

In the horseless carriage days in America in the late nineteenth century, it was far from certain that the car of the future would be driven by an internal-combustion engine. During the 1890's the electric automobile enjoyed a greater popularity than did the gasoline car. It was clean, quiet, and easy to operate. However, battery-powered electric autos had a limited range and low power, and enjoyed relatively low use after 1910.

Nevertheless, electric cars do have certain advantages: low noise, low maintenance costs, smooth acceleration, and flexibility in placement of the motor and batteries. For these reasons, and because there is no air pollution produced at the auto's location, electric autos are being reconsidered for general use in the U.S. The reason that electric automobiles lost in competition with cars using the internal-combustion engine is the limited top speed and driving range of a car based on acceptably priced batteries. The specific energy of a typical lead-acid storage battery is 20 watt-hours per pound for a 20-hour discharge time. If the energy is withdrawn more rapidly, the specific energy is closer to 10 watt-hours/pound.

The energy required to drive an auto may be written:

$$E = 1.64 \times 10^{-3} VDW, \tag{9.2}$$

where V is the constant velocity in miles/hour, D is the distance traveled in miles, and W is the weight of the car in pounds. Therefore, for a 2000-pound vehicle traveling 50 miles at 50 miles/hour, we have:

$$E = (1.64 \times 10^{-3})(50)(50)(2000)$$
$$= 8200 \text{ watt-hours.} \tag{9.3}$$

This figure is based on steady speeds on level ground. In actual driving there are periods of acceleration and hill-climbing, which require more energy. (Some of this energy might possibly be recovered during decelerations and downhill periods.) It is estimated that the additional energy needed for accelerations, even allowing for partial energy recovery, amounts to about 25 percent more in city conditions. Therefore we might estimate the actual requirement for energy as 10,000 watt-hours. If the battery supplied only 10 watt-hours/pound we would require 1000 pounds of batteries (or one-half the weight of the car). Clearly the range and speed of an electric car based on lead-acid batteries is quite limited. However, in urban commuting, speeds seldom exceed 35 mph, and the range required is less than 20 miles. For this case, the energy we require from the batteries is:

$$E = (1.64 \times 10^{-3})(35)(20)(2000) = 2296 \text{ watt-hours} \qquad (9.4)$$

for an auto of 2000 pounds. Allowing for a 25 percent increase in the requirement to account for starting and stopping, we require 2870 watt-hours, or 287 pounds of lead-acid batteries.

The specific power versus the specific energy for automotive power plants is shown in Fig. 9.3 for a 2000-pound vehicle, assuming the power plant equals 25 percent of the weight of the vehicle. Superimposed on the figure is a grid indicating the range in miles of the vehicle versus the steady driving speed. The figure shows that the lead-acid battery would provide only a 50-mile range at 20 mph for this 2000-pound vehicle, while the internal-combustion engine exceeds the requirements of range and speed (assuming a steady speed). The fuel cells generally meet the energy requirement but not the power-level requirements. However, the lithium-chlorine and sodium-sulfur batteries meet both the energy and power requirements and warrant intensive further development. The main disadvantage

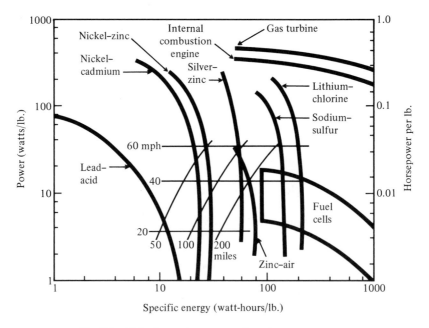

Fig. 9.3 *Vehicle requirements and motive power sources.*

of sodium-sulfur and lithium-chlorine batteries is that they must be maintained at high temperatures (250°C), and the sodium sulfide and chlorine are dangerous if they leak from the batteries. Rechargeable versions of the zinc-air cell have already achieved energy densities of over 60 watt-hours/pound. [8]

Despite the limitations of electric vehicles, many are used today. In Great Britain over 40,000 are used as milk delivery trucks; and in the U.S. and in Europe, battery-powered vehicles are used inside factories as fork-lift trucks.

The efficiency of an electric-powered automobile may be compared with a similar average gasoline vehicle. [9, 10] A schematic diagram of the propulsion efficiency of an electric car is shown in Fig. 9.4. The overall efficiency of this vehicle is the *product* of the efficiencies of each sequential portion of the system, or $\eta = 0.20$. A schematic diagram of a gasoline powered vehicle of the same weight as the electric car is shown in Fig. 9.5. The overall efficiency of this vehicle ranges from 0.07 to 0.13, where the higher efficiency is achieved for a constant speed while the lower value is obtained over a start-and-stop driving cycle. This savings of energy through the use of an electric vehicle is significant indeed. An electric car for urban commuter use, with a range of 40 miles and a top speed of 45 miles/hour, would be very useful. Such a vehicle might weigh 2000 pounds, and could be recharged overnight at the owner's home. One source estimates that there will be 5 million electric cars on the road in the U.S. by 1985. [10]

Whether or not the electric auto is less polluting than its internal-combustion-engine counterpart is very difficult to determine. The effect of the substitution would be an exchange of one form and magnitude of pollution for another form and magnitude of pollution. The present internal-combustion automobile releases significant amounts of hydrocarbons and oxides of nitrogen. It is a major contributor to urban smog. It does have the advantage of distributing its waste heat over a rather wide range. The pollution due to the electric automobile comes

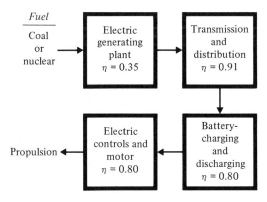

Fig. 9.4 *A schematic diagram of the propulsion efficiency of an electric car.*

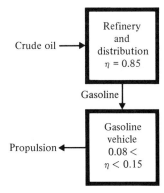

Fig. 9.5 *A schematic diagram of the propulsion efficiency of a gasoline powered car.*

Fig. 9.6 *The Copper Electric Town Car built by the Copper Development Association. The car has a range of 120 miles at a cruising speed of 40 mph and a top speed of 55 mph. It has a range of 75 miles in stop-and-go city driving, and energy costs are estimated between 2¢ and 3¢ per mile. The car weighs 2952 pounds, including the 1200 pounds of 18 batteries. The front-wheel drive uses a separately excited motor. (Photo courtesy of Copper Development Association.)*

primarily from the electric-power plant, where the emission of particulates and oxides of carbon, nitrogen, and sulfur occurs (for fossil-fuel plants). Of course, if an economical, nonpolluting electric-power source becomes available, then the electric car would grow rapidly in use. One version of an electric car is shown in Fig. 9.6.

9.4 AIRPLANE ENERGY CONSUMPTION

The history of aviation is essentially a 20th-century story. The Wright brothers recognized that the power of the internal-combustion engine, coupled with an efficient propeller, could generate sufficient propulsion to provide lift for an airplane. They achieved controlled flight in 1903. Subsequently the airplane became a weapon of war and, in both peace and war, a means of rapid transport.

Airplanes are the most energy-intensive mode of transport. They require 60 times the energy to move a specific amount of freight a given distance, compared to a train. Both freight movement and passenger travel by air have increased significantly over the decade 1960 to 1970. Over this period, freight and passenger transport increased more than 14 percent per year. The only pause in this previously uninterrupted growth came about as a result of the rapid rise in airplane fuel prices after 1974.

Direct fuel use for commercial airplane propulsion amounted to approximately 1×10^{15} Btu in 1971, or 6.3 percent of the total use of fuel for transportation.

Steps toward reduced consumption of petroleum for air transport include: (1) increasing the load factors, (2) reducing the number of flights of less than 200 miles, (3) increasing cruise altitude, and (4) towing airplanes to take-off positions. Increasing the load factors, or the percentage of seats occupied, on a plane is a very significant step toward increased efficiency, if the *number* of flights is correspondingly decreased. By decreasing the number of flights on overserviced routes, the airlines have been able to increase their load factor from 40 percent to 60 percent in many cases. The shortage of jet fuel in 1974 caused a reduction in the number of flights over U.S. routes, with an attendant reduction in the fuel expenditure per passenger-mile since the load factor was correspondingly increased.

REFERENCES

1. E. HIRST, *"Transportation Energy Use and Conservation Potential,"* M.I.T. Press, Cambridge, mass., 1977.

2. R. A. RICE, "Toward more transportation with less energy," *Technology Review*, Feb. 1974, pp. 45–53.

3. J. R. PIERCE, "The fuel consumption of automobiles," *Scientific American*, Jan. 1975, pp. 34–44.

4. J. K. TIEN, *et al.*, "Reducing the energy investment in automobiles," *Technology Review*, Feb. 1975, pp. 39–43.

5. J. T. KUMMER, "The automobile as an energy converter," *Technology Review*, Feb. 1975, pp. 27–35.

6. T. F. KIRKWOOD and A. D. LEE, "A generalized model for comparing automobile design approaches to improved fuel economy," *Rand Report 1562*, Jan. 1975.

7. S. WILDHORN, *et al.*, "How to save gasoline: Public policy alternatives for the automobile," *Rand Report 1560*, Oct. 1974.

8. N. VALERY, "A car to beat the fuel crisis?" *New Scientist*, Dec. 20, 1973.

9. T. J. HEALY, "The electric car: Will it really go?" *IEEE Spectrum*, April 1974, pp. 50–53.

10. J. T. SALIHI, "Kilowatthours vs. liters" *IEEE Spectrum*, March 1975, pp. 62–66.

11. T. J. HEALY, *Energy, Electric Power and Man*, Boyd and Fraser Pub. Co., San Francisco, 1974, Chapter 14.

12. E. HIRST, "Total energy use for commercial aviation in the U.S.," *Report EP–68*, Oak Ridge National Laboratory, April 1974.

13. R. F. Post and S. F. Post, "Flywheels," *Scientific American*, Dec. 1973, pp. 17–24.

14. J. G. Voeth, "The airship can meet the energy challenge," *Astronautics and Aeronautics*, Feb. 1974, pp. 25–30.

15. R. S. Shevell, "Technology, efficiency, and future transport aircraft," *Astronautics and Aeronautics*, Sept. 1975, pp. 37–42.

16. M. Chiusano, "Conserving transport energy," *Technology Review*, June 1974, pp. 54–55.

17. J. E. Ullmann, "Getting the rails back," *Environment*, Dec. 1975, pp. 32–36.

18. T. B. Reed and R. M. Lerner, "Methanol: A versatile fuel for immediate use," *Science*, Dec. 28, 1973, pp. 1299–1304.

19. F. P. Povinelli, *et al.*, "Improving aircraft energy efficiency," *Astronautics and Aeronautics*, Feb. 1976, pp. 18–31.

20. E. Hirst, "Transportation energy conservation policies," *Science*, April 2, 1976, pp. 15–20.

EXERCISES

9.1 Unfortunately, we seldom operate automobile engines at maximum power where we can obtain high thermal efficiency. We use engines that supply 100 horsepower at the rear wheels when needed, yet during constant driving, only 20 horsepower may be needed. [5]

If we use a smaller engine (lower maximum horsepower) for a given-sized vehicle with a constant ratio of engine speed to vehicle speed, we gain in fuel economy because the friction loss is smaller for the small engine. Also, we have to operate at higher intake manifold pressures for the same vehicle and engine speed, in order to obtain the required power output, and consequently incur lower pumping and heat losses. The vehicle with the small engine will, of course, have poorer acceleration performance because of lower peak power. Examine the horsepower of your auto (or a friend's) and estimate the horsepower required for normal driving.

9.2 Automobiles of lower weight consume considerably less energy per mile. Policies to promote new cars of lower weight might include: (1) vehicle weight taxes; (2) a graduated excise tax on vehicles with fuel economy less than 20 miles per gallon; (3) increased gasoline taxes; and (4) higher urban parking rates for large cars. Rate each of these alternatives in order of preference, and explain your preference.

9.3 Most regular domestic automobiles are designed to accelerate from 0 to 60 mph in 10 seconds. If this acceleration time was doubled to 20 seconds, the fuel economy might improve by 40%. Would this alteration be acceptable to you and your friends? Complete a survey of at least 10 people concerning this proposal.

9.4 The automobile fuel consumption for a trip length of 2 miles is 16,000 Btu while it is only 40,000 Btu for a trip of 10 miles. Compare the energy consumed per mile, and determine why autos are inefficient for trips under five miles.

9.5 The energy consumed by commercial airplanes is dependent upon the distance traveled and the type of aircraft. The average 100-mile flight consumes about 1.5 times more fuel per passenger-mile than does the average 1000-mile flight. Determine the reasons for this difference, and propose a policy regarding airline routes and intermediate stops.

9.6 An urban vehicle weighing 1500 pounds with 15 to 20 horsepower has been proposed. [2] This vehicle could achieve 30 miles/gallon in urban driving and 40 miles/gallon in country driving. Examine this concept and determine the parameters of an urban car you believe would be accepted by people today.

9.7 Electric trains are more efficient and can pull larger loads than diesel trains. Electric locomotives also offer superior acceleration characteristics and more easily controlled pollution. Of course, electric trains are not pollution-free, but electricity-generating stations are large centralized units and it is easier to clean their exhausts than to clean up the emissions from every individual diesel propulsion unit. Examine the benefits of shifting 50% of the railway system in your state to electric trains.

*9.8 The aerodynamic drag coefficients for several vehicles are given in Table E9.8. Calculate the drag force, in newtons, for several of these vehicles, and compare this force for the best and worst case when the vehicle is traveling at 50 miles/hour and the frontal area is 2 square meters.

Table E9.8
AERODYNAMIC DRAG COEFFICIENTS

Vehicle	Truck	Porsche 904	Station wagon	1972 Camero
C	0.7	0.35	0.6	0.5

*9.9 Consider a standard auto with $C_d = 0.5$ and $A = 2$ m². The rolling resistance due to tire friction with the road may be written as

$$F_r = \frac{M}{1000}[118 + 1.1v + 0.2(v - 20)^2],$$

when $v \geq 20$ m/sec, and M is the mass in kilograms.

Calculate the drag force and the rolling resistance for speeds of 15, 20, and 25 meters/sec, and prepare a table recording them. Calculate the constant-speed power required as $P = v(F_r + F_d)$ in kilowatts and horsepower, and add these values to the table.

9.10 The Anglo-French supersonic transport (SST) called the Concorde consumes fuel rapidly compared to a subsonic jet. The Concorde travels at 1400 miles/hour and takes only $3\frac{1}{3}$ hours from New York to London. The Concorde yields 19

---

* An asterisk indicates a relatively difficult or advanced exercise.

seat-miles per gallon of fuel, while the Boeing 747 yields 40 seat-miles per gallon. A fleet of 100 Concordes, each flying 12,000 miles per day, would consume 100,000 barrels of oil per day more than the same number of seat-miles with 747's. Examine the implications of this consumption rate, and state what you believe would be an appropriate policy regarding the introduction of the SST to intercontinental travel.

9.11 While the average power required by an auto is low, requirements for accelerating or hill-climbing require large power outputs for relatively short periods of time. An auto with stored energy, which can be drawn on to meet the peak demands, requires engines to meet only the lower *average* power demand. A flywheel, even one of steel, can reasonably provide the needed energy storage. Not only is the power required of the engine reduced, but with a proper coupling system, the engine can run more nearly at a steady speed. It is considerably easier to reduce the emissions of a fixed-speed engine than one which must go from idling to sudden high-speed power bursts; and the smaller constant-speed engine has a lower energy consumption. [13] Establish the general outline of such an auto, and determine the size of a suitable flywheel of fiber material.

9.12 Various options for saving crude oil have been proposed. Figure E9.12 shows the estimated percentage savings in fuel for several alternative options. [16] What

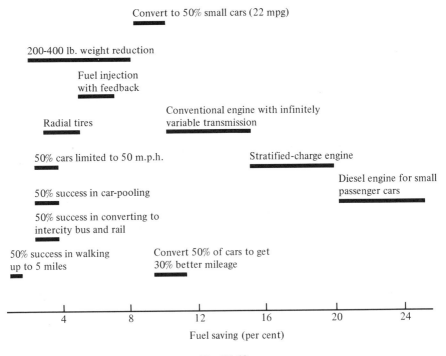

Fig. E9.12

options do you believe we should adopt during the next five years, and can you substantiate their feasibility?

9.13 The Copper Development Association Town Car incorporates several innovative features. For example, the car does not use any type of mechanical speed changing, thus eliminating the weight of a manual or automatic transmission. Explore the features of this and other electric autos, and suggest several innovations that should be incorporated in future experimental vehicles. Also consider the possibility of designing and marketing an electric motorcycle. Figure E9-13 shows the batteries in their tray as used in the Copper Electric Town Car.

Fig. E9.13

9.14 Cargo airships are being studied for use in moving large equipment and other loads. [14] Large dirigibles could be used to move million-pound payloads and could pick up the load without runways. Estimate the energy intensity of a dirigible (or other airships) and compare them with airplane freight service.

9.15 The fuel efficiency of airplanes depends upon many factors such as structural weight, wing-structure design, and engine design. For example, the B-747, DC-10, and L-1011 aircraft deliver 50% more seat-miles per gallon than the B-707 or DC-8. [15] Investigate the various factors of aircraft design affecting fuel efficiency, and describe a priority approach to aircraft design yielding significant energy savings in aircraft transport. [19]

9.16 a) Consider some specific automobile you are familiar with (your own, your family's, a friend's). Compare the purchase cost of the car itself with the cost of the fuel it will burn in a year, and in its lifetime (say, 10 years). Use estimates for your calculations.
   b) What kinds of energy consumption (other than gasoline consumed as fuel) are required to keep the automotive portion of American life thriving?

*9.17 The era of the electrified trolley car faded from U.S. cities by 1950. Nevertheless, the electric-powered trolleys provided convenient, inexpensive, energy-efficient and relatively nonpolluting transportation. Over 12 major U.S. cities had large electric trolley systems. For example, San Francisco had the Key system, Los Angeles had the Pacific Electric Company, and St. Louis an extensive system. [17] Examine the history of an electric trolley system in a city near your home or college, and determine why it was discontinued. Would you recommend the system be reinstituted?

*9.18 Methanol (which is also called wood alcohol or methyl alcohol) can be used as fuel for vehicles. [18] Examine the qualities of methanol as a fuel for an internal-combustion engine. Particularly examine the engine exhaust pollution, the efficiency of combustion, and the necessary design of the engine.

9.19 A conventional internal-combustion engine automobile loses 36% of its fuel energy to cooling water and 36% through exhaust gases. Engine friction and auto air resistance cause a loss of 5.6% and 7.1%, respectively. Overall, the energy used to accelerate a climb up a hill may be under 10%. Discuss several modifications you would propose for the purpose of increasing the overall energy efficiency of an auto.

*9.20 Coal, a relatively plentiful fuel in the U.S., was extensively used for railroads prior to the late 1940's. List several reasons why coal-fueled railroad engines were replaced by diesel engines. Explore the appropriateness of using coal to generate electricity to be used to run electrified railroads.

*9.21 We may define the specific power of a vehicle as the ratio of the maximum power available, to the gross weight of the vehicle. For example, the specific power of a merchant ship moving at a speed of 30 mph is about 1.5 hp/ton. Similarly, the specific power of an automobile is 20 hp/ton at a speed of 60 mph. Prepare a table listing the approximate specific power and speed of the following vehicles: (1) bicycle, (2) merchant ship, (3) automobile, (4) helicopter, (5) commercial airplane, (6) race car, and (7) jet fighter airplane, and any other vehicle you wish to add.

9.22 Automobiles are so poorly insulated that they require an air conditioner capable of cooling a small home. Explore and estimate the potential energy savings from (a) improved insulation and (b) the use of absorption air conditioners powered by waste heat from the car engine.

9.23 Design a Federal program, similar to the highway fund program, for restoring railroads to a more competitive position for freight and passenger traffic.

9.24 The supersonic transport Concorde is half as energy-efficient as subsonic air transportation. Calculate the added expenditure of fuel per seat-mile for Paris to New York. Is this expenditure worth the three-hour time savings? Should the Concorde be permitted to land in New York?

10

Agriculture

The objective of agriculture is to collect and store solar energy as food energy in plant and animal products, which are then distributed throughout the region (or the world) to serve as food for the population. To collect solar energy in plants and animals, farmers spend fossil-fuel energy and electric energy in tillage, fertilizers, irrigation, harvesting, and processing to help crops convert solar energy into food energy and to distribute it.

Energy has helped to make a revolution in farm life, farm work, and farm output since 1900. Only a hundred years ago in the U.S., we were a rural people: farmer, planter, trapper, and pioneer. Most work was done on farms, in the 19th century, by muscle power, human or animal. The chief sources of energy were wood and coal used for heating. Many farms were fortunate enough to have a windmill for pumping water. In 1900, one farm worker was able to supply the needs of about seven people; today, a farm worker, with the help of three other off-farm workers, supplies the needs of 50 people.

The first major contributions that energy made to farming were in the use of commercial fertilizer, an energy-intensive product, and in factory-made farm machinery, which required energy to produce. Motorized farm machinery, which also required energy for its operation, first became practical for the farmer around 1910 when farm tractors became available.

Table 10.1 shows the number of farms and the total acreage under production for the period 1900 to 1971. During this period the size of the average farm more than doubled while the farm population declined to one-third of its 1900 level. This change was made possible by the introduction of new technologies to the farming process, including the use of tractors, commercial fertilizers, motorized harvesters, and new scientific methods. The use of energy-intensive products in U.S. agriculture for the period 1900 to 1970 is shown in Table 10.2. By 1975 there were approximately 5 million tractors in use in the U.S. with a total capacity

Table 10.1

THE NUMBER OF FARMS, LAND IN FARMS AND THE
FARM POPULATION FOR 1900–1971

Year	Number of farms (Thousands)	Land in farms (Million acres)	Average acres per farm	Farm population Millions of people	Farm population Percent of total population
1900	5,737	839	146	29.4	38.7
1910	6,406	879	137	32.2	34.9
1920	6,518	956	147	32.0	30.1
1930	6,546	987	151	30.5	24.9
1940	6,350	1,061	167	30.5	23.2
1950	5,648	1,202	213	23.0	15.3
1959	4,105	1,183	288	16.6	9.4
1969	2,971	1,124	378	10.3	5.1
1971	2,876	1,117	389	9.4	4.6

Table 10.2

USE OF ENERGY-INTENSIVE PRODUCTS IN U.S. AGRICULTURE

Year	Tractors on farms (Thousands)	Commercial fertilizer used (Thousands of tons)	Value of farm implements and machinery (Millions of constant 1967 dollars)
1900	—	2,730	2,727
1910	1	5,547	3,543
1925	549	7,329	5,544
1940	1,545	8,336	7,345
1950	3,394	20,345	14,873
1970	4,619	39,775	28,928

of 250 million horsepower (186 gigawatts). Tractors consume 21 gallons of gasoline and 20 gallons of diesel fuel per capita in the U.S. The use of commercial fertilizer grew by a factor of approximately 14 during the period 1900 to 1970, and is a critical factor in our ability to increase crop yield per unit of land cultivated.

10.1 OUTPUT AND PRODUCTIVITY

The agricultural output and productivity in the U.S. is given in Table 10.3 for the period 1929 to 1972. During this period an index of crop production per acre approximately doubled while expenditures for energy increased by a factor of eight. During this period the population of the U.S. increased by 71 percent, so the growth in agricultural output more than matched population growth in the U.S. The net balance of agricultural product available (after supplying the U.S. population) was exported, and showed a growth of 3.75 percent per year over the period. The growth of annual expenditures is shown in Fig. 10.1. Energy

Table 10.3
AGRICULTURAL OUTPUT AND PRODUCTIVITY IN RELATION
TO INPUT RESOURCES FOR 1929–72

Variables in constant 1958 dollars	Year				Annual percent change 1929–1972
	1929	1948	1958	1972	
Gross farm product (Billion dollars)	17.0	19.0	20.8	24.7	1.0
Hired and family labor (Billion hours)	23.2	16.8	10.5	6.5	−3.0
Farm energy expenditures (Billion dollars)	0.5	1.2	1.6	1.7	4.09
U.S. population (Millions)	121.8	146.6	174.9	208.8	1.39
Per capita income (Dollars)	1236	1567	1831	2771	2.27
Agricultural exports (Billion dollars)	2.6	2.2	4.0	6.9	3.75
Persons supplied by one farm worker	9.5	15.0	24.8	49.2	
Index of U.S. crop production per acre (1967 = 100)	57	67	80	120	

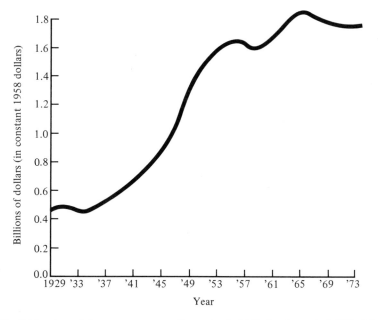

Fig. 10.1   *Annual farm energy expenditures in the U.S. for the period 1929–73.*

expenditures appear to have leveled off over the past several years. This leveling off may be due, in part, to the fact that most farms have attained the point of diminishing returns: increased energy expenditure cannot guarantee an increase in crop yield per acre. New technologies (and the energy required for these technologies) have permitted the U.S. to reap a bountiful harvest, with food sufficient for all its citizens, and a surplus for export. We have achieved this by substituting machines, and the energy to run the machinery, for human labor. We have released to crop production vast acreages formerly required to feed horses, mules, and oxen. Also, use of energy, in the form of fertilizers and in the form of pumped water for irrigation, has allowed production increases which otherwise would have required large increases in the amount of land under cultivation; thus energy was substituted for land.

Rural Electrification

The Rural Electrification Administration, established in 1935, brought about a rapid growth in electrical service to farms over the period 1935 to 1955, as shown in Fig. 10.2. While electricity was available to only 12.6 percent of the farms in 1935, the percentage grew to 96 percent by 1956. Farms with electricity used it for cooking, refrigerators, heating, lighting, and many kinds of farm machinery. Electric motors and lights in the barn and electric milking machines saved the farmer long hours. Electricity was used to pump the wells for irrigation and drinking water, and the windmills of the past went idle. The person who formerly had to stoke up a coal- or wood-fired stove at 5 A.M. had good reason to appreciate the advent of electricity.

By 1972, approximately 99 percent of U.S. farms had electrical service. The farmer's share of our total national electricity demand is about 3.5 percent, in-

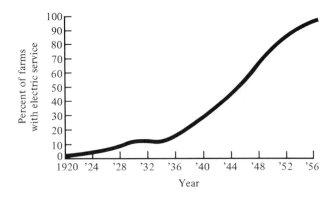

Fig. 10.2 *The growth of electric service to farms in the U.S. during the period 1920 to 1956.*

cluding household use. [15] Electric power, through machines such as milking machines, substitutes for manual labor. It is used for irrigation pumps, lifting and handling materials, drying crops, and hatching chicks, as well as continuously lighting the chicken houses to promote greater egg production.

10.2 ENERGY USE

Modern agriculture, based on the substitution of fuel energy for land and for muscular energy, has become a significant direct consumer of energy in the U.S. and other industrialized countries. It has been estimated that agriculture directly, on the farm, accounts for 3.5 percent of all U.S. energy use. [1] It is estimated that farm tractors consumed (as fuel) 1.1 percent and farmers purchased (as electricity) about 0.5 percent of the total national consumption. The production of fertilizers, chemicals, feeds, machinery, and other inputs purchased by farmers accounted for another 1.9 percent of the national energy consumption. Agricultural energy consumption on the farm is modest compared with that of transportation, which uses 25 percent of the nation's energy, or that of residential and commercial space heating, which uses about 18 percent of the nation's energy. Among U.S. industries, however, agriculture ranks third (after steel and petrochemicals) among the major industrial consumers of energy. Yet 3.5 percent is a modest commitment in terms of the output of food delivered to the nation and exported to the world.

The first input of energy to farming takes place during photosynthesis, using the sun's energy at an efficiency of approximately 1 percent. The maximum solar radiation captured by plants is about 5×10^3 kilocalories per hectare per year.* This energy of photosynthesis is then available for farming.

Table 10.4 shows seven categories of energy use on the farm, for 1950 and 1970. [2] The energy devoted to fertilizers and fuel increased significantly over the 20-year period.

The food-processing industry accounts for a large share of the total energy use for agriculture. The cost of processing, packaging, and transporting food is significant, as can be seen in Table 10.4, for the category labelled food-processing.

The final overall category in the food-distribution system in the U.S. is the energy use in commercial and home refrigeration and cooling, as shown in Table 10.4. This category is not insignificant, nor would most people be willing to lower the energy expenditures in this category.

The total energy use in the food system almost doubled, from 1134×10^{12} Calories to 2172×10^{12} Calories, in the period from 1950 to 1970. The growth of energy use in the food system is shown in Fig. 10.3 for the period 1940 to 1970. [2] This figure also draws a comparison between energy consumed and the caloric content of the food consumed during that period.

---

* For food and dietetic energy the unit kilocalories = Calorie is normally used, where one Calorie = 4184 joules. One hectare = 2.47 acres.

Table 10.4

ENERGY USE IN THE U.S. FOOD SYSTEM
(VALUES GIVEN IN 10^{12} KILOCALORIES)

	1950	1970
The farm		
Fuel	158.0	232.0
Electricity	32.9	63.8
Fertilizer	24.0	94.0
Agricultural steel	2.7	2.0
Machinery manufacture	30.0	80.0
Tractor manufacture	30.8	19.3
Irrigation	25.0	35.0
Subtotal	303.4	526.1
Food-processing industry		
Processing industry and machinery	197.0	314.0
Packaging and containers	105.0	207.0
Transport fuel and transport manufacture	151.5	320.9
Subtotal	453.5	841.9
Commercial and home		
Commercial refrigeration and cooking	150.0	263.0
Refrigeration machinery	25.0	61.0
Home refrigeration and cooking	202.3	480.0
Subtotal	377.3	804.0
Grand total	1134.2	2172.0

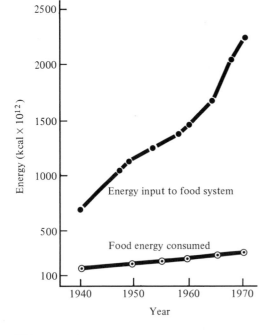

Energy input to food system

Food energy consumed

Fig. 10.3 Energy use in the U.S. food system for the period 1940 to 1970, and the annual caloric content of the food consumed during each year of that period. (Adapted from reference 2.)

The energy directly consumed on farms accounts for 3.5 percent of the nation's total energy consumption, as we noted earlier. However, the energy consumed directly on the farm accounts for only one-fourth of the total energy consumed in the food production, processing, and distribution system in the U.S. In fact, the energy dedicated to food-related uses in the U.S. is about 12 percent of the total national energy use. [3]

Expenditures for food account for 20 percent of the disposable personal income in the U.S., and totaled $132 billion in 1970. Therefore, it is not surprising that Americans use 12 percent of their total energy supply on the total food system (from seed to table).

Farmers process little of their own food, being dependent themselves upon the food-processing and retailing industries. For every farm worker, it is estimated that there are three food-system support workers. Thus, about 20 percent of the nation's work force and its industries are involved in supplying food.

Energy Use and Productivity

In the period 1910 to 1940, the introduction of energy-intensive farming changed our methods of farming and started to yield significant increases in farm output. In the period 1940 to 1960, the U.S. experienced an approximately linear growth of farm output as the energy input was linearly increased. (For example, the energy input to the farm essentially doubled during that period, while the farm output increased by a factor of 1.6.) However, the curve is now beginning to level off, and increasingly more energy is required in order to increase the farm output. New technologies will be required in order to maintain a useful increase in productivity for any increase in energy input. [4] We may be approaching the point of diminishing returns in this use of energy.

Energy Subsidies

The number of calories supplied to produce one calorie of food energy consumed by people has continued to grow since 1910. We might define this ratio as the Food–Energy Ratio (or FER), as follows:

$$\text{FER} = \frac{\text{Input energy}}{\text{Energy of food consumed}}. \tag{10.1}$$

This ratio was essentially 1 in 1900, 5 in 1945, and 10 in 1975. [2] There is some reason to believe that this ratio is beginning to level off, because of increasing fuel costs and limited increases in productivity for increases in energy outputs.

Figure 10.4 shows the food–energy ratio (FER) for various food crops, as well as the history of the U.S. food system [2] Many crops have progressed from a low-intensity method to a high-intensity, high-energy input method.

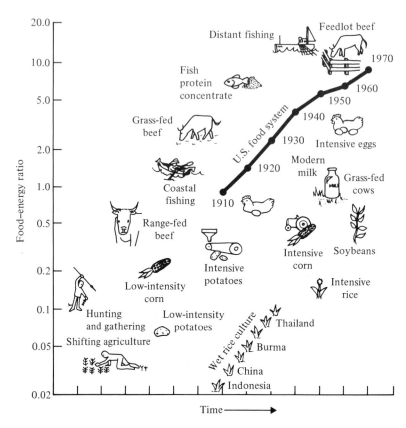

Fig. 10.4 *The food energy ratio, which is the input energy required per energy of food consumed for various food crops. Agricultural methods evolve over the centuries towards the right side of the graph. (Courtesy of J. Steinhart and the American Association for the Advancement of Science. Adapted from* Energy: Sources, Use and Role in Human Affairs, *Carol E. and John S. Steinhart, Duxbury Press, No. Scituate, 1974.)*

The Energy to Grow Corn

As an example of an analysis of the energy required to produce a crop, corn (maize) was chosen for study [5, 6], since it is intermediate in energy inputs and is one of the leading grain crops in the U.S. and the world. Corn ranks third in world production of food crops and has been the object of intensive farming methods. In the U.S., the corn yield per acre increased from 26 bushels/acre in 1909 to 87 bushels/acre in 1971. While corn yields increased about 240 percent from 1945 to 1970, the labor input per acre *decreased* more than 60 percent. Intense mechanization reduced the labor input and, in part, made possible the increased corn yield.

In 1970, fuel consumption per acre of U.S. corn production was 22 gallons, and 203 pounds of fertilizer was required—a total input of about 2.9 million

kilocalories. [5] From 1945 to 1970, mean corn yields increased from about
34 bushels per acre to 81 bushels per acre (2.4-fold); however, mean energy inputs
increased from 0.9 million kcal to 2.9 million kcal (3.1-fold).

Hence, the yield in corn calories decreased, from 3.7 kcal per one fuel kilocalorie input in 1945, to about 2.8 kcal, in the period of 1954 to 1970, a 24-percent
decrease.

In the U.S., in 1970, a total of $2,897 \times 10^3$ kilocalories was used as energy
input to the corn crop. The corn yield was $8,165 \times 10^3$ kilocalories of food
energy. Therefore, the Food–Energy Ratio is:

$$\text{FER} = \frac{2897}{8165} = 0.355. \tag{10.2}$$

This is a very favorable ratio compared to that for many other foods, as can be
seen by examining Fig. 10.4. Incidentally, the energy input from the sun, which
we do not include in our ratio, was 27 million kilocalories for the U.S. crop in 1970.

The energy input to the growing of corn from the use of nitrogen fertilizer is
about 32 percent of the total energy input. [5] Therefore, it is important to use
an optimum amount. Figure 10.5 shows the inverse of the Food–Energy Ratio
plotted against the input of nitrogen (the solid line). The yield in bushels per acre
has also been plotted against the input of nitrogen (the dotted line). The maximum
ratio (3.0) of output to input energy is obtained at about 120 pounds of nitrogen
per acre. It is clear that wise use of fertilizer energy would result when applications
of less than 160 pounds per acre are used.

A study in Britain showed that the Food–Energy Ratio for corn can be
improved (i.e., reduced) by several methods. Maize grain is used as the food energy
output in the U.S. while in Britain maize silage is used. Maize silage, where the

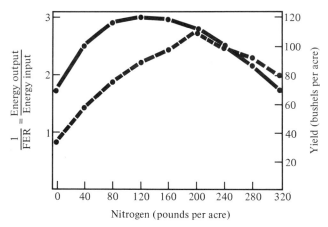

Fig. 10.5 *The energy output obtained divided by the energy input for corn, versus the amount
of nitrogen fertilizer applied (solid line). The dotted line shows the yield per acre of corn versus
the nitrogen fertilizer applied. (Adapted from reference 5.)*

whole corn plant is used, serves in Britain as a winter feed for both beef and dairy cattle. [6] Use of the whole plant doubles the consumable energy, while the energy inputs were reduced by using different methods, so that the Food–Energy Ratio was decreased to 0.112 (compared to the ratio of 0.355 computed earlier).

It was also shown that if livestock slurries are used as a substitute for all or part of the fertilizer, the Food–Energy Ratio can be further reduced to 0.0056. [6] Livestock slurry is a mixture of dung, urine, and water which accumulates on farms when animals are housed indoors, or in intensive collections of animals. An application of 20 tons/acre in the spring would come close to supplying the fertilizer requirements for silage maize. Some energy would be required for the

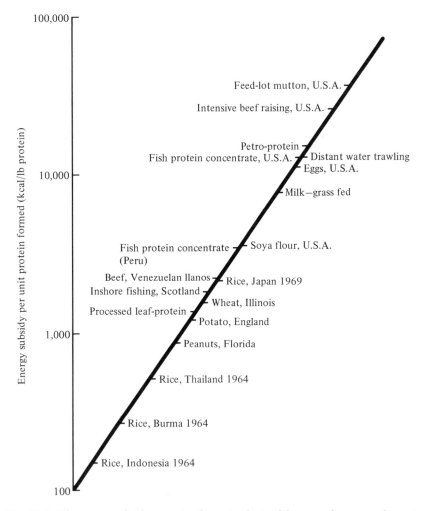

Fig. 10.6 *The energy subsidy per unit of protein obtained for several sources of protein.*

collection, storage, and disposal of the slurry whether or not it were used to fertilize a crop; so if it is available in the vicinity of the corn field, it can be used for fertilizer with very little additional energy expenditure.

Clearly, the production of forage or silage maize is much more efficient in terms of energy usage than the production of maize grain. Much of the energy required for growing a grain crop is not required for the production of maize silage. Furthermore, the gross energy/pound from maize silage is similar to that from maize grain, but the yield of maize silage is double that of maize grain. Here is an example of using an alternative method of agriculture in order to utilize a more energy-efficient process.

Energy Subsidy

The Food–Energy Ratio (defined as the ratio of the input energy to the energy of the food consumed) is greater than one for many foods. The difference between the actual input energy and the ideal value could be described as unnecessary loss, or as a *subsidy* to the food. [7, 8]

The energy subsidy for several approaches to modern protein production is shown in Fig. 10.6. The energy subsidy is expressed as the energy input (in kilocalories) required *per unit of protein obtained* (in pounds). Thus, for example, feed-lot beef require about ten times as much energy input per pound of consumable protein as does soya flour or rice.

If the purpose of crop and animal agriculture is the production of food energy, then direct consumption of plant products is a more efficient use of energy than is consumption of animals that have been raised on feed grains instead of roughages. Grains are fed to animals to produce meat that is about 17 percent protein, but soybeans that contain 37 percent protein can be produced with a substantially greater energy efficiency than that of meat. Indeed, soybeans clearly produce substantially more protein per unit of energy than do broilers, hogs, or cattle. Also, animal protein production requires more land than grains.

Energy Conservation in the Food System

The purpose of crop and animal agriculture is not limited to simply producing food energy, at least in the wealthier industrialized countries, but also to producing desirable and appetizing foods. Even so, the question remains, how can we minimize our dependence on fossil fuels in the whole food and agriculture process, from seed to table?

The efficiency of fertilizer use needs to be examined for each crop that is dependent upon heavy applications of fertilizer. Another heavy energy use is for irrigation. The net pumping requirements, for example, of the new California Water System, which brings water from Northern California to Southern California, are 8 billion kilowatt-hours of electricity per year. Also, the wise and limited use of pesticides needs to be continually considered.

The wise use of farm machinery and food-transportation equipment should be studied since the food system uses more gasoline and diesel fuel than any other single industry. Over 8 billion gallons of gasoline are used each year for farm machinery alone. New approaches to tillage and harvesting could produce beneficial reductions in fuel use. The direct application of power to the land by actuation of implements, rather than indirectly through drawn implements entailing heavily laden wheels, can be classed as a power-conservation technique; and we can envisage, in years to come, new implements as revolutionary in their way as was McCormick's reaper.

Since food processing, retailing, and home processing use a major portion of the energy in the food system, new energy-saving methods are important in this sector.

New plants which capture a greater percentage of the solar energy would result in an improved Food–Energy Ratio by working on the output side of the equation. Of the solar radiation reaching the earth, less than 10 percent is captured as human food in cropped areas.

10.3 AN APPROPRIATE TECHNOLOGY

The agricultural technology for the future and for less industrialized nations may not be the technology that already exists in Europe and the U.S.A. A technology appropriate to the village system of India and the rice fields of Asia should be developed, rather than concentrating on the introduction of energy-intensive methods. The appropriate technology for the agriculture of a country utilizes the skills, resources, climate, and cultural resources of the nation to produce food to feed its population.

As Schumacher notes in his important book, *Small is Beautiful*, what is needed is production by the masses, not mass production. This system, which he calls an intermediate technology, mobilizes the resources of the nation and supports them with first-class tools. [9, 16] This approach makes use of knowledge and experience, but is conducive to decentralization, compatible with ecology, and based on simple and efficient methods.

An example of an appropriate technology for agriculture in the future would be a system of less intensive energy use, which would encourage the development of smaller, less energy-intensive farms; the use of farming methods based on diversity; the planting of legumes such as soybeans, which add nitrogen to the soil and minimize the need for nitrogen fertilizers; and the use of biological pest-control methods as substitutes for the intensive use of chemicals. In addition, hand-spraying of pesticides and herbicides, rather than application by large machines and airplanes, may save significant energy, in the opinion of many. Diversity of this sort could be the starting point for the development of a sounder, more efficient agricultural system.

The packaging of food has passed beyond the point where it is necessary to hold and store considerable amounts of food in one place. Appropriate energy-saving containers and packages should be introduced into the system. The transportation of food can be examined for appropriate energy savings with little loss of food availability to the customer.

The use of chemical fertilizers can be reduced through the use of manure and crop rotation. The decentralization of energy-intensive animal feedlots would assist in this process.

Better use could be made of natural energy sources such as solar power. Solar crop-drying techniques are under development in many states, including Indiana, where 60 million gallons of liquid petroleum fuel and 1.5 billion cubic feet of natural gas are now used yearly to dry the corn crop. Windmills, instead of fossil-powered electrical pumps, can be used in many cases to pump water for agricultural irrigation.

The challenge to all nations in the future is to make the people's relationship to the soil a sustaining and natural relation, while utilizing all the tools for increasing their ability to feed the people of the world.

Nevertheless, it is important not to allow our concern for reducing the energy required to produce a food crop to degenerate into a mood of antitechnology. The appropriate use of machines and other tools is still required to produce enough food to feed the world's population. A return to subsistence farming is not energy-efficient since, if the human and animal energy and the feed necessary for the animals is accounted for, the Food–Energy Ratio is increased and may be worse than the case for farms using appropriate machinery, fertilizer, and other inputs. [10] An appropriate amount of irrigation may make a significant difference in raising the productivity of farms. In some cases, more or less centralization of agricultural crops and animals is appropriate, and each region's needs and resources must be examined.

REFERENCES

1. G. H. HEICHEL, "Energy needs and food yields," *Technology Review*, July/August 1974, pp. 19–25.

2. J. S. STEINHART and C. E. STEINHART, "Energy use in the U.S. food system," *Science*, April 19, 1974, pp. 307–315.

3. E. HIRST, "Food-related energy requirements," *Science*, April 12, 1974, pp. 134–138.

4. G. BORGSTROM, "Food, feed, and energy," *Ambio*, **2**, No. 6, 1973, pp. 213–219.

5. D. PIMENTEL, *et al.*, "Food production and the energy crisis," *Science*, Nov. 2, 1973, pp. 443–448.

6. B. PAIN and R. PHIPPS, "The energy to grow maize," *New Scientist*, May 15, 1975, pp. 394–396.

7. M. SLESSER, "Energy analysis in policy making," *New Scientist*, Nov. 1, 1973, pp. 328–330.

8. M. SLESSER, "Energy analysis in technology assessment," *Technology Assessment*, **2**, No. 3, 1974, pp. 201–208.

9. E. F. SCHUMACHER, *Small is Beautiful*. Harper and Row, Publishers, New York, 1973.

10. A MAKHIJANI and A. POOLE, "*Energy and Agriculture in the Third World*," Ballinger Press, Barton, 1975.

11. R. W. GRAHAM, "Fuels from crops: Renewable and clean," *Mechanical Engineering*, May 1975, pp. 27–31.

12. G. CHEDD, "Cellulose from sunlight," *New Scientist*, March 6, 1975, pp. 572–575.

13. M. KENWARD, "Food for thought," *New Scientist*, Aug. 21, 1975, pg. 436.

14. B. LEON, "Agriculture—A sacred cow," *Environment*, Dec. 1975, pp. 38–40.

15. G. H. HEICHEL, "Agricultural production and energy resources," *American Scientist*, Feb. 1976, pp. 64–72.

16. R. C. DORF and Y. HUNTER, *Appropriate Visions: Technology and the Individual*, Boyd and Fraser Pub. Co., San Francisco, 1978.

EXERCISES

10.1 Prepare a profile of energy use in agriculture in your state. Determine the number of farms in 1950 and the current yield per acre for a primary crop of your state. Determine the use of fertilizer, the number of tractors per farm, and the farm population for 1950 and this year. Prepare an analysis of the utilization of energy on farms for agriculture in your state.

10.2 Agricultural technology can become more efficient with an investment in research. Consider the following research objectives and rank them in priority order. State your reasons for selecting priority one and two.

Encourage the optimum use of manufactured fertilizers and animal manures in various cropping systems and on different soil types.
Develop more efficient methods of conserving and feeding grassland crops to animals.
Reduce cultivation operations and develop versatile implements.
Develop more efficient harvesting and transport systems.
Optimize the supplementary heating of glasshouses by fossil fuels.
Explore the possibilities of maintaining plant-foliage temperatures with lower air temperatures (by radiant heating, for example).
Assess policies for the use and replacement of farm machinery.
Improve the photosynthetic ability of growing crops.
Keep under review the exploitation of methane gas produced from animal wastes.
Encourage the use of straw as an animal feed, and explore its potential as a direct fuel, particularly in crop drying.

10.3 Waste hot water from electrical generating power plants can be used for agricultural purposes. By cycling the waste through a 2-inch-diameter pipe buried at a depth of 1 foot, the heat can be dissipated and also used as a soil-warming system for agricultural production. Consider the benefits and costs of this project.

10.4 The conversion of crops to fuels is being considered. [11] An agricultural crop would be used to collect solar energy and then the crop would be processed into a gaseous or liquid fuel. Cellulose in the form of plants or trees could be converted to methane or other fuels. Another approach is the growing of kelp offshore and then converting it to fuel. [12] Examine the technical and economic factors involved in the establishment and operation of an energy farm.

10.5 By 1972 the energy to provide the average protein consumption in Britain had risen as much as the use per capita in the developing world for all purposes. [13] If the whole world enjoyed the same standard of food consumption as Britain and the equivalent agricultural system was employed worldwide, the annual use for agriculture would be 40% of the total global fuel consumption in 1972. Energy use in agriculture in Britain includes a large commitment to the processing and transportation of food. Examine the agricultural system of Britain, Poland, and an African nation, and compare the suitability of the approaches.

10.6 The most energy-intensive food items in a kitchen are the throw-away aluminum canned beverages, plastic-containered milk, instant dinners, and frozen prepared foods. Examine the use of such items in your home or college cafeteria, and determine whether useful alternatives exist for the cook. What actions would you propose?

10.7 Fertilizer prices have increased by a factor of 3 or 4 since 1973. India's grain production fell ten million tons, despite good weather, when it could only afford to buy 3 million tons of fertilizer instead of the 30 million tons it wanted.

Attention is turning to making fertilizer from natural gas now burned off into the atmosphere at oil fields. A total of 4.5 trillion cubic feet is burned off every year in Venezuela, Nigeria, and the Persian Gulf region—enough to make twice the amount of nitrogen fertilizer now produced worldwide. What are the deterrents to using such natural gas to make fertilizers? Can you suggest some incentives to promote the development of such plants?

*10.8 In order to reduce the dependence of a farm on energy input, a farmer can reduce tillage, use less herbicide, improve the timing of irrigation, or change his method of drying crops. However, energy is such an integral part of the production system that the farmer's hands may be tied—at least over the short term. For example, a farmer who now burns fuel to dry corn to prevent spoilage in storage cannot simply let the corn dry in the fields as he once did. The high-yield corn varieties being planted require a longer growing season; later-maturing ears are deprived of the hot, pre-harvest weather that dries them in the field. In addition, instead of harvesting corn on the cob and storing it to dry, corn is now machine-harvested as grain (machines strip the kernels from the cob in the field) which usually includes forced evaporation. Thus, to avoid burning fuel for drying, a farmer might also have to use different genetic varieties, different harvesting machinery, and different storage facilities.

What changes in farming methods do you recommend and in what priority order?

---

* An asterisk indicates a relatively difficult or advanced exercise.

10.9 American agronomists have long made favorable comparisons of their cattle industry with the seemingly wasteful social practice that protects the Indian cattle population. In the past the predominantly Hindu population has not eaten the beef or condoned the killing of cattle. In contrast, the U.S. has a thriving cattle industry, which annually slaughters 30% of the cattle population to support the American people's beef-centered diet. It is difficult for many to understand why India does not use its cattle resource. In a recent study four types of utility for cattle and buffalo were identified: (1) milk production, (2) traction, (3) dung, and (4) meat and hides. [14] A measure of efficiency was defined as the sum of the outputs of cattle used by people divided by the sum of all inputs to cattle. Using this measure, the energy efficiency for cattle was 7% and 22%, for the U.S. and India, respectively. Explore this analysis and determine whether the U.S. could learn some useful methods from India and the Third World.

10.10 In order to reduce the amount of energy used in food production we might: (1) alter plant mechanisms to enhance yields, (2) develop new energy-efficient schemes for the nurturing of plants and animals, and (3) exploit underutilized plant byproducts as an energy resource. There is renewed interest in substituting animal manures for commercial fertilizer. About 50% of the domestic animal manure, about 10^9 tons annually, is readily accessible. [15] Investigate the usefulness of such an approach, and determine what percentage of reduction in fertilizer use could be achieved in the U.S. If manure was all used for corn acreage, what percentage of that acreage could be fertilized with manure?

10.11 Most large supermarkets use open-topped freezers for frozen foods and open coolers for dairy products. Visit a local supermarket and estimate the refrigeration load and the electric energy used. Estimate the possible savings if glass doors were used on all refrigerators.

Hydroelectric Power Generation

11.1 INTRODUCTION

Energy flows in the hydrologic cycle of evaporation, precipitation, and surface runoff of water, and can be harnessed to generate electric power. Worldwide, as we noted in Fig. 3.7, this flow is 4×10^{16} watts, and only a small fraction is used by man to turn water turbines and electric generators.

The earliest use of water power was in water mills several thousand years ago. By the 16th century, water mills had to be adapted to many processes. The use of water falling through a distance, in order to turn a turbine and later an electric generator, began after the general introduction of electric power in the U.S. and Europe at the end of the 19th century. In 1882 a hydroelectric station was placed in operation on the Fox River in Appleton, Wisconsin. A waterwheel drove two Edison-type generators with a total output of 25 kilowatts, which was sold to two paper mills and one residential customer for lighting incandescent lamps.

In 1895 the first large hydroelectric power plant was built at Niagara Falls, New York. A large central power station was used at the Falls and the power was distributed over a wide area. Large water turbines were used to harness the falling water. The height of the falls is about 200 feet and an estimated 210,000 cubic feet (5.95×10^6 liters) of water per second flows over it. Three million dollars of capital was raised to build the power plant, and it was decided to generate alternating-current electric power to be transmitted over a transmission line to Buffalo. Approximately 4 megawatts was generated at the Niagara power plant. Generators in the first Niagara Falls power plant in 1896 are shown in Fig. 11.1.

11.2 THE HYDROELECTRIC POWER PLANT

A simple schematic diagram of a hydroelectric plant is shown in Fig. 11.2. The water, often stored behind a dam, falls through a height (or head) of distance h. The water's potential energy is converted to kinetic energy, and the flowing water

Fig. 11.1 *Generators in the first Niagara Falls power station (photo, 1896). (Courtesy of the Niagara Mohawk Power Corporation.)*

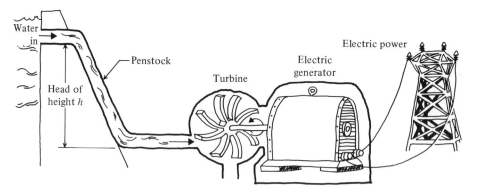

Fig. 11.2 *Schematic diagram of a hydroelectric generating plant.*

turns a water turbine. The rotating shaft of the turbine turns the electric generator, which yields the electricity.

The potential energy of the water is $P = mgh$. Let us calculate the power generated by water held by a dam with a head of 370 meters. If we assume, for this calculation, that all the potential energy is converted to electric energy, we have, for each liter of water:

$$E = mgh = 1 \times 9.8 \times 370 = 3630 \text{ joules/liter.} \qquad (11.1)$$

The speed of the water, v, as it enters the turbines is:

$$E = \tfrac{1}{2}mv^2 = \tfrac{1}{2} \times 1 \times v^2; \qquad (11.2)$$

therefore, the maximum attainable velocity is

$$v = (7260)^{1/2} = 85.2 \text{ meters/second.} \qquad (11.3)$$

Water turbines are highly efficient, converting most of the kinetic energy to rotational energy of the shaft. The outlet speed of the water leaving the turbine is sufficiently small so that the efficiency of the turbine is very high (95 percent). If the efficiency of the turbine is 95 percent and the efficiency of the electric generator is 95 percent, the overall efficiency of these two devices is $(0.95)^2 = 0.90$. Therefore, if we wish to generate 400 megawatts of power, we might use four turbine–generator sets. Then we have:

$$\eta \frac{E}{\Delta t} = 400 \text{ megawatts,}$$

or

$$\frac{0.9 \ mgh}{\Delta t} = 400 \times 10^6 \text{ watts.} \qquad (11.4)$$

Therefore, from Eq. 11.1 we have:

$$\frac{0.9 \ m \times 3630}{\Delta t} = 400 \times 10^6$$

or

$$\frac{m}{\Delta t} = 1.22 \times 10^5 \text{ kilograms/second.} \qquad (11.5)$$

Thus, we would need a mass of water of 1.22×10^5 kilograms moving through the turbines every second.

There are additional losses due to friction in the penstock and at the entrance to and exit from the penstock. Therefore, the actual total efficiency of a hydroelectric turbine, generator, and associated penstock may be 0.75 to 0.85.

A typical power plant with a high head is shown in Fig. 11.3. This plant uses two Pelton wheels on vertical shafts and generates 172,000 kilowatts.

Fig. 11.3 *A hydroelectric power plant using a head of 1,193 feet on the Pit River in California. The water descends from a reservoir on the McCloud River through 5500 feet of steel penstock. (Courtesy of Pacific Gas and Electric Company.)*

Hydraulic Turbines

The hydraulic turbines perform a continuous transformation of the potential and kinetic energy of the water into useful work. In an impulse turbine, the available head is converted into kinetic energy by a contracting nozzle. The resulting water jet drives the bucketlike structures on the shaft. A modern Pelton impulse-type turbine is shown in Fig. 11.4.

The other type of commonly used turbine is the reaction turbine, which uses a pressure head across the turbine as well as kinetic energy flow. A Francis reaction turbine is shown in Fig. 11.5. Pelton impulse turbines are normally used with heads exceeding 1000 feet, while reaction-type turbines are used for heads of 100 to 1000 feet. Fixed-blade propeller turbines are also used for heads up to 100 feet.

Fig. 11.4 *A Pelton turbine runner. This photograph, looking almost directly up from the pit, shows the fabricated bucket wheels of a vertical impulse-type turbine and some of the nozzles and jet deflectors. This 39,000-hp (29-megawatts) six-jet installation operates at 429 revolutions/minute under a 1233-foot head. (Courtesy of Allis-Chalmers Corporation.)*

Fig. 11.5 *A Francis-type turbine runner with a diameter of 19 feet. This runner is used in a 140,000-hp (104-megawatt) turbine operating at 100 revolutions/minute under a 162-foot head. (Courtesy of Allis-Chalmers Corporation.)*

11.3 HYDROELECTRIC POWER IN THE U.S.

Hydroelectric power is an important source of electric power in the U.S. and many Northern European nations. The total installed electrical generating capacity of the U.S. in 1972 was 418,000 megawatts, of which 54,000 megawatts, or 13 percent, was generated by hydroelectric generating plants. [1] In 1971, water power generated $269,580 \times 10^9$ watt-hours in the U.S.A.

Electrical power from a hydroelectric plant is directly usable as electric energy in the home or industry. The use of fossil fuel, such as gas or coal, to generate the electric power would require an input energy approximately equal to three times the output energy, since the average efficiency of a fossil-fuel power plant is 0.33. Therefore, in order to generate 270×10^{12} watt-hours (9.2×10^{14} Btu) it would have required 810 watt-hours (2.76×10^{15} Btu). The total energy input in 1971 was 69×10^{15} Btu; therefore, hydropower accounted for 4 percent of the necessary input energy.

The hydroelectric capacity of U.S. power stations is given in Table 11.1, which also shows the electric energy provided by hydroelectric power stations over the period 1920 to 1971. Both of these measures grew at an average rate of 5 percent per year over the past several decades. The growth of hydroelectric capacity and electric energy provided by these hydropower stations is shown graphically in Fig. 11.6.

Table 11.1
WATER POWER IN THE UNITED STATES

Year	Capacity (Megawatts)	Electric energy output provided (Gigawatt-hours)
1920		18,779
1925		25,496
1930		35,878
1935		42,727
1940		50,131
1945	15,892	84,747
1950	18,675	100,885
1955	25,742	116,236
1960	33,180	149,423
1965	42,948	196,984
1970	51,952	250,611
1971	53,404	269,580
1972	53,778	—
1975	60,000	289,000
1980*	74,000	357,000

* Projected

A capacity of 53,404 megawatts (in 1971 in the U.S.) could provide electric energy for 52 weeks a year at 24 hours a day. In practice, hydroplants are not utilized much more than half the time since the electrical demand fluctuates and the plant must also be shut down periodically for maintenance. If the capacity of

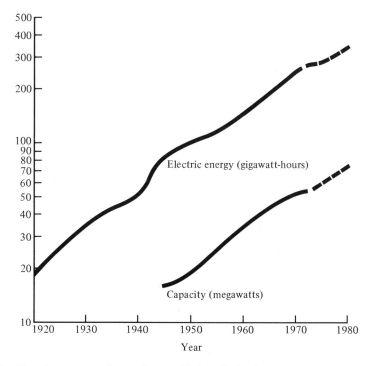

Fig. 11.6 *Electric energy and capacity provided by hydroelectric generating stations in the period 1920 to 1972 in the U.S.A.*

53,404 MW were used all the time, we would have:

$$E = 53{,}404 \times 10^6 \times 365 \text{ days} \times 24 \text{ hours/day}$$
$$= 467.8 \times 10^{12} \text{ watt-hours.} \tag{11.6}$$

The actual energy generated in 1971 (see Table 11.1) was 269.6×10^{12} watts. The *load factor* of an electric power plant is:

$$\text{L.F.} = \frac{\text{Actual energy generated}}{\text{Theoretically possible energy generated}} \tag{11.7}$$

for a given period, usually a year. For hydropower, in 1971, we have

$$\text{L.F.} = \frac{269.6 \times 10^{12} \text{ watt-hours}}{467.8 \times 10^{12} \text{ watt-hours}}$$

$$= 0.576. \tag{11.8}$$

The capacity and electric energy generated by hydropower plants will continue to grow, but at a lower rate over the next decade, as shown in Fig. 11.6. The principal forms of new development will be small sites of less than 200-MW capacity located in the western areas of the U.S.

There was a capacity of hydropower plants in January 1976 of 60,000 megawatts and the estimated undeveloped water power in the U.S. is equal to 126,078 megawatts. Seventy-one gigawatts of this undeveloped hydropower is located in California, Oregon, and Washington. There are currently under construction about 8 gigawatts of hydropower plants.

While there has been a fivefold increase in the installed hydroelectric generating capacity of the U.S. since the 1930's, this dramatic growth has been overshadowed by an even faster rate of expansion of steam-electric plant capacity. In the past, hydroelectric plants provided a third of the total electric generating capacity, while today they provide only 14 percent of the capacity. This decline is expected to continue because of the advent of nuclear power plants and the growing difficulty in developing additional hydroelectric plant sites.

11.4 HYDROELECTRIC GENERATING PLANTS IN THE U.S.

In 1972, there were 1511 hydroelectric power plants in the U.S. There are 129 plants with a capacity of over 100 megawatts.

The Grand Coulee Dam on the Columbia River in Washington will be the largest hydroelectric facility in the world when its ultimate capacity of 6 gigawatts is reached. There are three turbine generators currently approaching completion of construction, each at a rating of 700 megawatts, which will add to the existing capacity of 2100 megawatts.

The Hoover Dam Reservoir on the Colorado River was completed in 1936 to provide flood control and electric power. Lake Mead, behind the Dam, provides irrigation water to the U.S. and Mexico, and water to residents of Southern California. The dam is 726 feet high and is 660 feet thick at the bottom and 45 feet thick at the top. The Lake is 110 miles long and holds 26×10^6 acre-feet (3.2×10^{13} liters) of water. The Hoover Dam and generating station is shown in Fig. 11.7.

The hydroelectric plant has a capacity of 1345 megawatts which is generated by means of 17 generators rated at between 40 and 95 megawatts each. Hoover Dam and its associated power plant cost 175 million dollars to construct, which resulted in a cost of \$130 per kilowatt of capacity. The power from the plant is sold to Arizona, Nevada, and the Los Angeles area. The power plant is operated by the Los Angeles Department of Water and Power and by the Southern California Edison Company, under Federal license. The operating income for the power plant was 11 million dollars in 1971.

11.5 HYDROELECTRIC GENERATING CAPACITY IN THE WORLD

Hydroelectric power is used throughout the world. In 1969, there were 274.7 gigawatts installed in the world, and North America accounts for only 30 percent of the total. The installed capacity in several regions of the world is given in Table 11.2. It can be noted that hydroelectric capacity is an important share of the total installed electric generating capacity in many regions of the world.

Fig. 11.7   *Hoover Dam and electric power plant. (Courtesy of Los Angeles Department of Water and Power.)*

Table 11.2

INSTALLED HYDROPOWER CAPACITY OF THE WORLD IN 1969

Region	1969 installed hydroelectric capacity, 10^3 MW	Percentage of total installed electrical generating capacity
North America	83.9	22.1
Central America	0.5	41.7
West Indies	0.2	4.5
South America	13.4	50.0
Europe, including U.S.S.R.	126.4	26.4
Africa	6.1	29.7
Asia	37.6	32.1
Oceania	6.6	35.4
World	274.7	26.3

The Aswan High Dam is an example of a very important development in Africa and the Middle East. The dam, used for flood protection and to provide irrigation water in the Nile Valley, will store one year's river flow in order to provide water in times of drought. The dam is 111 meters above the river bed and drives an installed electric capacity of 2.1 gigawatts, larger than Hoover Dam. There are 12 generators installed and over 10^7 megawatt-hours are provided by the plant in one year.

In Chile, half of the installed electric capacity of 2140 megawatts is provided by 15 hydroelectric stations. It is planned to install an additional 1400 megawatts of hydropower by 1984. [2] It is not difficult to understand how important hydropower is to a developing nation.

The world's largest single-site hydroelectric power plant commenced total operation in 1975 at Churchill Falls, Labrador, Canada. This plant produces 5225 megawatts of electricity.

11.6 PUMPED-STORAGE PLANTS

Water can be pumped up to a reservoir by electric pumps and then allowed to fall through a turbine to generate electricity when it is needed. On the eastern shore of Lake Michigan is the biggest pumped-storage plant in the world. The Ludington facility takes water from Lake Michigan and stores it in a manmade reservoir 1 mile wide by 2 miles long. At times of peak electrical power demand, water is released from the reservoir and drives turbines generating 1.9 gigawatts of electricity. Reversible turbines are used as pumps and driven turbine-generators. Water is usually pumped up at off-peak hours, usually at night and on weekends. The Ludington pumped-storage system is shown in Fig. 11.8.

It is estimated that the world's annual growth rate for pumped storage up to 1980 will be about 15 percent. In Europe alone it is likely that some 26 gigawatts of pumped-storage capacity will be in service by 1980.

A pumped-storage system uses power available at off-peak hours to pump the water to the reservoir at a pump efficiency η_p. The efficiency of the turbine and electric generator is η_{tg}, which is approximately 0.80. If the pump efficiency is $\eta_p = 0.90$, then the efficiency of the pumped-storage system is:

$$\eta = \eta_p \eta_{tg} = 0.72. \tag{11.9}$$

Therefore, approximately one-fourth of the originally available electric energy is lost in the process. Nevertheless, it is a useful process since it uses energy generated by existing power plants at off-peak hours when they would otherwise be idle.

11.7 ECONOMICS OF HYDROELECTRIC POWER PLANTS

The capital investment required to construct a new hydroelectric project varies widely according to the size and location of the project and the cost of relocating buildings or other items in the path of the project. The average cost of construction

Fig. 11.8 *The Ludington pumped-storage project in Michigan. This facility can yield 1.9 gigawatts maximum power and 15,000 megawatt-hours of stored energy. The Ludington plant uses Lake Michigan as the lower reservoir while the upper reservoir is a manmade lake measuring about 3 km by 1.6 km. It cost $340 million to construct. The upper reservoir is 105 meters above Lake Michigan. The Ludington facility is the largest pumped-storage project in the world. (Photo courtesy of Ebasco Services, Incorporated.)*

of a hydroelectric plant per kilowatt installed is higher than for a thermal electric plant. However, the hydroelectric plant requires no fuel, and therefore is much less expensive to operate, compared to a thermal plant that uses a fossil fuel. Hydroelectric power is the least expensive available. The capital investment costs for hydroelectric plants range from $100 to $400 per kilowatt, depending upon what portion of the project is allocated to electrical generation and what portion is allocated to irrigation and flood control. Hoover Dam was built for $130/kilowatt in 1936. As the costs of fossil fuels rise, the low costs of operating hydroelectric power plants will become even more advantageous.

11.8   ENVIRONMENTAL EFFECTS OF HYDROPOWER PLANTS

Hydroelectric projects often are combined with flood control and irrigation projects and therefore must involve the evaluation of the total environmental effects of the project. Furthermore, the dams often provide recreational areas with a lake and beaches. Undesirable effects include upstream flooding of river valleys, downstream water-flow reduction, impact on the area required for the lake, and the effects of long electric transmission lines from the project site to the area where the electric power is used. The environmental impacts of a hydro-electric project must be thoroughly analyzed since, after it is completed, they are essentially irreversible.

For example, the Aswan High Dam reduces the supply of silt that maintains the delta of the Nile, thus threatening coastal agriculture. Therefore, as the Nile plain is irrigated, the delta loses its necessary supply of silt. The balance of all these factors must be included in the decision to construct a dam.

In Table 11.3, an attempt is made to summarize a comparison of the adverse effects of hydro, nuclear, and fossil-fuel power plants. In general, hydropower projects are valuable and economic alternatives to fossil-fuel or nuclear-power plants.

Table 11.3

COMPARISON OF ADVERSE EFFECTS ON SEVEN ENVIRONMENTAL PARAMETERS
OF HYDRO, NUCLEAR, AND FOSSIL-FUEL POWER PLANTS

Adverse effects	Hydro	Nuclear	Fossil-fuel steam
Land use	Greatest effect	Great effect	Least effect
Major river regulation	Yes	No	No
Thermal pollution	No	Yes	Yes
Air pollution	No	Yes	Yes
Radioactive waste	No	Yes	No
Consumes natural resources	No	Yes	Yes
Sensitivity to earthquakes	Moderate	Greatest	Least

Hydroelectric power projects use a renewable resource, are able to make rapid changes in power output, have low operating costs, and are well adapted for serving peak-load demands. The hydroelectric plant will remain an important part of the world's source of electric energy for many decades in the future.

REFERENCES

1. *Energy Facts*, Science Policy Research Division, Library of Congress, U.S. Government Printing Office, Washington, D.C., 1973.

2. M. AGUILAR, "Three important hydro projects in Chile," *Water Power*, May 1973, pp. pp. 161–165.

3. F. BASSLER, "New proposals to develop the Qattara Depression," *Water Power*, August 1975, pp. 291–296.

EXERCISES

11.1 A large hydroelectric project is being pursued on the Congo (now the Zaire) River in the country of Zaire. The project is being built in three stages. The first stage, which has been built, will supply power to Kinshasa, capital of Zaire, and to a copper smelter. Its output will be 2400 million kilowatt-hours each year. The second stage, which recently came into production, provides an additional 9600 million kilowatt-hours. The third stage would have a potential of 19,700 million kilowatt-hours. Compare this project with the Columbia River plants in the state of Washington. Examine the potential for rapid industrialization for Zaire as a result of this hydroelectric development. Can you identify other potential hydroelectric sites in developing countries and estimate their potential?

11.2 The first generator began producing power at Hoover Dam on October 26, 1936. With the installation of the final generating unit in 1962, there are now 17 turbines, with a total rated capacity of 1344 megawatts. The power generation has averaged 4.3 billion kilowatt-hours annually. The water falls through 600 feet to the turbines. Calculate the load factor of the power plant.

11.3 Lester Allen Pelton came to California in 1850 seeking his fortune. While he never struck gold, he did invent what is now called the Pelton wheel. The waterwheels that furnished the motive power for the local mines and machinery fascinated him. Little had changed since the time of the Pharaohs. The wheel was powered by a stream of water on buckets around its rim—but efficiency was impaired because water would splash back and impede the progress of the next bucket. His idea was to split the stream of water so it would be deflected from the path of the next bucket. It was in 1878 that Pelton introduced his wheel with its radical principle of twin cup-shaped buckets. He built a number of buckets of differing shapes until, in 1880, he applied for a patent on his invention. Draw a common shape of Pelton's wheel and investigate the increased efficiency of the Pelton wheel compared to the old waterwheel with rotating buckets. Estimate the percentage increase in efficiency.

11.4 The Tennessee Valley Authority (TVA) brought electric power to the citizens of an 80,000-square-mile area of seven Southeastern states in the late 1930's. TVA is capable now of delivering 110 billion kilowatt-hours each year by means of its generating plants at over 35 dams on the Cumberland and Tennessee Rivers. Investigate the growth of TVA and its service to the region. Recently, some environmental concerns have been raised about TVA's expansion plans for hydroelectric plants and coal-fired plants. Explore these concerns, and summarize the costs and benefits of TVA's planned developments.

11.5 A pumped-storage facility can be started rapidly and will offer low operating and maintenance costs. Determine the operating costs and initial costs of one pumped-storage system and the associated usefulness of the system. One such project, the Cabin Creek system, of the Public Service Company of Colorado provides 300 megawatts, using a head of 1199 feet. Another project is the Raccoon Mountain project of the TVA near Chattanooga, Tennessee, with a maximum capacity of 1.5 gigawatts.

11.6 The Bonneville Dam was built in 1937 on the Columbia River of Washington and Oregon. The dam is 58 meters high, and the generating plant has a capacity of 518,400 kilowatts. The Bonneville Power Administration (BPA) has built over 20 hydroelectric projects since 1937, and has plans for additional projects. Trace the growth of generating capacity of BPA and describe its planned growth. What are the advantages of BPA to the residents of the Northwestern U.S.? Are there offsetting disadvantages?

*11.7 A pumped-storage scheme has been proposed for the Qattara depression in Egypt. The scheme uses a canal to bring the water from the Mediterranean Sea to the storage basin. The basin would have a head of 205 meters and would be located at Deir Kirayim on the upper rim of the depression. The pumps would operate using the surplus off-peak electricity developed from a hydro-solar power plant in the depression. [3] Investigate the costs, benefits, and details of the proposed system, and estimate the electrical capacity of the pumped-storage system.

11.8 In Fig. E11.8 the Pit 7 Powerhouse of Pacific Gas and Electric Company on the Pit River, Shasta County, California, is shown. The powerhouse has two Francis or reaction-type water turbines on vertical shafts, utilizing a large volume

Fig. E11.8 *The Pit 7 Powerhouse on the Pit River, California. (Courtesy Pacific Gas and Electric Company.)*

---

* An asterisk indicates a relatively difficult or advanced exercise.

of water under low head, in contrast with the small-volume, high-head operation of the James B. Black Powerhouse, which uses Pelton impulse waterwheels driven by high-velocity jets of water.

Locate a low-head powerhouse in your area and determine the characteristics of the turbines.

11.9 The most recent group of three hydraulic turbines for the Grand Coulee power plant are the most powerful units in the world. The maximum output of each of 3 identical machines is 700 megawatts when operating under a head of 285 feet at a speed of 86 revolutions/minute. Determine the total power generation at Grand Coulee and the total energy production for the most recent year. Determine what agency paid for the latest installation, its cost, and the price of electric power from Grand Coulee. (Contact U.S. Bureau of Reclamation.)

*11.10 Large hydroelectric power projects are under construction throughout the world. The Bratsk Hydro-power Station on the Angara River in Eastern Siberia is shown in Fig. E11.11. Investigate the details of financing, construction, and utilization of a hydroelectric plant in another nation. Provide a brief report on these details, as well as the plant's importance to the region it is located within.

11.11 The ultimate hydroelectric capacity in the U.S. is estimated to be 180 gigawatts. Assume that 95% of this potential will be realized by 2020. Draw a curve of the installed capacity in the U.S. for the period 1900 to 2040, assuming that the installed capacity in 1900 was less than one gigawatt.

Fig. E11.11 *The Bratsk Hydroelectric power station on the Angara River in Eastern Siberia (Soviet Union). This first portion of this project was completed in 1961. By 1975 the project had a capacity of 4.15 gigawatts. The plant produced 27 billion kWh of energy in 1974. (Courtesy of Novosti Press Agency.)*

12

Nuclear Fission Power

Of all the major new sources of energy, nuclear fission has received the most financial support since 1945 and is correspondingly the best developed technology. The development of nuclear fission power dates back to the 1940's and the achievement of the first nuclear explosion at Alamogordo in New Mexico in July, 1945. After World War II, the U.S. formed the Atomic Energy Commission (A.E.C.) in 1946 to oversee the development of nuclear power reactors as well as nuclear weapons. The A.E.C. was divided in 1975 into the Energy Research and Development Administration (ERDA) and the new Nuclear Regulatory Commission (see Chapter 23), to which was assigned the regulatory portion of the A.E.C.'s responsibility. The U.S. budget of ERDA assigned to nuclear-fission research and development was 717 million dollars in 1977.

The recent ERDA energy plan calls for a goal of 6×10^{15} Btu (1.8×10^{12} kWh) in 1985 generated by nuclear-fission reactors. If this goal is achieved, nuclear-fission energy could account for one-half of the electric-energy production in 1985 (see Table 8.1). If the U.S. must have sources of energy other than fossil fuels by 1985, the only source that can make a major contribution by that time is nuclear fission. [1] Nevertheless, there is a growing concern in the nation regarding the extent to which research funds are concentrated on nuclear-fission reactors and the consequences of large-scale use of nuclear fission as a source of electric power.

These concerns include operating hazards, particularly the chances of a serious reactor accident, the difficulties of safeguarding the fissionable materials used as reactor fuels, and the still unresolved problem of long-term storage for radioactive wastes. The possibility of technological failures, earthquakes, and other unforeseen natural disasters, and human actions ranging from carelessness to deliberate sabotage appear to be particularly significant with nuclear power systems. Because of the consequences to human health (and to the environment) of any large release of radioactive substances, nuclear fission is viewed by many people as a hazardous process for providing electric power. Extremely safe and reliable systems for control of the reactor are required in order to ensure public acceptance of nuclear-fission reactors.

Nevertheless, nuclear fission has substantial advantages over traditional sources of energy. Nuclear plants do not emit particulates or sulfur oxides, as do fossil-fuel plants. The fuel requires less mining and thus results in less disruption of the environment.

In a recent survey in 1975, about 63 percent of the people interviewed favored the use of nuclear power compared with 19 percent opposed. [2] However, support is wavering as many outspoken opponents have articulately provided their views to a public anxious about achieving a sufficient supply of energy, but at the same time not significantly disturbing the environment, to the detriment of human beings.

12.1 THE FISSION PROCESS

Essentially all of the atom's mass is in its nucleus, which is composed of protons and neutrons, collectively called *nucleons*. The number of nucleons is indicated by a superscript to the left of the symbol of an element as in ^{235}U. Uranium has several nuclear species with a different number of neutrons in each; for example: ^{233}U, ^{235}U, and ^{238}U. These different species are called *isotopes* of uranium.

The source of nuclear energy lies in the equivalence of mass and energy, expressed by Einstein as:

$$E = mc^2, \qquad (12.1)$$

where m is the mass of the particle and c is the speed of light ($c = 3 \times 10^8$ m/sec). A number of heavy nuclei with masses of about 240 amu* and about 90 protons can undergo a process of fission. The process of fission is exemplified by ^{235}U absorbing a neutron, splitting into barium and krypton, and releasing 3 more neutrons, as follows:

$$^{235}U + n \rightarrow {}^{92}Kr + {}^{141}Ba + 3n. \qquad (12.2)$$

Accounting for the total mass on both sides of the equation, the loss of mass after splitting ^{235}U is 0.215 amu (3.57×10^{-28} kg). Thus the energy released is

$$E = (3.57 \times 10^{-28} \text{ kg}) \times 9 \times 10^{16} \text{ m}^2/\text{sec}^2$$
$$= 3.2 \times 10^{-11} \text{ joule}. \qquad (12.3)$$

This is the energy produced from one fission. If 1 percent of the 2.5×10^{24} atoms in a kilogram of uranium-235 were to fission over a very short time, then the total

---

* amu = atomic mass unit = 1.66×10^{-27} kg.

energy would be:

$$\text{Total } E = (0.01)(2.5 \times 10^{24})(3.2 \times 10^{-11} \text{ joule})$$
$$= 8 \times 10^{11} \text{ joule}. \tag{12.4}$$

The energy content of a ton of coal is about 3×10^{10} joule. Thus, about $\frac{1}{27}$ of a kg of ^{235}U would be an energy source equivalent to a ton of coal.

In order to achieve nuclear fission, the free neutrons must strike other nuclei, causing them to fission. This process is called a *chain reaction*; the atoms must be packed close enough together to ensure a constant rate of fission. The amount of material and the arrangement required for this is called the *critical mass*.

If the neutrons do not strike a ^{235}U nucleus, the process will slow down. Natural uranium cannot support a chain reaction in most types of reactors since the neutrons strike a ^{238}U nucleus, which usually does not fission. Thus, regular uranium is processed as a fuel to increase the proportion of ^{235}U. The enriched fuel contains about 3 percent ^{235}U.

The products of a fission are unstable and may emit electromagnetic radiation as well as alpha and beta radiation. This process is called *radioactive decay*, and the substances are said to be radioactive. The time for one-half of the radioactive atoms of an element to decay is known as the half-life.

12.2 NUCLEAR REACTORS

A nuclear-reactor power plant essentially consists of a uranium fuel core, a converter loop, a turbine–generator, and a condenser. The critical size and arrangement of the reactor depends upon several factors, including: (1) the geometry of the fuel elements, (2) the purity of the fissionable material, (3) the converter for extracting the energy from the reaction, and (4) the control mechanism.

Most fission reactors include the following components: the core, a coolant, control rods, and a moderator. The core is made up of bundles of fuel rods, which contain uranium oxide pellets. When a number of bundles of rods are assembled, a critical mass is reached and the chain reaction starts. Individual fuel rods do not contain sufficient fuel for a critical mass.

The coolant, either gas or liquid, flows over the fuel core, removing heat from the fission process. The coolant does not come in contact with the actual fuel, since the radioactive material itself is sealed within the fuel rods.

The control rods are made of material that readily absorbs neutrons. These rods are usually cylinders or sheets of metal (boron steel or cadmium), positioned inside the fuel assembly. If the rods are pulled out of the bundle, more neutrons are available to cause fissioning of the fuel, so the rate of reaction increases. If the rods are inserted into the fuel bundle, they act as a neutron sponge, so that there are fewer neutrons available to the fuel. Thus the chain reaction slows, or may be stopped completely. This makes it possible to produce heat at a desired rate, or to shut down the reactor completely.

The moderator is a material in the core which serves to slow down the neutrons as they emerge from the fissioning atoms. This is necessary because neutrons travelling too fast are less readily captured and do not cause more fissions. Graphite, water, or heavy water are commonly used moderators.

A reactor is usually housed within a containment vessel and includes a shielding arrangement. The shielding surrounds portions of the reactor to prevent radiation leakage into the environment.

A schematic diagram of a water–steam circuit in a boiling-water reactor is shown in Fig. 12.1. About one-half of today's reactors are boiling-water reactors (BWR). In a BWR, water is used as the coolant and serves also as the moderator. The water is brought into the reactor and allowed to boil. It then is taken out as pressurized steam and used to drive the turbine. Typically, the BWR operates at 7×10^5 kg/m^2 (1000 lbs/in^2) pressure and a temperature of 300°C. The BWR has the advantage of simplicity, but suffers from the disadvantage that there is the potential of radiation passing to the turbine via the water. Also, the efficiency of this system is only about 34 percent.

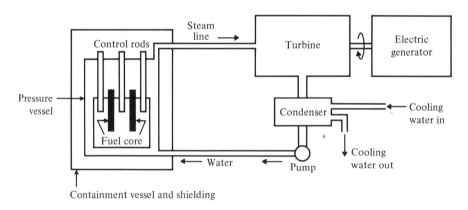

Fig. 12.1 *A schematic diagram of a water-steam circuit in a boiling-water reactor.*

A schematic diagram of a two-loop system using a heat exchanger is shown in Fig. 12.2. The fluid or gas in the primary loop may be water, helium gas, or carbon dioxide.

When water is used in the primary loop, it is pumped over the fuel core under pressure and removed from the top as a heated liquid. The water is kept under sufficient pressure to keep it from boiling. This system is called a pressurized water reactor (PWR). The water is then circulated through a heat exchanger, in which steam is produced in the secondary loop and used to drive the turbine. The PWR operates at pressures of 1.4×10^6 kg/m^2 (2000 lbs/in^2) at about 330°C. The PWR has several advantages similar to those of the BWR. The coolant used

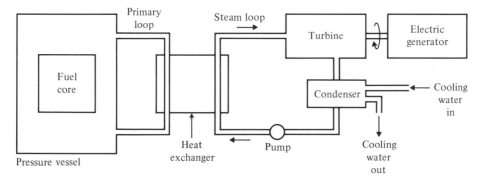

Fig. 12.2 *A schematic diagram of a two-loop system using a steam generator to provide the steam to drive the turbine. The fluid or gas in the primary loop may be water, helium gas or carbon dioxide gas.*

at the reactor core does not directly contact the turbine; thus the turbine area remains uncontaminated with radioactive materials. The higher pressure allows more efficient heat transfer and requires a smaller surface area for the core. The PWR, however, requires higher operating pressures and additional heat exchangers, which lowers its efficiency. The high temperature increases the corrosion of the fuel rods, the cladding (the walls of the fuel elements), and the vessel.

In the high-temperature gas-cooled reactor (HTGR), the core is cooled by passing a gas over it and through the primary loop. Usually, purified carbon dioxide or helium is employed as the gas. This type of reactor has a low fuel-consumption rate because very few neutrons are captured by the coolant. A United States HTGR uses helium as a coolant and graphite as a moderator. The coolant operates at about 700°C compared with only 330°C for a PWR. It may be possible to raise temperatures to 1000°C, thus increasing the efficiency. The HTGR is a significantly safer plant than the BWR and it consumes less fuel per kWh produced. The efficiency of a HTGR is about 40 percent, about equal to new fossil-fuel plants. The first HTGR was installed at Peach Bottom, Pennsylvania, in 1967 and has a capacity of 40 megawatts. A 350-MW unit has recently gone into operation at Fort St. Vrain for the Public Service Company of Colorado.

Most new nuclear-fission power plants are built with a capacity of about 1000 megawatts. A plant in California was opened at Rancho Seco, near Sacramento, in October 1974 with a capacity of 913 megawatts. A unit of 850 megawatts was built about 65 miles northwest of Little Rock and completed in 1976 for Arkansas Power Company. A 1250-MW plant is under construction for Mississippi Power Company near Grand Gulf. The first 1250-MW unit is scheduled for completion in 1979. This unit, when completed, will be the largest nuclear power plant in the U.S. The Yankee Atomic power plant is shown in Fig. 12.3. This plant was built in 1960 at a total cost of $44 million. This PWR plant produces 185 megawatts and has over 16 years of safe operating experience to its credit.

Fig. 12.3   The Yankee Atomic Power Plant at Rowe, Massachusetts. The containment vessel can be seen behind the electric generating building. (Photo courtesy of Yankee Atomic Electric Company.)

Nuclear Power Production

The first nuclear power plant went into operation in 1957 at Shippingport, Pennsylvania. The Shippingport plant is a 90-megawatt unit. By the end of 1975, the U.S. had 58 operating nuclear power plants, with a combined capacity of 39,900 megawatts, which is about 9 percent of the total electric generating capacity of the U.S. In addition, 69 reactors were in various stages of construction, with a capacity of 72,000 MW.

The growth of nuclear generating capacity in the U.S. is shown in Table 12.1. The capacity increased by a factor of eight from 1970 to 1975, and the number of nuclear plants grew to 58. It is projected that the nuclear capacity will be 85 gigawatts in 1980 and 160 gigawatts in 1985. These projections are sensitive, however, to the delays introduced by opposition from environmental groups and concerned citizens, and shortages of capital.

The production of energy by nuclear power is shown in Table 12.2. The actual production of energy by nuclear plants has become significant only in the past few years.

Table 12.1

NUCLEAR POWER CAPACITY
IN THE U.S.

Year	Capacity (Gigawatts)	Number of plants
1957	0.09	1
1960	0.290	2
1965	0.891	7
1970	4.9	14
1972	14	
1973	21	
1975	39.9	58
1980*	85	100
1985*	160	

* Projected.

Table 12.2

NUCLEAR ENERGY
PRODUCTION IN THE U.S.

Year	Production	
	($\times 10^{12}$ Btu)	(10^6 kWh)
1957	1	10
1960	6	518
1965	38	3,657
1970	229	21,801
1971	391	37,899

The Federal Power Commission has recorded a scheduled completion of 155 gigawatts of nuclear power-plant capacity for the period 1975–84. [3] The number of units scheduled to yield this power is 147 new plants. Thus, the new plants scheduled for this period average 1050 megawatts per unit.

Uranium Resources

Uranium is an element occurring widely at low concentrations on land and in seawater. Limitations on its use are set by the cost of mining and concentrating the ore. From ores containing about 0.1 to 0.5 percent of U_3O_8 by weight, the cost is below $20/kg. However, it would cost about $200/kg it if was necessary to tap the enormous reserves in the sea.

Estimates of the world's assured reserves of uranium vary, but for ore at a cost of $20/kg or less, about one million metric tons can be obtained. World uranium resources used in BWR or PWR will yield about 2×10^{21} joules (2×10^{18} Btu). Current annual world consumption of energy is approximately 2×10^{20} joules, so the uranium resource could provide 10 years of the world's supply of energy. Assuming that nuclear energy supplies 25 percent of the world's energy by the year 2000, the uranium resource might be sufficient to last another 40 to 60 years, depending upon the growth of energy consumption in the world.

Annual production of uranium ore in the U.S. and the world is given in Table 12.3. Production in the U.S. peaked in 1960 and then declined over the period 1960 to 1970, when it again rose to 17,000 tons in 1976. If a plant capacity of 85 gigawatts is to be supplied in 1980, then U.S. uranium production will need to increase to 30,000 tons.

Table 12.3
ANNUAL PRODUCTION OF
URANIUM ORE

| Year | Production (thousand tons) | |
	United States	World
1960	17.8	43
1965	10.3	19
1970	12.9	
1973	13.2	25
1976	17.0	
1980*	30.3	80
1985*	45.0	135

* Projected.

Table 12.4
DOMESTIC RESOURCES OF URANIUM
IN THE U.S. (THOUSANDS OF TONS)

Cost	Reasonably assured	Estimated additional	Total
Up to $20/kg	315	960	1,275
Up to $30/kg	420	1,500	1,950
Up to $60/kg	600	2,900	3,500

The resources of U.S. uranium ore are given in Table 12.4. The reasonably assured resource for a cost up to $30/kg is about 420,000 tons. [4] This resource, at an annual production of 30,000 tons in 1980, could supply fuel sufficient for 14 years. Clearly, much of the estimated additional resource will have to be discovered and economically mined.

One estimate of the total uranium required in the U.S. over the period 1975 to 2000 is 1.5 million tons. [5] This would require most of the U.S. resources at a price up to $30/kg. This assumes that there will be 200 nuclear plants by 1985 and about 800 by 2000. These plants would require two million more tons of uranium during their lifetimes. Thus, a grand total of 3.5 million tons of uranium would be required if the number of plants was increased to 800 by the year 2000 and these plants continued to operate while new plants were built. This uranium requirement could exhaust the U.S. domestic resource of U_3O_8. However, it is exceedingly unlikely that the number of nuclear plants will approach 800 in number by the year 2000.

Most of U.S. uranium comes from sandstone, predominantly in New Mexico. As further reserves are exploited, uranium from shale may have to be developed. At some point there is *no net energy* release from uranium since the energy required for mining and processing is equal to that eventually released.

Nuclear fuel proceeds from the mine, through a conversion process to the reactor and then is recycled, as shown in Fig. 12.4. The enrichment process increases the ^{235}U content from 0.7 percent to 2 or 3 percent for ultimate use in a reactor. An enrichment plant costs about one billion dollars to build. A full fuel load for one new 1000-MW reactor costs about $38 million, and replenishment every year or so costs another $11 million, or about $400 million over the life of the plant.

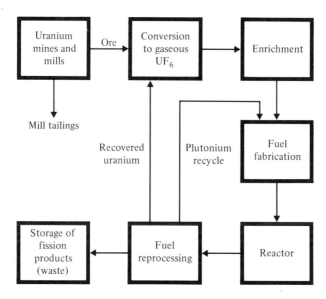

Fig. 12.4 *The fuel cycle for reactors using water as a coolant.*

12.3 THE BREEDER REACTOR

Manmade plutonium-239 is suitable as a fuel in a nuclear reactor. However, it is only through the use of reactor-produced neutrons that substantial quantities may be obtained. A breeder reactor starts with fissionable fuel for the initial fuel and, besides producing energy, it produces more fuel than is consumed in the process.

Plutonium-239 can be produced from ^{238}U. Thus, although ^{238}U is not readily fissionable, it converts to ^{239}Pu, which is fissionable. This conversion to plutonium also takes place in conventional light-water reactors but not at as high a rate as in *breeder reactors*. In a breeder reactor, for each atom of fissionable material consumed, more than one atom of fertile material is converted to fissionable material.

The Liquid Metal Cooled Fast Breeder Reactor (LMFBR), it is hoped, will be able to produce more ^{239}Pu from the fertile material ^{238}U than it would consume. It is called a *fast reactor* since it contains no moderator material to cause a slowdown of the neutrons. At these higher neutron velocities, there is a greater probability that the neutrons not needed to maintain the chain reaction will be captured by the fertile ^{238}U than by reactor core components.

The LMFBR uses a liquid metal, sodium, as the reactor coolant. An inert gas, argon, is used to blanket the sodium. The fuel used in the LMFBR is a mixture of uranium and plutonium. The primary loop coolant, sodium, conveys the heat to the heat exchanger where it is passed on to a secondary loop and eventually makes steam for the turbine.

Liquid metal fast breeder reactors have the potential of greater efficiency than light water reactors. Sodium is considerably more efficient than water in transferring heat from the core. Also, the reactor core can be operated at a higher temperature without pressurization, since sodium has a much higher boiling point than water. As a consequence, the thermal efficiency of such a power plant will be 34 percent or more, compared to 31- to 33-percent efficiency for light-water reactors. This means a decrease in waste heat.

The breeding action is accomplished by placing a blanket of ^{238}U around the core. The LMFBR is designed so that, for every atom fissioned, about 1.2 atoms of ^{239}Pu is created. Thus plutonium can be removed by fuel reprocessing and used to refuel the core, while the excess can be used to fuel another LMFBR or LWR.

The LMFBR can utilize over 200,000 tons of discarded uranium now stored at Oak Ridge, Tennessee, and other fuel-processing sites. Essentially, the LMFBR will use about 70 percent of the potential energy present in the uranium ore. By contrast, only 1 percent is utilized in light-water reactors. A 1000-MW LMFBR will consume about one tonne of uranium annually, so the present stockpile of discarded uranium would supply 400 plants for about 500 years. If 1 million tonnes of the U.S. uranium resources could be utilized over the next several centuries, this resource could supply 800 LMFBR for over ten centuries. Using the LMFBR, uranium ore is essentially an inexhaustible resource.

The LMFBR has a high research and development priority in Europe and has operated successfully in France, Great Britain, and the Soviet Union. The Prototype Fast Reactor (PFR) at Dounreay in Scotland started generating power in late 1974. It is designed to deliver a net of 250 MW using a sodium coolant. It was constructed for about $110 million.

The French 250-MW Phénix breeder reactor went into operation in 1973. The breeding ratio of this reactor is about 1.2 neutrons for each neutron used in fission. The breeding ratio is defined as the number of fissible atoms produced per fissible fuel atom consumed. Since 1974 the Phénix has provided 250 MW with a load factor of 80 percent. The Phénix reactor is shown in Fig. 12.5. The French are now planning a 1000-MW breeder reactor, called the Super-Phénix and scheduled to come into operation in the early 1980's.

Fig. 12.5   *The Phénix breeder reactor located on the Rhone River near Marcoule, France. (Courtesy of the French Information Office.)*

The Soviet Union has a prototype breeder plant of 350-MW capacity at Shevchenko on the Caspian Sea. [6]

The U.S. is planning the construction of a breeder plant on the Clinch River near Oak Ridge, Tennessee, and it may be completed by the late 1980's. This plant will be 350-MW and will be used to experiment with LMFBR commercial development. ERDA is supporting the cost of this experimentation and it is estimated that the total cost of this program through 1990 is $10 billion. However, in April 1977 President Carter suspended construction plans for the Clinch River LMFBR.

Meanwhile, Admiral Hyman Rickover of the Navy's nuclear reactor program is pursuing the conversion of the light-water cooled reactor at Shippingport, Pennsylvania, to a breeder reactor. This reactor will use ^{233}U as a fuel; the neutrons emitted will bombard a thorium blanket placed around the core. This thorium will yield 1.02 uranium-233 atoms for every atom consumed with slowly breeding new fuel. Whether the light-water breeder reactor will be economically feasible is not assured.

The primary concerns with LMFBR's and LWR's are their cost and environ-mental and safety hazards. The LMFBR is more difficult to control than a light-water reactor because an accidental loss of the sodium coolant from the core may result in an increase in reactor power. The opposite effect occurs in light-water reactors. The cost of building a LMFBR will be considerably greater than that of light-water reactors, primarily because of the more exacting specifications and closer tolerances required. Finally, plutonium is an extremely hazardous material and can be diverted to weapons use, if not properly safeguarded.

12.4 NET ENERGY FROM A REACTOR

The average load factor of a nuclear power plant is about 62 percent. That is, a plant on the average actually produces 62 percent of the energy it is capable of producing if it operated all of the time. It is interesting to examine how long it takes a nuclear reactor operating with a load factor of 62 percent to yield output energy equal to the energy invested in its construction and used to fuel the reactor.

Chapman recently completed a study of a 1000-MW reactor, which required 10^{10} kWh of input energy, assuming the fuel was obtained from ore consisting of 0.3 percent fissionable uranium-235. [7] The net electrical capacity of the reactor, after losses and refueling energy are subtracted, is 523 MW (electrical), assuming a load factor of 62 percent. This plant provides 4.6×10^9 kWh of electrical energy in one year. If one calculates the number of years that it takes for the electrical output to equal the thermal input, one obtains a period of 2.2 years. As Chapman points out, if a rapid growth in the number of reactors were to occur, it would take a large number of years for the net output to become positive. For example, if the growth rate of reactors was 14 percent per year, it could take 13 years to yield a positive net output.

As Wright and Syrett have pointed out, electrical energy is in a more valuable and useful form than the original thermal energy. [8] If we assume it is 3 times superior, then it would take only 0.7 years to yield a positive net output, since the plant output of 4.6×10^9 kWh (electrical) is equivalent to 13.8×10^9 kWh (thermal). A reasonable estimate is that it takes about a year to yield a positive net energy output from a nuclear reactor. Nevertheless, it is interesting to note that if either the rate of growth of the number of nuclear power stations is too fast, or if the grade of uranium ore is too low, then nuclear power programs could consume more energy in the form of fossil fuel than they produce in the form of electricity.

12.5 ECONOMICS OF NUCLEAR ENERGY

It is about $100/kW more expensive to build a nuclear power plant than a coal- or oil-fired plant. However, when it comes to fuel costs, the position is reversed. It costs about three times as much per kilowatt-hour output for fuel for a fossil-fuel plant as for a nuclear plant. In 1975 it cost about $3/thousand kWh to fuel a

coal-fired electric power plant, while it took about $1/thousand kWh for fuel for a
nuclear reactor. The cost for fuel for an oil-fired plant is about twice that of a
coal-fired plant, or approximately six times that of a nuclear-fission plant.

The cost of building a nuclear plant has increased over the past several years,
to about $600/kW for plants put in operation in 1976. It is estimated that the cost
of a nuclear plant going on-line in 1985 will be $1100/kW to build (in 1985 dollars).
As shown in Table 12.5, the cost of nuclear reactors has increased at a rate of
about 20 percent per year. In addition, the *planned* cost of a reactor, usually
calculated 8 to 10 years before the first operation of a plant, is at least $100 lower
than the actual cost incurred.

Table 12.5

CAPITAL COSTS OF A NUCLEAR
FISSION REACTOR

Year of first operation	Planned or budgeted cost* ($/kW)	Actual cost ($/kW)
1966	$120	$120
1968	$165	$165
1970	$150	$220
1972	$240	$400
1974	$310	$520
1975	$420	$610
1980	$700	
1985	$1100	

* Cost is given in dollars of the year indicated.

The capital cost of a nuclear reactor coming into operation 8 to 10 years from
now is difficult to calculate because of several unknown factors. Inflation, construction delays, new regulatory requirements, environmental and safety studies,
and changes in reactor design result in increasing costs. [10] Therefore, a
1000-MW plant coming into operation in 1980 may have a capital cost of about
$700 million.

The cost of the nuclear fuel is low compared to the fuel cost for oil- or coal-
fired plants, and thus nuclear power is competitive with fossil-fuel plants. In 1975,
the total cost of producing electricity was 1.2¢/kWh for a nuclear plant compared
with 1.6¢/kWh for a coal-fired plant. [11] It was estimated that a nuclear plant
going into operation in 1985 will produce electricity at a cost of 4.0¢/kWh, compared to the cost of 6.5¢/kWh for a coal-fired plant in that same year. A cost
advantage of 0.5¢/kWh for the nuclear plant will translate to an annual savings
of nearly $40 million in the cost of power generation.

However, as capital costs of nuclear plants increase, it becomes more difficult
for nuclear plants to compete. In 1966 the cost of fuel represented 34 percent of
the cost of nuclear power. In 1975 nuclear fuel costs accounted for only 18 percent
of the cost of the nuclear energy. Engineering estimates of reactor capital cost are

insufficient guides for predicting actual future costs. However, as the fuel costs of coal-fired plants increase, utilities may increasingly be willing to take the risks associated with planning a nuclear-reactor plant, recognizing that the capital costs are difficult to estimate for a plant coming into line some ten years after planning commences. One possibility for reducing the escalation of costs of nuclear reactors and the lead time required for a nuclear plant is to develop several standard plant designs. These designs, once accepted by the Nuclear Regulatory Commission and proven by industry, could be more rapidly approved for site location, licensing, and financing.

The LMFBR is an unknown factor in the U.S. nuclear-reactor industry. The first demonstration plant at Clinch River is estimated to cost as much as $2 billion, up from the original estimate of $700 million. It is estimated that breeders will cost about $100/kW more than standard light-water fission reactors. However, it is estimated that the LMFBR can be commercially built for $800/kW (in 1980 dollars), compared with $700/kW for the LWR for the same year. A commercial LMFBR could not be available in the U.S. before 1990, so it is exceeding difficult to estimate the actual costs of such an approach, since many technological changes would impact the capital cost of such a plant. It is reasonably safe to state that the economics of the LMFBR are yet to be established with any reasonable security.

12.6 RADIATION

With the use of nuclear fission reactors to provide electric power, an examination of the attendant risks and environmental effects is required. Radioactive pollution from a reactor is created by fission in the reactor core. Each fission usually produces two fission fragments. In a 1000-megawatt (electric) nuclear power plant, about 10^{20} new fragments are produced every second. This radioactive material is contained within the reactor, but the question arises: The escape from containment of *how much* radioactive pollution should be tolerated within the plant?

Radioactive fragments include numerous isotopes such as strontium-90 and cesium-137, as well as xenon. Fragments often emit neutrons as well as a succession of alpha and beta particles and gamma rays. The decay of the isotopes may take minutes or several years. Strontium (^{90}Sr) has a half-life of 28 years, for example. (The half-life of an isotope is the time it takes for the intensity to decay to half of its original radiation intensity.) As a radioactive material passes through its decay chain, it may emit alpha and beta particles and gamma rays.

The most penetrating of these radiations are gamma rays. High-energy gamma rays can penetrate a concrete block. Beta radiations, which are high-energy electrons, are capable of penetrating a person's skin. An alpha particle is a positively charged particle made up of two protons and two neutrons bound together. Low-energy alpha particles can be stopped by a piece of paper.

Energetic particles (α-particles and β-particles) from radioactive decay lose energy as they pass through matter, by the process of ionization. As the particle passes through matter, it draws an electron away from the molecules of the material. The energy lost by the fast charged particle in creating the ion pair is that required to break the electron away, plus the kinetic energy imparted to the electron, which is about 30 eV. Thus a 3-MeV alpha particle will create about 10^5 ion pairs as it slows down.

There are several units for measuring exposure to radiation. The *roentgen*, R, is the unit of exposure related to the number of ion pairs produced in air by gamma rays. It is the amount of radiation required to produce ions carrying a standard charge and therefore, 1 R = 2.58×10^{-4} coulombs/kg in air.

The *radiation absorbed dose*, rad, indicates the amount of energy deposited in a material by radiation. One rad is equal to 100 erg/gm (10^{-2} joule/kgm). The rad and the roentgen are about equal when the absorbing substance is tissue.

The *roentgen equivalent man*, rem, is the unit of dose equivalent of radiation. The rem is that dose which has the same biological effect on a human being as 1 R of γ radiation.

The *curie* of any radioactive substance is the amount of substance which will give rise to 3.7×10^{10} decays per second. One gram of radium is equal to about one curie.

Beta particles are hazardous if ingested or inhaled by humans. Gamma rays may have energies in the 1- to 5-MeV range and travel 300 meters in air. Thus they may be harmful from a source external to the human body. Beta particles can be harmful if ingested or received at short distances (10 m).

The normal functions of living cells of human beings depend upon genetic information which is stored in deoxyribonucleic acid (DNA). Ionizing radiation may cause damage to the DNA, resulting in a defect showing up in a deleterious mutation. When genetic damage occurs, an offspring is damaged or experiences a mutation effect.

If the level of radiation is great enough, acute somatic damage occurs. At a level of 500 rem or greater, death occurs for more than half of the persons experiencing this effect. At a level of 250 rem, severe radiation sickness occurs for the majority of individuals.

Chronic somatic damage occurs as a result of repeated small dosages of radiation, which result in cumulative effects. It is assumed that there is no known threshold of radiation below which there is no effect, and that all dosages are additive. This is the basis of the linear theory of radiation effects. Thus, it is assumed that every small dosage of radiation has its effect.

The average dosage of radiation per person in the U.S. is about 200 millirem per year. An individual receives cosmic radiation, radium in the water, and radiation from television sets, building materials, x-ray diagnostic tests. Undoubtedly, these radiation doses cause some number of mutations and undesirable consequences such as leukemia and cancers in all age groups.

The fear is that an increase in the radiation level will result in an increased incidence of cancer in the population. Thus, it is estimated that 100 rad increases the naturally occurring cancer rate by 100 percent. If so, then 1 rad would increase the rate by 1 percent (if the linear theory holds). If the entire population of the U.S. were exposed to 0.17 rad/year, according to this theory, there would be about 30,000 additional cases of cancer per year in the U.S. Other studies place the number in the range of 3000 to 15,000.

The radiation standards for reactors in the U.S. (at the reactor site) are 0.005 rem/yr, reduced from the earlier standard of 0.17 rem/yr for the total population.

Since a reactor producing 1000 MW of electric power involves 10^{20} fissions/sec, we have 7×10^{20} decays/sec, or 2×10^{10} curies of radiation. In a year a 1000-MW reactor produces 3×10^6 curies of ^{90}Sr and 4×10^6 curies of ^{137}Cs. [11] This radioactive material is more than enough to contaminate one of the Great Lakes. Clearly, the radioactive material of a reactor must be severely contained.

Nuclear Reactor Wastes

Essentially all of the radioactive substances must be contained within the nuclear-reactor containment system. The small radioactive emissions permitted from a reactor are 1 percent of the natural background. However, the spent fuel and surrounding fission products are radioactive, and remain so for hundreds of years. Thus, the nuclear-power industry is faced with the problem of processing and storage of nuclear-reactor wastes.

The nuclear-fuel cycle is shown in Fig. 12.4. The process of fuel reprocessing and the storage of fission products has an associated risk of radioactive release. "We nuclear people have made a Faustian bargain with society," physicist Alvin Weinberg wrote in 1972. "On the one hand, we offer . . . an inexhaustible source of energy. But the price that we demand for this magical energy is both a vigilance and a longevity of our social institutions that we are quite unaccustomed to . . ."

Radioactive wastes are created when spent nuclear fuel is removed from commercial or military reactors. The material is dissolved in nitric acid and the reusable uranium and plutonium are reclaimed. What remains is a brew of liquid wastes containing strontium-90, cesium-137, and other toxic and long-lived substances. Both strontium and cesium in this liquid waste take 600 years to decay to harmless levels, and in the process of decay they emit harmful radiation that must be contained.

Plutonium, deemed hazardous for 250,000 years, is harmful if inhaled or ingested and is either used as a nuclear-reactor fuel or held at storage sites for later use in nuclear breeder reactors.

The U.S. government has several sites for the storage of nuclear waste products and has already accumulated 81 million gallons of waste from its weapons and military reactor programs. About 8 million gallons of waste are added annually from military sources. It is estimated that 60 million gallons of waste will be added from commercial reactors by the year 2000.

The U.S. is currently storing its nuclear waste at sites in Idaho, Georgia, and at Hanford, Washington. At Hanford, in 1973, some 430,000 gallons leaked into the soil from the storage tanks. Officials claim that none of the liquids have penetrated to the water table, but with the long life of the radioactive wastes, local water may eventually be contaminated. To prevent further leaks, wastes are now required to be solidified before long-term storage. In the proposed design, the solidified wastes are sealed in 10-foot-high, 2-ton steel canisters and then embedded in concrete casks. Then these casks would be placed in a barren portion of the desert.

The half-life of cesium-137 is 30 years; for strontium-90 it is 28 years. Thus, in order to decay to a factor of one thousandth of its original intensity, it would require 300 years. However, in the case of plutonium-239, its half-life is 24,400 years. Thus, to reach one-thousandth of its original intensity, it may require 250,000 years. If all the electric power used by one person in his lifetime were generated by a nuclear reactor, the high-level radioactive wastes would fill a 16-oz beer can. For the 200 million U.S. citizens over the next several decades, the waste generated would then be about 90,000 metric tons.

In a typical reprocessing plant, the spent fuel elements are received and put in a storage pool. When ready they are chopped and dissolved in nitric acid. The tubes which contain the fuel pellets are not dissolved by the acid, but are buried. An organic solvent dissolves the uranium and plutonium, which are later separated, purified, and further converted into plutonium oxide and uranium hexafluoride. The nitric acid containing the fission-product solution goes to an evaporator where the acid, along with some fission products, is boiled off. The remaining high-level waste is condensed and put into storage tanks where it is cooled.

The wastes at Hanford, Washington, are divided into three categories; those termed low-level in terms of their radioactivity are piped directly into surface ponds on the site. Intermediate-level wastes are treated more cautiously, being emptied into concrete-covered trenches known as cribs. The cribs are open to the soil at the bottom, and the water in the wastes gradually seeps downwards, taking the radioactive isotopes with it. The most radioactive wastes, known as high-level, are buried in steel-lined concrete tanks in the ground.

In Europe, the low-level wastes are placed in drums and dropped into a northeastern section of the Atlantic Ocean. The ocean is 2000 meters deep in that region, and over 40,000 tonnes has been dropped in the area centered on a point 46°15′ north and 17°25′ west. [12]

The storage of the long-term radioactive wastes is the major challenge to the nuclear industry. Any storage approach must meet the following criteria:

1. The wastes must be isolated for at least 250,000 years;
2. The storage sites must be proof against sabotage or theft;
3. The sites must be safe from the effects of natural disasters such as earthquakes and hurricanes;
4. The sites must be geologically stable; and
5. Handling and transportation methods must be fail-safe.

Candidates for methods of long-term storage are (a) disposal in salt formations and mines, (b) dumping into oceans, and (c) shooting into outer space. The disposal of waste containers in oceans is unsafe, and in outer space it is dangerous because of potential failure of the rocket (criterion 5). Thus, long-term disposal of wastes is currently planned for mines. Of all the techniques considered, storage in a salt mine has received the most thorough investigation. West Germany has established such a facility, and the U.S. had investigated several salt mines near Lyons, Kansas, and in southeastern New Mexico, as potential sites. It is estimated that salt-mine disposal would add about 0.5 percent to the cost of generating electric power by nuclear reactors. The long term safety of salt-mine storage depends upon preventing the intrusion of water into the salt beds by any means. [13]

Another scheme is to bury the canisters at moderate depth in Antarctic rocks. To a depth of about 1 km, all ground water is frozen in the Antarctic. The question is, however, the durability of the ice sheet over 250,000 years. One method that could be explored is to place the canisters on the ice sheet and allow them to slowly melt down through the sheet to the bedrock. Another possibility is to anchor an emplacement of the canisters within the ice sheet.

The problem of nuclear-reactor wastes remains unsolved at present. As Frank Zarb, former head of the Federal Energy Administration, has said, "A single aspirin tablet has the same volume as the waste produced in generating seven thousand kilowatt-hours, which is about one person's share of the country's electric output for an entire year. Compared to large quantities of other harmful materials, the volume of nuclear waste is miniscule." What Mr. Zarb fails to mention is that, unlike a real aspirin tablet, his radioactive one is sufficiently toxic to kill a hundred people. Furthermore, nuclear wastes are persistent in their danger. As the number of nuclear plants increases, the world is going to face the unresolved issue of nuclear wastes.

Diversion of Nuclear Fuels and Wastes

Atomic bombs are fabricated from enriched uranium or plutonium. Thus, the danger of the diversion of these substances from their peaceful use to use in nuclear weapons must be considered. Even small amounts of plutonium in the hands of terrorists would expose a nation to intense danger. Plutonium might be stolen in the fuel-processing plant or in the transportation system between the processing plant and the reactor.

The transportation of nuclear fuels and wastes will require extremely secure methods. The loss of plutonium in "materials unaccounted for" (MUF) is difficult to monitor in a fuel-processing plant and during transportation. Dr. Theodore Taylor has stated that it would be relatively easy for a terrorist group to fabricate a bomb from a small amount of stolen plutonium. In a March, 1975, episode of the public television program *Nova*, the development of a method of building a nuclear bomb by a college student was described. [14] The bomb was designed to be built from stolen plutonium and parts from a hardware store. It is known that two kilograms of plutonium would be sufficient to fabricate a bomb.

If a relatively large number of LMFBR breeder reactors are operating in the late 1980's throughout the world, a significant amount of plutonium fuel and waste will be circulating from processing plant to reactor. It is this potential that will require a vast security system for safeguarding nuclear fuel. [15, 16] Some authorities believe that the present security system is inadequate and unable to routinely determine MUF to within an accuracy of 0.5 percent. Security systems must be designed to address the problems of detection, prevention, and recovery of any diversion of nuclear fuel materials.

If commercial development of LMFBR continues throughout the world, by the year 2000 there may be enough plutonium created each year to make thousands of atomic weapons. [17] The most vulnerable link in the system is the transportation of the nuclear fuels. Extensive communications systems and extremely secure trucks or train cars will be required to monitor and maintain secure transportation. A national police force for securing this plutonium may even be required. All this danger and concern with the diversion of plutonium may weigh against any extensive development of LMFBR in the U.S. Nevertheless, if several nations in the world proceed to build LMFBR's, the potential still exists for the diversion of plutonium from one nation to terrorist use in another nation.

Reactor Safeguards

In order to assure reactor safety, many features are built as an integral part of a reactor facility. The safeguards include: (1) monitoring of reactor neutron intensity in order to maintain it within a prescribed range, (2) reactor control systems, (3) reactor safety circuit instrumentation, (4) electric power maintenance, and (5) an emergency core-cooling system.

The emergency core-cooling systems, ECCS, are intended to prevent serious damage to the reactor in the event of a loss-of-coolant accident (LOCA). This type of accident may occur as a result of a break in the pipe system of the primary coolant system. After such an accident, the peak temperature in the reactor core is predicted to be 2000°C after one or two minutes.

After such an accident, the coolant water rushes out of the pipes and the control rods are programmed to rush into the core, damping the reactors. The ECCS then proceeds, according to plan, to release cool water stored in accumulators for such emergencies. Under pressure from nitrogen gas, the water flows over the core, flooding the reactor and cooling the core.

Several critics of the ECCS point out that, if the temperature rises too high before the stored water is released, it will be turned to steam and vented or blocked from flowing to the core.

The flooding of the core by the ECCS is estimated to be at a rate of one inch per second. Since the system is not able to be tested under real conditions, it has not been proved that it will function as planned. Currently, the ECCS has been analyzed by means of computer models. A loss-of-fluid test (LOFT) program in Idaho, using a 55-megawatt test reactor, will be used to examine the ECCS. It is designed to provide a verification of the analytical methods used for prediction of behavior in the case of a loss-of-coolant accident.

The LOCA is the worst sort of accident that could befall a reactor. If the fission process is turned off, and that happens automatically if the water moderator is removed, the core still generates a significant percentage of heat. The inevitable rise in temperature of the fuel rods during the period when the core is without cooling water will weaken them sufficiently so that the cladding on many could rupture before emergency cooling water refloods the core. This means that a substantial fraction of the gaseous fission products could escape from the reactor pressure vessel along with the cooling water. These radioactive gases carry more than 10 percent of the core's total radioactivity of about 10 billion curies. It is essential that these radioactive gases not be allowed to escape into the environment.

If the ECCS were to fail, the core might melt and some of the radioactive core could conceivably leak from the pressure vessel. The pressure vessel is normally constructed of six-inch-thick steel. This pressure vessel, if breached, is contained within a reinforced concrete containment structure over two feet thick.

The ECCS must work within a minute, or the high temperature may generate steam which drives the ECCS cooling water away. The greatest fear may be not that a natural accident may occur but that a deliberate accident—sabotage or terrorism—may result in a loss of coolant and the disablement of the ECCS. This is essentially the worst case conceivable. This would result in a core meltdown and eventually a breaching of the containment structure.

Another issue concerns the potential dynamic reaction of the molten material plunged into a large volume of cool liquid. If the core experiences a meltdown and drops into the rapidly entering ECCS cooling water, it may result in the development of a large pressure wave, fracturing the containment vessel. [18] All these results are hypothetical, since a LOCA has fortunately not been experienced in practice.

Previous Nuclear Reactor Accidents

The likelihood of a significant nuclear-reactor accident can, in part, be judged in terms of the safety history of the nuclear power industry. There has been a number of significant (but nonfatal) nuclear-reactor accidents in the world since 1950. These accidents are typically due to carelessness, design error, or mechanical failure.

The first accident of significance was at the Chalk River, Ontario, Canada, facility in 1952. The accident occurred as a result of several human operator errors. As a result, in this case, the uranium core began to melt and the top of the reactor lifted off releasing radiation. Nevertheless, the accident at this low-power reactor did not result in extensive injury or any fatalities.

Another nuclear-reactor accident occurred at Windscale in England in 1957. This reactor accident was due to the failure of the procedure used in the reactor process. Radioactivity did leak out from the reactor, which was not enclosed in a containment vessel, and covered a 200-sq-mile area of the surrounding region. This experimental plutonium production reactor was a type used in the U.S.

A significant reactor accident occurred at the Fermi 300-MW fast breeder reactor near Detroit. [19] The accident took place, after an interim shutdown, during the startup procedure in 1966. The fuel began to melt due to stoppage of the coolant flow by a broken plate. Fortunately, the containment vessel was not breached and no radiation was released.

An accident occurred in March 1975 at the Brown's Ferry 1000-MW nuclear plant near Decatur, Alabama. Below the plant's control room, an engineering aide and an electrician inadvertently started a fire at 12:20 P.M. They tried to extinguish the fire for 15 minutes before turning in a fire alarm. Six minutes after the alarm sounded, the operator in the control room noted that the ECCS started automatically. [2] At 12:51 P.M. the decision was made to shut down the reactor. The ECCS began to function to avoid meltdown but at 12:55 P.M. the electrical supply failed due to the fire. However, by improvising, the core was kept covered by water and the fire eventually extinguished, seven hours later. At least 7 of the plant's 12 safety systems had failed. The accident was an indictment of the limited training of the nuclear operators and technicians and a reminder of the risk of nuclear plants.

Nonetheless, there never has been a nuclear-reactor-related accident in an American reactor which has caused injury to any worker, much less the public at large.

Insurance for Nuclear Accidents

Acknowledging the considerable risks of nuclear power plants, the U.S. Congress passed, in 1957, the Price–Anderson Act, which authorizes the government to insure against the potential costs of a reactor accident, up to the sum of $560 million. When the reactors were first built in the 1950's, private insurance companies were unable or unwilling to provide that amount of liability insurance.

However, private insurance companies do now provide a share of the total. Private companies now provide $125 million of the total $560 million insurance. Current discussion of an extension of the Price–Anderson bill involves the phasing out of the federal government as the primary insurer against person and property-damage claims in the event of an accident. Also under consideration is the possibility of raising the limit on the liability, perhaps reaching $1 billion by 1980. Many critics of the Price–Anderson Act believe that the current liability ceiling is too low for adequate coverage of potential damage.

The Price–Anderson Act provides for a liability framework intended to encourage participation in and development of the nuclear-power industry. It has also provided a base of liability insurance for the citizens of a region surrounding a nuclear reactor.

Nuclear Safety

Current concern over the safety of nuclear reactors and the possible consequences of an accident revolves around an accident that fortunately has never occurred. The consequences of such an accident could be serious, but the probability of its occurring are very small. This risk situation is similar to the risk of a dam breaking. It is very rare that a dam will fail, but if it does, the consequences to the persons living in the river valley below the dam would be catastrophic indeed. What is the risk of a 747 passenger aircraft crashing into a stadium filled with football fans? It is difficult to assign a probability of occurrence to that possibility, and undoubtedly it is very small. In a similar way, the likelihood of a nuclear reactor failing to the point where meltdown occurs and the containment vessel is breached is very small, but it is difficult to estimate. No such meltdown has occurred in nearly 2000 reactor-years of operation involving commercial and military light-water reactors in the U.S. Nevertheless, one may ask: Exactly how likely is such an accident?

Reactors are designed so that, in the case of component failure, the reactor can still be safely shut down. However, when *several* components fail, as well as the safety system, an accident could occur. The probability of such accidents occurring can be calculated using the event-tree and fault-tree methods. [21] A research group, under the direction of Professor Norman Rasmussen of the Massachusetts Institute of Technology, spent three years on a project with two objectives: (1) to identify those accident sequences with potential public consequences and estimate their probability of occurrence, and (2) to calculate the consequences of the release of radioactivity, in varying amounts, resulting from the identified potential accidents.

The event-tree method was used to attempt to identify the possible accident sequences. However, many believe that it is unlikely that all the possible sequences were identified, and it is difficult to predict absolute probabilities of each event. Finally, the contribution of human error is difficult to quantify accurately.

In a study of probabilities, an estimate is made of the probability of the failure of one important reactor element, and it is then assumed that failures of two *different* elements are independent, so that the probability of simultaneous failure of the two is the *product* of the individual probabilities. This, however, is not always true. There can be common-mode failures, where one event triggers two or three failures of essential elements of the reactor; in that case, the probability is not the product of the two individual probabilities. As an example, consider the reactor shutdown system which the Rasmussen study found to be triply redundant. Each individual system had a failure probability of 10^{-3}. Thus, if independent, the probability of all three systems failing would be $(10^{-3})^3 = 10^{-9}$. However, including the common-mode factors, the study estimated the probability of failure to be 10^{-5}.

The Rasmussen group estimated the probability of the worst accident as a result of a series of events occurring, with the most stable weather conditions, with

the wind blowing toward the population center. The probability per year is
then: [31]

Probability of worst consequence		Probability of initiating event	Probability of safety system failure	Probability of worst weather	Probability of contain- ment failure	Probability of highest population density
P_{wc}	$=$	P_{ie}	P_{ssf}	P_{ww}	P_{cf}	P_{pd}
	$=$	$[10^{-3}]$ $\times$	$[10^{-2}]$ $\times$	$[10^{-1}]$ $\times$	$[10^{-2}]$ $\times$	$[10^{-2}]$
	$=$	10^{-10}.				

The Rasmussen group found that while the probability of meltdown accidents
are not as small as some would have hoped, these meltdown accidents were very
unlikely to cause much damage outside the nuclear installation itself. [22] The
group calculated the probability of a meltdown to be about 10^{-4} per reactor year.
With 100 reactors operating in the 1980's in the U.S., this would imply a probability
of one percent per year that a reactor would experience a meltdown. The chance of
a meltdown occurring in a radiation release, however, was calculated to be 10^{-5}
per reactor year. Thus the overall chance of a catastrophic accident was estimated
to be only 10^{-9} per reactor year.

The consequences of such a low-probability accident are difficult to estimate.
The number of deaths due to radiation-induced cancer from such an accident are
estimated to be 3300 by the Rasmussen group, while a similar study estimates
about 10,000 deaths. The Rasmussen study estimates that there would be 45,000
cases of radiation illness and $14 billion of property damage due to contamination.
Long-term health effects from the same accident are estimated to be 1500 latent
cancer fatalities per year for up to ten years. These results are summarized in
Table 12.6. This is the classic case of a very unlikely accident with very large
consequences.

Table 12.6

THE WORST-CASE NUCLEAR ACCIDENT

Probability $= 10^{-10}$ per reactor year
Early cancer fatalities $= 3300$
Radiation illness $= 45,000$ persons
Property damage $= $14 billion$
Delayed cancer fatalities $= 1500$ per year for up to 10 years

The expected fatalities for 100 nuclear-power plants are shown in Fig. 12.6
for man-caused events and in Fig. 12.7 for natural events. [23] These figures
indicate that the risk of a large number of fatalities due to a reactor accident is
about one-thousandth of that due to the failure of a dam. The frequency of
occurrence of 1000 fatalities is comparable to the probability of a meteor hitting
the center of population density and much less than the probability of an
earthquake.

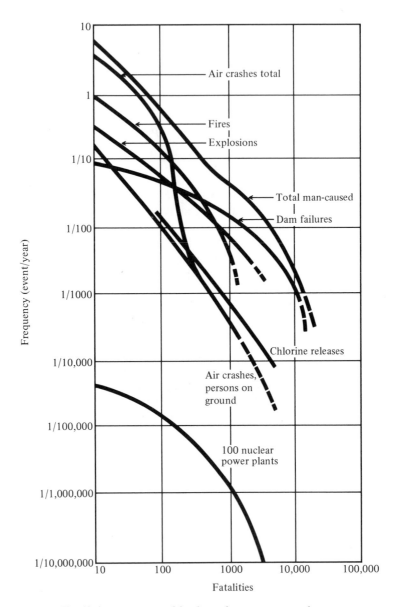

Fig. 12.6 *Frequency of fatalities due to man-caused events.*

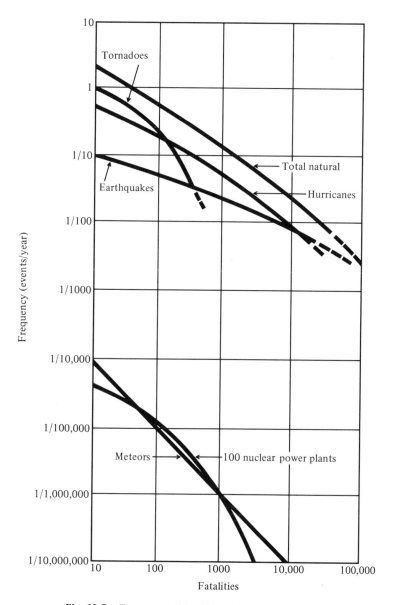

Fig. 12.7 *Frequency of fatalities due to natural events.*

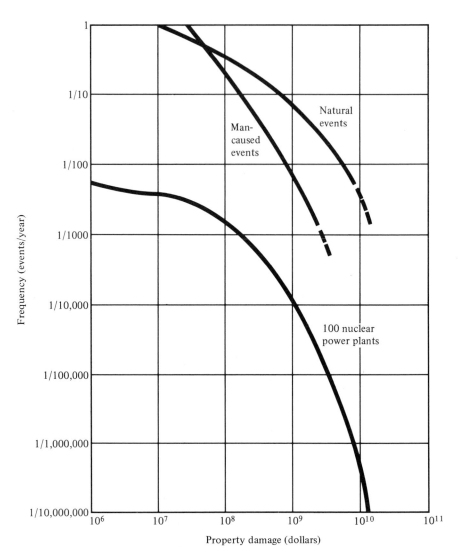

Fig. 12.8 *Frequency of property damage due to natural and man-caused events.*

The frequency of property damage due to a nuclear accident is compared to the effects of man-caused and natural events in Fig. 12.8. The likelihood of an event causing more than one billion dollars of damage is about one-hundreth of other man-caused events (such as dam failures).

The risk of an individual being fatally injured for various possible accidents is listed in Table 12.7. A person's chance of being killed in an air-travel accident is one in 100,000, while it is one in 5×10^9 for a nuclear accident.

Table 12.7

AVERAGE RISK OF FATALITY BY VARIOUS CAUSES FOR
1969 IN THE U.S.

Accident type	Total number	Individual chance per year
Motor vehicle	55,791	1 in 4,000
Falls	17,827	1 in 10,000
Fires and hot substances	7,451	1 in 25,000
Drowning	6,181	1 in 30,000
Firearms	2,309	1 in 100,000
Air travel	1,778	1 in 100,000
Falling objects	1,271	1 in 160,000
Electrocution	1,148	1 in 160,000
Lightning	160	1 in 2,000,000
Tornadoes	91	1 in 2,500,000
Hurricans	93	1 in 2,500,000
All accidents	111,992	1 in 1,600
Nuclear-reactor accidents (100 plants)	—	1 in 5,000,000,000

Critics of the Rasmussen study point out that the charts (such as Fig. 12.6) depict only the early fatalities that would occur within a short time after the accident, while ignoring delayed fatalities caused by cancer in the area downwind of a reactor accident.

It is difficult to accurately estimate the probable consequences of a rare accident. One critique of the Rasmussen study indicates that the probable consequences of a catastrophe or worst-case failure would be similar to that obtained for dam failures. [24] The debate continues as to whether nuclear reactors are sufficiently safe. In the interim, it would be worthwhile to continue to develop improved containment-vessel designs and to develop strategies for investigating the effects of a worst-case accident. Examples of the latter include the distribution of iodide pills to block the thyroid's uptake of radioiodides, and the development of practicable decontamination techniques to reduce long-term population doses from the 30-year half-life isotope cesium-137. [24]

12.7 THE NATIONAL NUCLEAR DEBATE

The debate over the safe and economic use of nuclear-power reactors in the U.S. has intensified over the past several years. Several advocacy groups of scientists, engineers, and environmentalists have issued statements calling for unbridled development of nuclear power or for a drastic reduction in new starts on construction of nuclear plants.

During 1975 and 1976, several groups affiliated with Ralph Nader have come together under the slogan "Stop nuclear power." [25] One of the outcomes of these group efforts is an attempt to develop a "nuclear safeguards" proposition for the ballot in many state elections. Twenty-two states permit the initiative process, and efforts are underway to qualify nuclear-safety initiatives in 16 states during the next several years.

In California, a nuclear-safeguards initiative was qualified for the June 1976 primary election. Earlier, in 1972, a vote on a proposed 5-year moratorium on the building of nuclear power plants lost, 2 to 1. This earlier initiative and the 1976 initiative had the support and assistance of the Los Angeles-based People's Lobby and its director the late Ed Koupal. This group was effective in the passing of the Coastal Zone initiative in California in 1972.

California has three nuclear power plants in operation, and 28 more are planned over the remainder of the century. The California Nuclear Safeguards initiative would have delayed this schedule, and eventually put a ban on nuclear plants if its conditions were not met. As a result, a shift to another source of power would have been required.

The requirements of the Nuclear Safeguards Initiative are as follows:

1. The federally-imposed limitations on insurance liability for nuclear accidents must be removed within a year.
2. The effectiveness of all safety systems, including the emergency core-cooling system, must be demonstrated by the testing in actual operation of substantially similar physical systems.
3. Radioactive wastes must be stored with no likelihood of escape.
4. Conditions 2 and 3 above must be demonstrated to the satisfaction of the state legislature, as expressed by a two-thirds vote in each house.

Opponents of this initiative said it was tantamount to shutdown, since it was a foregone conclusion that the U.S. Congress was not going to remove the insurance liability limit. Also, they pointed out, the obtaining of a two-thirds vote on such a difficult issue would be exceedingly difficult.

The so-called Western Bloc of 19 states is seeking to qualify nuclear-safeguards issues in their own states. A fundamental question, however, is whether a state has the right to reject or restrict nuclear reactors on its own authority, or whether such powers have largely been assigned to the federal government. Oregon and Vermont have already adopted laws which the nuclear industry believes will restrict or prohibit nuclear development. These nuclear initiatives are primarily based on a concern about the radiological safety of nuclear power plants, an issue reserved for federal regulation. [26]

The supporters of the nuclear-initiative movement maintain that they will force public review of nuclear power by the legislation, and require the provision of full compensation for accidents. The opponents maintain that the result of the passage of an initiative will be a moratorium on nuclear power and a shift to more fossil-fuel plants during the next two decades. Since nuclear-power plants require a large capital investment, it is clear that, with the passing of an initiative, utilities will be reluctant to invest in a nuclear plant and will shift to the planning of new fossil-fuel plants.

The California Nuclear Safeguards initiative was defeated by a vote margin of two to one in the June 1976 election. Nevertheless, the vote indicates that a

Table 12.8
RESULTS OF A 1975 SURVEY OF OPINIONS
REGARDING NUCLEAR POWER

	Total public (Percent)
1. Attitude toward building more nuclear power plants in the U.S.	
Favor	63
Oppose	19
Not sure	18
2. Attitude toward having nuclear power as main source of electric power in your community.	
Favor	54
Oppose	24
Not sure	22
3. How safe are nuclear power plants?	
Very safe	26
Somewhat safe	38
Not so safe	13
Dangerous	5
Not sure	18

significant number of persons are highly concerned about the development of nuclear power in California.

The Louis Harris organization surveyed the general public in 1975. [27] This opinion poll showed that Americans feel nuclear power is safe. The public surveyed felt that the advantages of nuclear power were that nuclear energy is cheaper and cleaner than fossil-fuel plants. The results of this survey are summarized in Table 12.8. In summary, most persons believe that nuclear power plants are safe and should be used as a source of electric power. However, as the nuclear debate continues, these attitudes may change.

Offshore Nuclear Power Plants

In order to improve the safety and reduce the environmental effects of nuclear-power plants it may be wise to place the plants underground or off the coast. Underground plants could be built to contain any radiation leakage due to a meltdown, but they would be relatively expensive to construct. The construction of offshore power plants is being pursued, and one company is planning the sale of such plants.

The concept of producing floating nuclear plants (FNP's) by assembly-line methods and putting them offshore has been responded to with significant interest; and it is possible that the first plant may be delivered in the early 1980's. The FNP concept emphasizes standardization in the design and manufacture of power plants. The FNP could be standardized since no special adaptation to a given site is required (as would be the case on land).

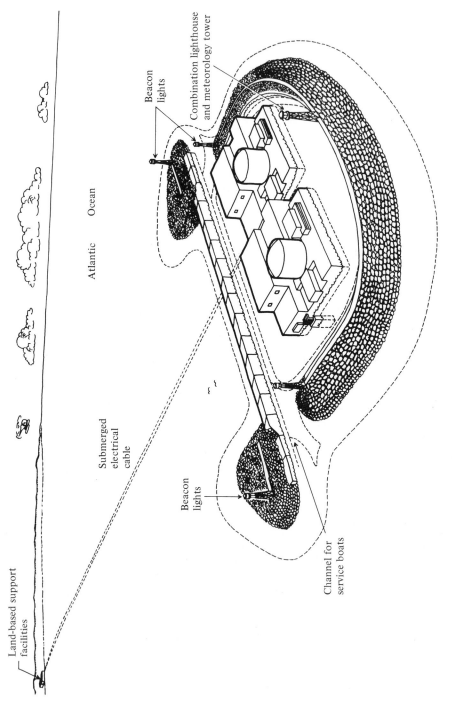

Land-based support facilities

Atlantic   Ocean

Submerged electrical cable

Beacon lights

Combination lighthouse and meteorology tower

Beacon lights

Channel for service boats

Fig. 12.9   A drawing of the floating nuclear plant concept.

Nevertheless, the consequences of a meltdown accident are in question. The consequence of the meltdown material contacting the sea water could release a large quantity of radioactivity into the ocean.

The FPN is planned to be 120 meters square, have a 10-meter draft, and rise 50 meters above the waterline. It would contain quarters for 100 operating personnel and would supply 1150 megawatts. Normally, two FNP's would be moored together, in water from 13 to 21 meters deep. They would be protected by a massive breakwater, from storms and collision by ships. The breakwater would cover an area of 100 acres (40 hectares, or 4×10^5 sq meters), which is about one-fourth of the area required by land-based reactors.

With the FNP, the once-through cooling water can be seawater and can be employed with minimal environmental effect. The cost of one FNP is estimated to be \$500/kW; thus a total site with two FNP's would cost one billion dollars.

It is hoped to place the first FNP facility off the coast of New Jersey, if all the licensing issues can be resolved. The advantages of the standardized plant would be (1) repetitive licensing, (2) reduced time and costs for construction, and (3) improved quality assurance. The plants could operate under hurricane conditions and provide 345 kilovolts of electricity to the nearby coastal cities. A conceptual drawing of a FNP is shown in Fig. 12.9.

12.8 NUCLEAR POWER IN THE WORLD

The commitment to nuclear power in other nations of the world is as great as or greater than in the U.S. In 1974 there were 149 commercial reactors in operation in 16 countries. The electrical capacity of these reactors was over 57,000 MW. By 1980 it is estimated that 350 reactors will be in operation, with about 200,000 MW of capacity, on line in 24 nations. The number of nuclear power plants and their capacity is given in Table 12.9 for several nations in the world. It is clear from the projections for 1980 that several nations expect to significantly increase their commitment to nuclear power.

Table 12.9

NUCLEAR POWER IN THE WORLD

Nation	1974 Capacity (gigawatts)	1974 Number of plants*	1980 Capacity (gigawatts)	1980 Number of plants*
U.S.	38.0	58	86	100
U.S.S.R.	2.6	9	15	25
Japan	3.1	7	32	40
Canada	2.5	6	7.5	13
Great Britain	5.6	14	12	19
France	2.8	11	15.0	23
West Germany	2.3	11	20	30
World total	57.0	134	200	450

* A plant may contain one or more reactors.

By 1980 it is estimated that about 25 percent of West Germany's electric-power production will be from nuclear reactors. In Italy, by 1980 the total nuclear power output may reach 6 gigawatts, up from 0.6 gigawatts in 1974. In Japan, the 1974 nuclear capacity was 3.1 gigawatts. By 1980 Japan plans to have increased its nuclear capacity by a factor of ten.

In Great Britain, the capacity is expected to double, from 6 gigawatts in 1974 to 12 gigawatts in 1980. Plans are being reviewed that call for the nuclear generating capacity in Great Britain to be over 100 gigawatts by the year 2000. Britain has decided to standardize its reactor program, basing it on the Steam Generating Heavy Water Reactor (SGHWR).

France currently imports about 70 percent of its energy requirements, primarily in oil from the Middle East and Africa. As a result of increased ore prices, France has embarked on an accelerated nuclear-power program. The French Pressurized Water Reactor (PWR) is built by Framatome, a company owned partially by Westinghouse Corporation. Framatome is building reactors of 900 MW and 1200 MW. France plans to have 15 gigawatts of nuclear capacity by 1980 and 50 gigawatts by the late 1980's. France is also embarking on the development of a breeder reactor and is operating the 250-MW Phénix near Marcoule. At current energy prices in France, nuclear energy is about one-half as expensive as all other forms of energy. [28]

As the use of nuclear reactors spreads through the world, an associated concern is the possible proliferation of the capacity to build nuclear weapons. India used the waste from a nuclear power plant purchased from Canada to detonate an atomic weapon in 1974. Since then, Pakistan and Brazil have been negotiating with France and West Germany to purchase nuclear reactors. In order to avoid the proliferation of atomic weapons, the nations providing nuclear reactors need to agree on the following points: [29]

1. Any country buying nuclear technology from them must accept inspection and supervision by the International Atomic Energy Agency;
2. Purchasers must give assurances not to use any imported nuclear facilities to make nuclear weapons, and
3. Those purchasers will employ tight safeguards against the theft or diversion of nuclear materials.

Forecasts are that only 5 percent of the world's nuclear reactor capacity will be in the less developed nations by 1980. This is due to a problem of size. Nuclear power stations of less than 500 MW have not proved generally economically competitive, yet plants larger than this would be too big for the power needs of most developing countries.

REFERENCES

1. H. A. BETHE, "The necessity of fission power," *Scientific American*, January 1976, pp. 21–31.
2. F. C. OLDS, "Nuclear power engineering," *Power Engineering*, December 1975, pg. 16.

3. *FPC News*, Dec. 12, 1975, pg. 3.

4. M. C. DAY, "Nuclear energy: A second round of questions," *Bulletin of the Atomic Scientists*, Dec. 1975, pp. 52–59.

5. R. E. LAPP, "We may find ourselves short of uranium, too," *Fortune*, October 1975, pp. 151–152, 194.

6. W. D. METZ, "European breeders: France leads the way," *Science*, Dec. 26, 1975, pp. 1279–1281.

7. P. CHAPMAN, "The ins and outs of nuclear power," *New Scientist*, December 19, 1974, pp. 866–869.

8. J. WRIGHT and J. SYRETT, "Energy analysis of nuclear power," *New Scientist*, Jan. 9, 1975, pp. 66–67.

9. L. BROOKES, "The growth of a myth," *New Scientist*, Oct. 16, 1975, pp. 143–147.

10. B. COMMONER, "Energy," *New Yorker*, Feb. 9, 1976.

11. E. H. THORNDIKE, *Energy and the Environment*, Addison-Wesley, Reading, Mass. 1976, Chapter 6.

12. N. TURNER, "Nuclear waste drop in the ocean," *New Scientist*, Oct. 30, 1975, pg. 290.

13. A. S. KUBO and D. J. ROSE, "Disposal of nuclear wastes," *Science*, Dec, 21, 1973, pp. 1205–1211.

14. D. F. SALISBURY, "Can we protect ourselves against plutonium?" *Technology Review*, Dec. 1975, pp. 6–7.

15. J. McPHEE, *The Curse of Binding Energy.* Farrar, Straus and Giroux, New York, 1974.

16. M. WILLRICH and T. B. TAYLOR, *Nuclear Theft: Risks and Safeguards.* Ballinger Publishing Co., Cambridge, Mass., 1974.

17. D. F. SALISBURY, "Quarantining plutonium," *Technology Review*, January 1976, pp. 4–5.

18. K. P. SHEA, "An explosive reactor possibility," *Environment*, Feb. 1976, pp. 6–11.

19. J. G. FULLER, *We Almost Lost Detroit*, Readers Digest Press, New York, 1975.

20. "Incident at Brown's Ferry," *Newsweek*, Oct. 20, 1975, pp. 113–114.

21. H. RAIFFA, *Introductory Lectures on Making Choices.* Addison-Wesley Pub. Co., Reading, Mass. 1968.

22. J. PRIMACH, "Nuclear reactor safety," *Bulletin of Atomic Scientists*, Sept. 1975, pp. 15–20.

23. *Reactor Safety Study*, U.S. Nuclear Regulatory Commission, WASH-1400, October 1975, Washington, D.C.

24. F. VON HIPPEL, "A perspective on the debate," *Bulletin of Atomic Scientists*, Sept. 1975, pp. 37–41.

25. J. WALSH, "Opposition to nuclear power: Raising the question at the polls," *Science*, Dec. 5, 1975, pp. 964–966.

26. P. M. BOFFEY, "Nuclear energy: Do states lack power to block proliferation of reactors?" *Science*, January 30, 1976, pp. 360–361.

27. "Ebasco survey: Public backs nuclear plants," *Ebasco News*, Aug. 1975, pp. 5–8.

28. E. BONER, *et al*, "Nuclear energy: A fateful choice for France," *Bulletin of Atomic Scientists*, January 1976, pp. 37–41.

29. A. DEMING, "Diplomacy: Bombs away?" *Newsweek*, March 8, 1976, pg. 51.

30. J. R. LAMARSH, *Introduction to Nuclear Engineering*. Addison-Wesley Publishing Co., Reading, Mass., 1975.

31. N. C. RASMUSSEN, "Rasmussen on nuclear safety," *IEEE Spectrum*, August, 1975, pp. 46–55.

32. S. S. PENNER, *Energy: Nuclear Energy and Energy Policies*, Vol. III. Addison-Wesley Publishing Co., Reading, Mass, 1976.

33. G. HARDIN, "Living with the Faustian bargain," *Bulletin of the Atomic Scientists*, Nov. 1976, pp. 25–29.

34. B. L. COHEN, "Impacts of the nuclear-energy industry on human health and safety," *American Scientist*, Oct. 1976, pp. 550–559.

35. I. C. BUPP, JR., "The status of nuclear power," *Ambio*, Vol. 5, No. 3, 1976, pp. 119–123.

36. G. I. ROCHLIN, "Nuclear waste disposal," *Science*, Jan. 7, 1977, pp. 23–30.

37. S. NOVICK, *The Electric War: The Fight over Nuclear Power*, Sierra Club Books, San Francisco, 1977.

38. F. VON HIPPEL, "Looking back on the Rasmussen Report," *Bulletin of Atomic Scientists*, Feb. 1977, pp. 42–47.

39. H. A. BETHE *et al*., "Six views of atomic energy," *Bulletin of Atomic Scientists*, March 1977, pp. 59–69.

40. V. GILINSKY, "Plutonium, proliferation and policy," *Technology Review*, Feb. 1977, pp. 58–65.

EXERCISES

12.1 Show that the change of mass in the fission process of Eq. (12.2) is equal to 0.215 amu.

12.2 If the half-life of cesium-137 is 30 years, show that it will take 300 years for the radioactive intensity of the material to be reduced to one-thousandth of its original level. What is its level of intensity after 600 years?

12.3 One author estimates that in 1986 the cost of electricity produced by nuclear plants will be 2.92¢/kWh compared to 2.74¢/kWh for coal-fired plants. [10] With the increasing costs of capital construction for nuclear plants, Professor Commoner predicts "The entire nuclear program is headed for extinction." Do you agree? Prepare a list of supportive and opposing arguments for this conclusion.

12.4 Nuclear power plants can be cooled by using water from a lake or a river. Other methods are the use of a wet cooling tower or a dry cooling tower. Wet cooling towers at the Rancho Seco 913-megawatt plant in Northern California use water from a nearby canal. Water enters the two 425-foot-high towers and is dispersed over 1 million asbestos sheets, cooling the water from 103 to 75°F. These towers cool about $\frac{1}{2}$ million gallons of water per minute by evaporation. An alternate dry cooling tower (nonevaporative) does not use water, but rather uses forced-draft air to cool the condenser. Examine the advantages and disadvantages of dry and wet cooling towers for nuclear power plants. Obtain information on the cooling method used for a nuclear power plant in your state or a neighboring state.

*12.5 Find the activity (in curies) of 10^{16} atoms if the half-life is (a) two hours, and (b) 100 years.

12.6 Examine the details of a nuclear-safeguards initiative in your state or in a nearby state. Do you find that it is establishing requirements tantamount to a shutdown of nuclear plants? List the arguments for and against the initiative. (One source: Nuclear Safeguards, 405 Shrader St., San Francisco 94117.)

12.7 It is proposed to use a HTGR for the purpose of steel-making. This plant would use an electric furnace as well as a preheater for the reducing gas. Examine the process, and list the advantages and disadvantages of such a scheme.

12.8 How much energy is deposited in an adult body when it receives a dose of 100 rad?

12.9 List several schemes for mitigating the possible consequences of a radiation release from a nuclear reactor.

12.10 Complete a survey of a random sample of persons in your community regarding nuclear safety and their willingness to have a reactor constructed 50 miles from the city. Also determine whether they estimate the risk of a dam failure is greater than the risk of a nuclear power-plant radiation release.

12.11 Examine several methods of nuclear waste disposal and list the advantages and disadvantages of each.

*12.12 The nuclear debate is active in Europe. West Germany and France are committed to the construction of many new nuclear reactors over the next decade. However, organized opposition groups are active in both countries. In 1976, several French anti-nuclear groups attempted to halt the nuclear power program. In West Germany a group of 20,000 residents is opposing a 4-reactor plant in the Ruhr Valley city of Whyl. Examine the situation in West Germany and France at this time, and assess the levels of opposition to nuclear power. Estimate whether these nations will be able to meet their projected development of nuclear power by 1985.

---

* An asterisk indicates a relatively difficult or advanced exercise.

13

Nuclear Fusion Power

With nuclear fission reactors slowly growing in number, and under scrutiny regarding their environmental safety, attention is turning to the potential for nuclear fusion. Whereas nuclear fission gets its energy from large nuclei as they break up, thermonuclear-fusion reactions yield energy when two light nuclei fuse together to form a heavier one. It is fusion that powers the sun and the stars, and the challenge is to establish a workable fusion reactor in practice.

Two light nuclei are brought together to yield energy upon fusion within a high-temperature gas. The very hot gas at a relatively low density is known as a *plasma*. Within the plasma are free electrons and ions that flow like a current.

Research and development in the U.S. and Russia on the physics of fusion has increased over the past several years. The U.S. Energy Research and Development Administration provided funds to researchers in the amount of $75 million and $322 million, in 1973 and 1977, respectively. It is estimated that the U.S.S.R. and the U.S.A. will each spend over 1.5 billion dollars over the period 1975 to 1980 on fusion research.

Fusion reactors may possibly be commercially available by the year 2000. The fusion reactor, if achieved, has many advantages. The first of these is the ready and inexpensive availability of fuel. The fusion reactor is expected to present fewer environmental hazards than do the fission reactors. No long-lived heavy-element isotopes are produced, and there is no production or handling of the biologically dangerous and radioactive plutonium. There is no need to transport radioactive fuel elements, and all processing would be done on site. Also, there is probably less danger of accident. Estimates for the efficiency of the reactor range from 40 to 60 percent.

13.1 THE FUSION REACTION

There are several basic fusion reactions that can be pursued in a fusion reactor experiment. The four most common reactions are listed in Table 13.1. The heavy isotopes of hydrogen ^{2}H and ^{3}H are called *deuterium* and *tritium*, respectively, and are usually represented by D and T respectively. Deuterium is a stable isotope of hydrogen; and its nucleus consists of one proton and one neutron. It is sometimes called *heavy hydrogen*. Thus, the primary fuel for fusion is D, which can be readily obtained from common seawater. Deuterium occurs as one out of every 6500 atoms in seawater. The energy obtainable by the fusion of all the D in a cubic meter of seawater is 12×10^{12} joules. This amount of energy can be compared with the energy in a barrel of oil, which is 6×10^9 joules. A cubic meter of seawater, then, corresponds to 2×10^3 bbl of oil, and a cubic kilometer to 2000×10^9 bbl. This last number is, coincidentally, nearly equal to recent estimates of the earth's total oil reserves. The oceanic volume is approximately 1.5×10^9 km^3, so there is more than one billion times the energy content of the world's oil reserve available to use through deuterium fusion.

Table 13.1

COMMON FUSION REACTIONS

1) ^{2}H + ^{2}H → ^{3}H + ^{1}H + 4.03 MeV*
2) ^{2}H + ^{2}H → ^{3}He + n + 3.27 MeV
3) ^{2}H + ^{3}H → ^{4}He + n + 17.59 MeV
4) ^{2}H + ^{3}He → ^{4}He + ^{1}H + 18.35 MeV

* 1 MeV = 1.60×10^{-13} joule

Attention has tended to focus on the third of the fusion equations in Table 13.1. Because tritium does not occur naturally, the reaction must be supplemented by one using lithium to reproduce (or breed) the T fuel. The two reactions are:

$$D + T \rightarrow {}^4He + n + 17.6 \text{ MeV}; \qquad (13.1)$$

$$n + {}^6Li \rightarrow {}^4He + T + 4.8 \text{ MeV}. \qquad (13.2)$$

The neutron, n, from Eq. (13.1) combines with lithium in Eq. (13.2) to yield helium and tritium, plus released energy. The tritium from Eq. (13.2) then is used again in the process of Eq. (13.1), thus recycling itself. Combining Eq. (13.1) and (13.2), we have the net reaction as:

$$D + {}^6Li \rightarrow 2\,{}^4He + 22.4 \text{ MeV}. \qquad (13.3)$$

The D–T reactor requires a second reaction (Eq. (13.2)) to take place outside the plasma, and this is technologically difficult. Nevertheless, the conditions needed to achieve a positive net energy for the D–T reaction are less demanding than for the others of Table 13.1.

The energy release of entry 3 in Table 13.1 is equal to 17.59 MeV. This is equivalent to 2.8×10^{-12} joules for one nuclear fusion. For the D–T reaction of Table

13.1, the equivalent energy per gram mass of reacting nuclei is 98,000 kWh/gm. [1] At this level, the amount of primary fuel, deuterium, needed for world electrical power would be small. The total U.S. electrical power could be supplied by an input of about 10 kilograms of D per hour, which could be readily obtained from seawater.

Fusion reactions can take place only when the nuclei of the fuel atoms are brought close enough together to fuse. The nuclei are positively charged and repel each other. Thus the nuclei must have sufficient kinetic energy to overcome this repulsion force; and plasmas of interest must have an average energy per particle in excess of 5 keV. A collection of particles with average energy of 5 keV has an effective temperature of nearly 40 million kelvins (4×10^7 K). At these temperatures, the gas is completely dissociated into its constituent, positively charged nuclei and free electrons. The electrical charge density is such that the behavior of the collection of particles is completely dominated by electromagnetic phenomena.

Plasma at high temperature cannot be confined by material walls but can be confined by electromagnetic forces. The central problem in the search for controlled fusion energy has been the design of a magnetic field configuration that allows for a stable containment. In a magnetic field, such a force balance can occur only in the presence of currents flowing within the plasma, with a nonzero component perpendicular to the field lines. This requirement is stated through the local equilibrium force-balance equation

$$\nabla \cdot \mathbf{P} = \frac{1}{c}(\mathbf{J} \times \mathbf{B}), \tag{13.4}$$

where $\mathbf{P}$ is the pressure tensor for the plasma, $\mathbf{J}$ is the current density in the plasma, and $\mathbf{B}$ is the local value of the magnetic-field intensity. The strongly coupled nature of the problem is evident when we note that $\mathbf{B}$ is itself a function, not only of currents in conductors outside the plasma, but of the current $\mathbf{J}$ within the plasma itself.

13.2 ACHIEVING POSITIVE NET ENERGY

Fusion fuel must be heated to a sufficiently high kinetic temperature so that the fuel nuclei collide with each other with sufficient energy. Every fusion reaction produces 22.4 MeV if the process of Eq. (13.3) is used; but a large number of such reactions must occur each second in order to generate more total energy than is required to produce, maintain, and heat the plasma in the first place. In order to achieve a net positive release of fusion energy, the product of particle density, n (in particles per cubic centimeter), and confinement time, τ (in seconds), must exceed about 10^{14}. This is called the *Lawson criterion*, and is written as

$$n\tau > 10^{14} \text{ atoms/sec/cm}^3. \tag{13.5}$$

The twin requirements of high kinetic temperature and adequate confinement times define the challenge to fusion research. Reasonable goals for a steady-state reactor are: plasma density n = 5 × 10^{14} atoms/cm^3; pulse duration τ = 0.5 second; and plasma temperature T = 10 keV.

The Tokamak Confinement Scheme

The Tokamak confinement scheme is an example of a toroidal confinement system. The Tokamak scheme was first proposed by Artsimovich of the Kurchatov Institute, U.S.S.R., in 1968. One special feature that characterizes the Tokamak concept is a toroidal vacuum chamber (shaped like a doughnut), as shown in Fig. 13.1. The plasma is held within a vacuum chamber; and magnetic field lines are set up by the current I_θ flowing in the coil wrapped around the vacuum shell. The ions and electrons spiral along the field lines, B_ϕ, and remain confined to the chamber.

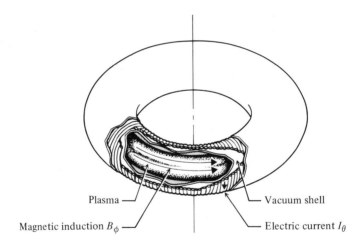

Plasma ——

Magnetic induction B_ϕ ——

Vacuum shell ——

Electric current I_θ

Fig. 13.1   *A toroidal magnetic field B_ϕ made by the current I_θ.*

The complete Tokamak scheme is shown in Fig. 13.2. The current induced in the plasma, I_θ, sets up a magnetic field B_θ. This current is induced by the external pulse I_ϕ. This induced current serves also to heat the plasma by resistive heating $(I_\phi{}^2R)$. The combination of B_θ and B_ϕ serve to confine the plasma within the toroid.

The largest Tokamak operating at present is the T-10 facility at the Kurchatov Institute in Moscow. An American experiment based on the same principle, the Princeton Large Torus, also started experimentation in 1975. The purpose of the Tokamak T-10 and the Large Torus at Princeton is to experimentally investigate

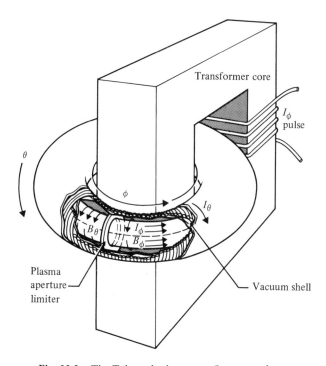

Fig. 13.2   *The Tokamak plasma-confinement scheme.*

the confinement density and the time for a stable confinement. The T-10 has a major radius of 150 cm. [2] The earlier Russian Tokamak T-3 yielded a plasma density n = 2 × 10^{13} ions/cm^3, a temperature of 0.6 keV, and a confinement time τ = 0.025 s.

The Stellerator developed at Princeton University is a form of toroidal confinement scheme with additional helical windings that impose a helical structure on the magnetic field. The Stellerator is a steady-state device, while the Tokamak is a cyclic device.

The conceptual design study of a possible future Tokamak reactor done by Oak Ridge Laboratory has a major radius of 10.5 meters. The toroidal magnetic field in the plasma is 60 kilograms and its toroidal current is 20 megamperes. This concept is planned to yield 518 megawatts of net power, produced at an efficiency of 56 percent. [1]

Magnetic Mirror Devices

A second device for magnetic containment is based on the principle that the charged particles in a plasma will be reflected at regions of very strong magnetic field. Such a device is called a *magnetic mirror* confinement system. A schematic diagram of a simple magnetic mirror is shown in Fig. 13.3. This so-called open-ended confinement system relies on the repelling effect of extra-strong magnetic field regions (the mirrors) on helically moving particles. By locating magnetic mirrors at both ends of a confinement region, charged particles can be trapped between these mirror regions and reflected back and forth for a long enough time to permit fusion reactions. Problems associated with this scheme are the containment at the end of the regions and instabilities in the mid-region.

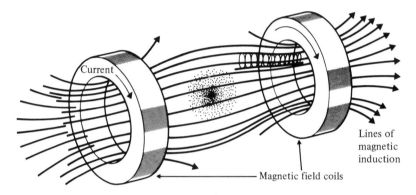

Fig. 13.3 *Magnetic-mirror plasma-confinement configuration.*

The Theta-Pinch Confinement Scheme

Magnetic compression of a plasma may be achieved by means of a confining field in the θ-axis, as shown in Fig. 13.4. This confinement scheme is called the *theta-pinch* plasma-confinement method. The plasma is placed inside a single-turn coil,

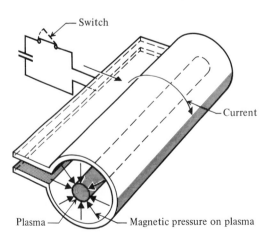

Fig. 13.4 *Theta-pinch plasma-confinement scheme.*

to which current is suddenly fed from a capacitor bank. This rapidly fills the coil with a magnetic field parallel to its axis. During the dynamic (shock-heating) phase, the surface of the plasma is driven rapidly inward by the axial field, heating the ions and electrons. Later there is a quiescent (adiabatic compression) phase after the magnetic field is built up to a steady value in the coil. The plasma is then held in a cigar shape by the steady magnetic field, gradually being lost out its ends along magnetic lines. The objective of this experiment is to achieve a very high-density plasma. Capacitors storing nearly a megajoule of energy at 50 kV are discharged to generate fields of 100 kilogauss.

The theta-pinch method is incorporated in the Scylla machines at Los Alamos Scientific Laboratory, New Mexico. Plasma densities of 5×10^{16} ions/cm^3 are achieved at temperatures of 3.2 keV, but insufficient plasma-confinement time τ is achieved ($\tau = 10^{-5}$ seconds). The three confinement schemes are summarized in Table 13.2, in terms of their performance in comparison to the Lawson criterion. [3]

Table 13.2
PERFORMANCE OF CONFINEMENT SCHEMES

	Tokamak (T-3)	Mirror	Theta-pinch (Scylla)	Lawson criterion
Plasma density (ions/cm^3)	2×10^{13}	5×10^{13}	5×10^{16}	5×10^{14}
Plasma temperature (keV)	0.6	8	3.2	10
Confinement time, τ (msec)	25	0.4	0.01	200

13.3 LASER FUSION

While research projects are pursuing the development of a magnetic confinement scheme, a different approach is also being investigated. Instead of containing the fusion fuel within a magnetic field configuration, a pellet of these materials is heated very rapidly with a very powerful laser beam. A laser beam impinging from many directions on the pellet of deuterium and tritium compresses the pellet to a very dense state, where fusion takes place.

The occurrence of laser-induced fusion was first demonstrated in 1968 at the Lebeder Institute in Moscow. A laser that, through the use of lenses and mirrors, delivers multiple high-powered beams, can deliver 1000 joules for a nanosecond and is a potential for laser-fusion experiments.

A laser that is useful for high-power experiments is the neodymium-doped glass laser, which is made from a special glass to which the rare-earth element neodymium has been added. Other lasers of recent interest are the carbon-dioxide laser and the high-pressure xenon-gas laser.

Experiments are only beginning to explore the potential of laser-induced fusion. The challenge is to develop pulsed lasers with sufficiently high power, energy, frequency, and efficiency to yield a laser-fusion reactor with a net energy gain.

13.4 RADIOACTIVITY

The only radioactive substance that would be released during routine operation of a fusion power plant is tritium. Thus, the essential design of a power plant is to minimize the release of tritium. An essential point is that tritium fuel has to be transported only once, for reactor start-up. The biological-hazard potential of a fusion reactor is ten to one hundred times less than the use of the fission reactor. [1] Finally, the diversion of tritium or deuterium is not sufficient to construct an atomic weapon. Thus, the environmental and safety aspects of the fusion reactor are more benign than for the fission reactor.

13.5 THE PACER POWER SYSTEM

An *in situ* power system, using an underground cavern to contain atomic explosions and generate steam to power, has been proposed as a Pacer system. [4] Pacer would use a fission reaction to ignite fusion weapons in an underground cavity to produce steam. This proposal has been made by the Los Alamos Laboratory to the Energy Research and Development Administration and has not, as yet, been adopted for experimentation.

A Pacer system capable of supporting a 2000-megawatt-capacity electrical plant would require two explosions of 50-kiloton magnitude each day, or about 700 such explosions each year. A small amount of uranium-235 would be used as a fission trigger to ignite deuterium fuel. These explosions would be contained within an underground cavity filled with water, which the explosions would vaporize for use in the steam electric generator. The cavity temperature would be held at about 500°C. Salt formations could be used for the underground cavity.

The safety and economy of such a system is uncertain; and a fuel study would be required before experimentation could proceed. The cost of one 50-kiloton device could be in the range of 100 to 400 thousand dollars. Thus, the cost of the devices above could be $70 million per year. Nevertheless, the cost per kWh generated could be reasonably economic.

The primary objection to the pursuit of the Pacer program or other peaceful uses of nuclear explosives is the reasonable fear of radioactive leakage from the underground cavity and the danger of diversion of the 700 atomic devices used in such a system each year. ERDA has not yet chosen to fund further studies, and the Pacer proposal may remain an untried method for tapping fusion power.

13.6 ENGINEERING OF A FUSION REACTOR

Fusion power is only in the research stage at present and therefore, it is difficult to know what engineering techniques may actually be required. However, it will most likely take until the year 2000 to bring a laboratory reactor to full commercial utilization.

A fusion reactor will probably consist of a plasma contained within a vacuum containment wall. In addition, a lithium–tritium regeneration region must be provided outside the vacuum wall. Finally, the total system should be surrounded with magnetic field coils, providing the magnetic confinement field.

The vacuum wall surrounding the plasma must be thin enough to pass neutrons readily but strong enough to withstand high plasma temperatures. Metals such as niobium and molybdenum may be candidates for the vacuum container walls. These walls must withstand embrittlement under neutron bombardment.

The magnet coils must be able to produce large fields, typically larger than 100 kilogauss. The most likely technique for producing large fields will be superconducting magnets.

Other engineering problems include fuel injection, electrical power generation, and the availability of high-temperature materials. As the fusion reactor is developed in the research laboratory over the next several decades, a full-scale engineering effort will be required to bring fusion power to commercial realization. The energy yield from the fusion reaction must be picked up in the lithium blanket region as heat and transferred to the secondary system by an intermediate heat exchanger. [5] The engineering tasks to tap fusion power are challenges for the future.

The economics of fusion power are very difficult to define prior to the engineering development of an actual prototype device. One estimate for a 2020-megawatt plant predicts a capital cost of $1215 million for a cost of $600/kW. [6] Since fuel costs are essentially negligible, the cost of fusion electricity, according to this study, would be 1.5¢/kWh. More likely the capital cost of a fusion plant would be about the same as that of a fission plant. The cost of a Tokamak reactor could be over $1000 per kW in 1976 dollars.

13.7 ENVIRONMENTAL EFFECTS

Since the fusion reactor will utilize deuterium, which is abundantly available from seawater, there will be no significant environmental impacts associated with obtaining the fuel for the reactor. The end products of the fusion process are helium and hydrogen isotopes. Thus, no long-lived, heavy-element isotopes are produced.

There is no need to transport radioactive-fuel elements, and all processing would be done on site. Because the fusion process readily shuts itself off, there is probably less danger of an accident.

A recent report confirms that radioactive tritium poses the most serious hazard to public health from fusion reactors. [7] Neutron-induced radioactivity in the material used in the reactor is the highest-intensity hazard to the operating staff. The tritium will be bred by neutron reactions in a blanket of liquid lithium or a lithium salt surrounding the reactor. Both the lithium/lithium salt and the "ash" produced during fusion reactions—the mixture of unburned fuel and fusion products left after thermonuclear reactions—will have to be processed to remove tritium.

The induced radioactivity in the reactor components will be the only true nuclear waste produced in a fusion reactor. In principle, none of this waste need leave the reactor site (where it can remain in suitable storage silos) until it has decayed to safe levels of radioactivity.

Further study will be required regarding the potential release of tritium by an accident. Also, leakage of tritium radiation is possible, although it is estimated that the radiation dose would be less than one-hundredth of the recommended levels, even if fusion reactors provided all the world's electric needs.

Fusion may be the safest source of energy for the future with a minimal environmental impact. Nevertheless, it will probably be another two or three decades before a commercially useful fusion reactor is available. Fusion power may be a source of energy for the 21st century. The advantages of fusion power are chiefly environmental and social ones, but present developments are not sufficient to enable us to determine whether fusion power is a dream or will become a reality. [8]

REFERENCES

1. R. F. POST and F. L. RIBE, "Fusion reactors as future energy sources," *Science*, November 1, 1974, pp. 397–407.

2. B. B. KADOMSTEV and T. K. FOWLER, "Fusion reactors," *Physics Today*, November 1975, pp. 36–41.

3. L. M. LIDSKY, "The quest for fusion power," *Technology Review*, January 1972, pp. 10–21.

4. W. D. METZ, "Energy: Washington gets a new proposal for using H-bombs," *Science*, April 11, 1975, pp. 136–137.

5. F. C. OLDS, "Fusion power teams begin to tackle the engineering tasks," *Power Engineering*, October 1974, pp. 33–41.

6. R. G. MILLS, "Problems and promises of controlled fusion power," *Mechanical Engineering*, Sept. 1975, pp. 20–25.

7. *Culham Study Group Report on Fusion Reactors and the Environment*, United Kingdom Atomic Energy Commission, Her Majesty's Stationery Office, 1975.

8. W. D. METZ, "Fusion research: What is the program buying the country?" *Science*, June 25, 1976, pp. 1320–1323.

9. S. S. PENNER, *Energy: Nuclear Energy and Energy Policies*, Addison-Wesley Publishing Co., Reading, Mass, 1976.

10. G. KAPLAN, "Europe: tilting toward fusion," *IEEE Spectrum*, December 1976, pp. 36–40.

11. D. J. ROSE, "The prospect for fusion," *Technology Review*, December 1976, pp. 20–43.

EXERCISES

13.1 Fusion research is being carried out in many laboratories in the U.S. Examine recent laboratory reports in the literature and prepare an analysis of when commercially available fusion power will become a reality.

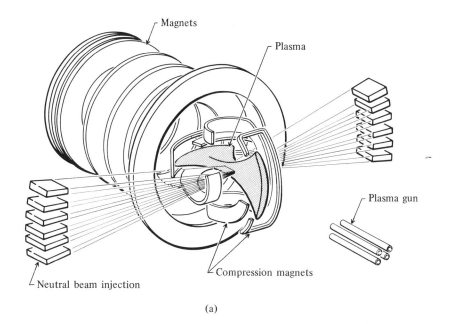

(a)

Fig. E13.8 *(a) The schematic diagram of 2XIIB. (b) The actual experiment. At lower left are the plasma guns. The domed tanks house the fusion fuel injection system. (Photo courtesy of Lawrence Livermore Laboratory, Univ. of Calif., Livermore, Calif.)*

13.2 The containment of the plasma is one of the critical problems of fusion reactors. Examine the various schemes, and select the one with the most potential. Justify your selection.

13.3 Contact the Princeton University Fusion Research Laboratory, and obtain the latest results of their experiments. Compare their results with the Lawson criterion.

13.4 Contact the Lawrence Livermore Laboratory, Livermore, California, concerning its work on laser fusion. Explore the potential for commercially available laser fusion.

13.5 Provide your written recommendation to ERDA on whether the Pacer project should be funded.

13.6 Weigh the potential environmental effects of a fusion reactor in contrast to other sources of electrical power. Consider fission reactors, coal-fired power plants, and natural-gas plants.

***13.7** A plasma is confined by a magnetic field. An electric charge travels a helical path along a field line. Calculate the radius of the path of travel, and show that a stronger magnetic field will hold a charge in a dense plasma, thus confining it.

***13.8** The Lawrence Livermore Laboratory is pursuing experiments on the magnetic mirror approach. The 2XIIB is shown in Fig. E13.8(a) and (b). Plasma guns fire a burst of preheated fuel into the device, where the magnets contain and squeeze the fuel. A neutral beam of 250 million degrees F is then injected. Explore the advantages of the magnetic-mirror concept and of the success of the LLL experiment (LLL, Livermore, Calif. 94550).

---

* An asterisk indicates a relatively difficult or advanced exercise.

Energy from the Wind and Oceans

14.1 HISTORY OF THE USE OF WIND ENERGY

A large amount of power is contained in the movement of air in the form of the wind. The Persians built the first known windmills as early as 250 B.C., several of which are still in use today. Their sails revolve around a vertical axis in the manner of carousels. One view of a Persian windmill is shown in Fig. 14.1. The wind is caught by sails rigged between the spokes radiating from the top and bottom of the shaft. [1]

Use of the horizontal windmill, whose sails rotate around a horizontal axis, spread throughout the Islamic world after the Arab conquest of Iran. The windmill in Europe in the 11th century was also a horizontal-axis type; and by the 17th century the Netherlands had become the world's most industrialized nation through extensive use of wind power in ships and the familiar Dutch windmill. A sketch of a Dutch windmill is shown in Fig. 14.2.

In the U.S., the first windmill was constructed a few miles from Jamestown in Virginia. Dutch settlers lined the shore of Manhattan Island with handmade windmills. Two of the most powerful Dutch windmills ever built stand on a bluff overlooking the Pacific Ocean in San Francisco's Golden Gate Park, where they once pumped water. [2]

During the 18th century, many important improvements were made in horizontal windmills. Windmills were built with 8 to 12 sails constructed from sailcloth stretched over a wooden frame. Late 18th-century windmills were characterized by 4 to 6 arms that supported long rectangular sails. In 1750, the Netherlands had 8000 windmills in operation. In Germany there were over 10,000 windmills by the mid-19th century. The designs of windmills continued to be improved and refined. Sails gave way to hinged shutters, like those of Venetian blinds, and the shutters in turn gave way to the propeller or airscrew.

The use of wind power for sailing ships has prevailed for several thousand years. However, the use of sailing ships for commercial trade began to decrease in

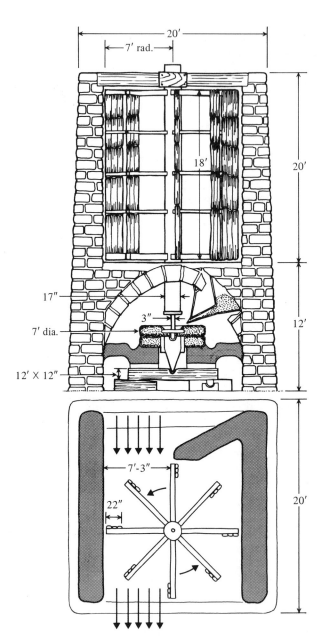

Fig. 14.1 *Persian vertical windmill with grindstone in sectional views. This version has bundles of reeds instead of cloth for its sails.* [*From* Traditional Crafts of Persia, *Hans Wulff (MIT Press, Cambridge, Mass., 1966)*]

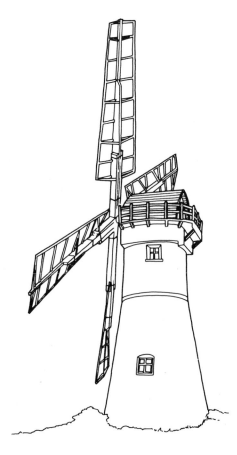

Fig. 14.2 *A sketch of a horizontal Dutch windmill. The blades rotate around a horizontal axis with the plane of the vanes essentially perpendicular to the wind-velocity vector.*

the early 1800's. The demise of the clipper ship era was begun by the introduction of the *S.S. Savannah*, which in 1819 was the first steamship to cross the Atlantic Ocean. Actually, the *Savannah* was a sailing vessel with substantial auxiliary steam power, which was used to supplement failing winds; but this advantage proved to be important. The 320-ton *Savannah* could travel at 5 knots with its 72-horsepower (54-kilowatt) engine, or at 7 to 9 knots under sail. The advantages of the steam-engine ship were its ability to maintain a reliable schedule for commerce, and the smaller crew required, compared to the sailing ship. These economic factors still prevail today as deterrents to the use of wind power to power ships on the seas.

14.2 WINDMILLS IN THE U.S.A.

An American, Daniel Halladay, constructed a self-governing windmill in Connecticut in 1854. This improvement and those developed by others resulted in a windmill with many short sails and self-governing speed. Factories throughout the country produced and sold millions of such machines in the next half-century. Wealthy men bought them to provide running water in their homes. The Union Pacific used them to fill trackside tanks with water for the first locomotives that chugged across the continent. And cattlemen and homesteaders bought them to help turn the great American plains into the source of much of our food today.

By 1850, the use of windmills in the U.S. provided about 1×10^9 kWh each year. [3] By 1870, the amount of power provided by windmills began to decrease as the steam engine became predominant. Up until the 1930's windmills provided power for water pumping and small (2-kilowatt) electric generators for rural locations. These electric generators of the automobile type charge lead–acid storage batteries at 32 volts for lighting farms. This system could not compete, however, even in its restricted application, with the central-station commercial service that expanded into the rural areas in the 1930's, principally under the aegis of the Rural Electrification Administration. Central-station service was not only more economical for the consumer, but it also offered increased reliability and capacity for heavier-duty applications. These old rural lighting systems, however, aside from economics and the capacity requirements of the users, illustrate the always present need for energy storage and backup facilities when one tries to match the variable wind energy available with the variable consumer demand for electric energy. Nevertheless, up until 1950, some 50,000 small windmills converted wind energy into electricity in the midwestern U.S.

American windmills worth visiting are those in Golden Gate Park in San Francisco; Williamsburg, Virginia; Newport, Rhode Island; and at Sandwich on Cape Cod. [1] In addition, still standing on the farms of America, many of them in disrepair, there may be up to 150,000 water-pumping windmills.

The Putnam Wind Generator

A four-vaned windmill was developed in Denmark by LaCour in 1890 to drive an electric generator. The sails of this mill were constructed of hinged wooden shutters that were operated towards the wind by turn fantails. The sails, hoisted on a steel tower, were 8.25 feet wide and 75 feet long from tip to tip. The structure was similar to an airplane propeller, with a continuously changing angle of twist. This windmill was used to drive two 9-kilowatt generators.

During the 1930's, impelled by a growing public interest in using windmills to generate larger amounts of electricity, Palmer Putnam explored the feasibility of constructing a large wind-power machine. A wind turbine, with propeller blades 100 feet in diameter, was constructed in 1931 on a bluff near Yalta, Russia, on the Black Sea. This windmill drove a 100-kilowatt generator to serve the city of Sevastopol, 20 miles away. Despite limitations in the equipment (the blade skins were made from roofing metal and the main gears from wood), the unit reportedly produced 279,000 kilowatt-hours of electrical energy in one year.

Putnam interested many engineers and manufacturers in the construction of a prototype, two-bladed propeller windmill. It was agreed that the unit would be financed and built by S. Morgan Smith Company of Pennsylvania and then be operated by the Central Vermont Public Service Corporation. A 2000-ft peak in Vermont, dubbed "Grandpa's knob," was chosen for the construction of the wind machine. Putnam engaged the assistance of Thomas S. Knight of the General Electric Company and J. P. Hartog of Harvard University, as well as John B. Wilbur of M. I. T.

The wind turbine had to be very large in order to economically tap the wind power and, therefore, the two-blade propeller was 175 feet from tip to tip. Each blade weighed 8 tons and was made of stainless steel. The tower was 110 feet high. A photo of the Smith–Putnam wind generator is shown in Fig. 14.3. The machine was rated to produce 1250 kilowatts, and was actuated for the first time on August 29, 1941. The blade pitch was adjustable, so that a constant rotor speed of 28.7 rpm could be maintained. This rotational speed was maintained in wind speeds as high as 70 to 75 mph. At higher wind speeds, the blades were feathered and the machine was brought to a stop. Electricity was generated and fed into the utility transmission network commencing on October 18, 1941.

Fig. 14.3 *The Smith–Putnam wind machine on Grandpa's Knob in Vermont. (Photo courtesy of National Aeronautics and Space Administration.)*

In 1942, a total of 179,000 kilowatt-hours was provided to the Central Vermont Public Utility. However, in 1943 a main bearing failed. Since World War II was in progress and this project had low priority, it took several years to obtain a new main bearing, which was finally installed early in 1945. Following the installation, the machine was operated only a few months when one blade failed as a result of overstress after being idle for two years; the machine was subsequently dismantled. The project, which had cost 1.25 million dollars, had been founded on the premise that, if several similar machines could be produced by Smith–Putnam, the cost could be reduced to $200 per kilowatt for construction. This, of course, was never realized since the Putnam project ended in 1945.

Percy Thomas of the Federal Power Commission devoted twelve years after 1942 to analyzing the Smith–Putnam machine and providing improved designs for wind-powered electric generators. [4] He described a twin-wheeled, two-bladed propeller design for a 7500-kilowatt unit, and a three-bladed design for a 7500-kilowatt unit. Thomas's calculations showed that wind machines were economical in the range of 5,000 to 10,000 kilowatts for utility operation. His 1951 design called for a tower height of 475 feet and a rotor diameter of 200 feet. Thomas estimated the capital costs for this 6500-kilowatt machine at $75 per kilowatt.

14.3   WIND-POWER ELECTRIC GENERATORS IN EUROPE

During the past 45 years, several experimental wind-powered electric generators have been built in Europe. In addition to the Yalta plant built by the Russians in 1931, a three-bladed machine rated at 300 kilowatts was built at Nogent Le Roi, France, in 1959. The diameter of the blade rotation was 31 meters (102 feet). A two-bladed, 100-kilowatt unit was built in West Germany in the period 1961–1966. Designed by U. Hutter in cooperation with the Allgaierwerks of Württemburg it used a blade diameter of 35 meters.

The Danes built a wind-turbine machine with three blades in 1957. This Gedser machine operated at 200 kilowatts in a 34-mph wind. It was connected to the Danish public power system and produced 400 megawatt-hours per year. The tower was 85 feet high and the rotor 79 feet in diameter. The installation cost was approximately $205 per kilowatt in 1957. This machine ran until 1968.

Although wind-turbine systems have been built and tested in a number of countries around the world, these systems have been dismantled after running for a while, because the installation cost per kilowatt has proved to be too high when compared with those of other methods of producing electric power. In addition, because of wind variability, it is not sufficient to have a wind turbine alone, and some form of energy storage must be used, thus adding to the cost of the total system.

14.4   ADVANTAGES AND THE POTENTIAL FOR WIND-POWER MACHINES

The advantages of using wind energy are (1) it does not deplete natural resources; (2) it is nonpolluting, making no demands upon the environment beyond the comparatively modest use of land area; and (3) it uses cost-free fuel. These advantages must be weighed against the disadvantages: (1) that wind is a variable source of energy and (2) that total system costs are high when a storage system is included.

The total potential production of electric energy from the wind in the U.S. is estimated to be 1.536×10^{12} kilowatt-hours per year. [5] Of course, to realize this potential by the year 2000, a vast capital investment would be required. For example, along the 100-meter contour offshore of the Eastern seaboard, it is estimated that a potential 283×10^9 kilowatt-hours could be realized if plants were constructed by the year 2000. Obviously, the technical and financial aspects of realizing such a potential are staggering. It is currently planned to put forth considerable research and development effort over the next decade towards the development and construction of several prototype wind-generator units.

Efficiency of a Horizontal-Axis Windmill

The efficiency of a windmill is defined to be the ratio of the power extracted from the wind to the power contained in the wind that passes through the swept area of the blades. The theoretical efficiency of a windmill can be calculated from one-dimensional aerodynamic theory. It is assumed, in the calculation, that the rotor is replaced by a disk containing innumerable blades that produce a uniform change in the velocity of the air passing through the disk.

The velocity of the air entering the rotor is v_f, the free-stream velocity, while the velocity of the air acting on the rotor is v_r. The velocity of the air after it has left the rotor is v_w, the wake velocity. [6] In the process of extracting power from the wind, the wind velocity acting on the rotor is less than the input air velocity; that is,

$$v_r = v_f(1 - a), \tag{14.1}$$

where a is called the axial interference factor. Then it may be shown that

$$P = 2\pi r^2 \rho v_f^3 a(1 - a)^2, \tag{14.2}$$

where ρ = density of the air and r is the radius of the rotating blades. Equating $dP/da = 0$, we find that the maximum power is developed when $a = \frac{1}{3}$. Substituting $a = \frac{1}{3}$ in Eq. (14.2), we have:

$$P_{max} = (\tfrac{8}{27})\pi \rho r^2 v_f^3. \tag{14.3}$$

The kinetic energy passing through the rotor per unit of time is:

$$\frac{dE}{dt} = \frac{d}{dt}(\tfrac{1}{2}mv_f^2)$$

$$= \frac{v_f^2}{2}\frac{dm}{dt}$$

$$= \frac{v_f^2}{2}\frac{d}{dt}(\pi r^2 \rho x)$$

$$= \tfrac{1}{2}\pi r^2 \rho v_f^3, \tag{14.4}$$

where x = length of a cylinder of air. Therefore, the maximum theoretical efficiency should be:

$$\eta_{max} = \frac{P_{out}}{P_{in}} = \frac{P_{max}}{dE/dt} = \tfrac{16}{27} = 0.593. \tag{14.5}$$

The actual efficiencies achieved by windmills are about half of the theoretical maximum. The actual power supplied by a windmill is then:

$$P_{actual} = \tfrac{1}{2}P_{max}$$
$$= (\tfrac{4}{27})\pi\rho r^2 v_f{}^3. \tag{14.6}$$

Clearly, the actual power increases as the radius of the blades increases and as the velocity of the wind increases. Therefore, it is most important to build wind machines with large-diameter propellers and in locations where the machine experiences high-velocity winds.

Let us calculate the maximum power available from a windmill with a rotor radius of 10 meters and experiencing a wind velocity of 10 meters/second (22.4 miles/hour). Then, using $\rho = 1.288$ kg/m^3 for the density of air, we have:

$$P_{max} = (\tfrac{8}{27})\pi \times 1.288 \times (10)^2 \times (10)^3$$
$$= 120 \text{ kilowatts.} \tag{14.7}$$

Since P_{actual} may be assumed to be half of the maximum power, we might expect a power output for these conditions of 60 kilowatts.

For a unit experiencing a wind velocity of 15 meters/second, we can calculate the required rotor radius in order to achieve an actual power output of 2000 kilowatts. We use Eq. (14.6) as follows:

$$2 \times 10^6 = (\tfrac{4}{27})\pi\rho r^2(15)^3; \tag{14.8}$$

thus the required r is 31.4 meters (103 feet). Recall that the Smith–Putnam wind machine generated 1200 kilowatts with a blade radius of 83 feet and a similar wind velocity.

We can conclude that 50 wind units with blades of 31-meter radius, and subject to a wind velocity of 15 meters/second (33.6 miles/hour), would be required to produce 1000 megawatts, the production of a modern large nuclear or fossil-fuel plant.

Wind turbines are mounted in high locations because wind velocity increases with height (air flow is not retarded by ground-level friction). Wind is most prevalent at higher velocities between 30 and 60 degrees latitude in the northern and southern hemispheres. The northern land areas include Europe, the Soviet Union, and North America. The World Meteorological Organization concludes that there are some 20×10^9 kilowatts of wind power available at favorable turbine sites throughout the world. If each unit were able to provide 1000 kilowatts, some 20 million units would be needed in order to realize this potential. Even if this figure proves to have been exaggerated by a factor of 100, it is obvious that the potential for wind power-energy use is essentially untapped.

Storage of Wind-Generated Electric Power

One inevitable disadvantage of wind-generated power is the variability of the wind. The storage of wind-generated power for use at times of peak demand is an important part of the overall system of wind-generated energy. In a home-sized wind-power plant producing direct-current electricity, storage is accomplished by charging lead–acid batteries. In a large system designed to feed electricity into the power lines of a utility, energy can be stored mechanically or hydraulically by pumping air or water into a reservoir, and then releasing it later to drive a generator when it is needed. In the case of hydroelectric generation, this would mean building up a water supply in the reservoir while the wind turbine is running. The degree to which this build-up would bring about an overall increase in generating capacity, with consequent increase in electrical energy available for sale, would depend upon whether the hydroelectric system had enough waterwheels and sufficient storage space in the reservoirs to take advantage of the supplementary electric energy. If the hydroelectric network were already working at full capacity, for example, new waterwheels or reservoir space would have to be added to achieve an increase in overall generating capacity. The costs of the new installations would have to be considered, as well as the cost of the wind turbine, to determine whether the wind machine would be profitable.

Another storage device is a flywheel, which can be spun by the kinetic energy of the windmill shaft. Then, when a demand for energy occurs, the flywheel is connected to an electric generator to provide the necessary power output. A properly designed flywheel, made perhaps of pressed bamboo and enclosed in a vacuum chamber to minimize friction (the main energy loss in the system), is not only much more efficient than battery storage, but is potentially cheaper. Compared with lead–acid batteries, which store only 8 to 10 watt-hours of energy per pound of material, the envisioned bamboo flywheel would store more than 20 watt-hours per pound.

Another storage approach is the production of a synthetic fuel, hydrogen, which can be produced by the electrolysis of ordinary water. The hydrogen would then be compressed and stored for later use as fuel. In the 1890's, Paul La Cour produced hydrogen by using wind-generated electricity and used it to illuminate the school where he taught.

14.5   NEW VERSIONS OF THE VERTICAL WINDMILL

The common approach to extracting wind energy by windmills, since the days of the first Dutch windmill, has been by means of horizontal-axis windmills. The Persian windmill was a vertical-axis device, and about 50 years ago two inventors patented new forms of the vertical windmill. The great advantage of a vertical-axis rotor is that it operates independently of the wind direction and thus has a potential for high efficiency in spite of changing winds. By contrast, horizontal-axis machines must reorient the blades so that they are perpendicular to the wind direction.

The first device, patented by a French inventor, G. J. M. Darrieus, in 1927, is shown in Fig. 14.4. The second device, invented by S. J. Savonius, a Finnish engineer, in 1929, is shown in Fig. 14.5(a). Both of these devices can operate equally well regardless of the direction of the wind.

The Savonius rotor was formed by cutting a cylinder into two semicylindrical surfaces, moving these surfaces sideways to form an S-shaped rotor, and then placing a vertical shaft in the center. End plates are used to provide structural strength.

The Savonius rotor (S-rotor) operates as a two-stage turbine whenever the wind impinging on the concave side is circulated to the center of the rotor towards the back of the convex side. The air flow is illustrated in Fig. 14.5(b). The S-rotor is particularly adaptable to remote areas and locations where the wind direction is variable. Electro GMBH of Switzerland produces a 6-kilowatt version of an S-rotor electrical generator. The efficiency of S-rotor devices is estimated to be about 31 percent. Currently there is some research being carried out on varying the parameters of the device. For example, one could examine the benefits of

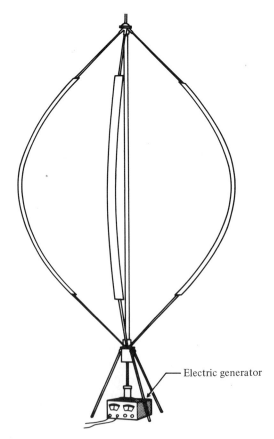

Electric generator

Fig. 14.4 *The Darrieus vertical-axis device. The shape of the blades is the natural shape a flexible cable would assume if spun about a fixed vertical axis.*

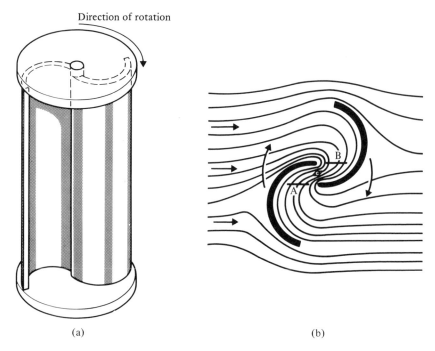

(a)   (b)

Fig. 14.5 *The Savonius rotor. (a) The two semicylindrical vanes joined together on a vertical shaft. (b) A top view showing the flow of air and the rotation of the device around a vertical shaft.*

varying the aspect ratio (height/diameter), the number of vanes, the shape of the vanes, and the rotor configuration.

The Darrieus device shown in Fig. 14.4 consists of two thin airfoils with one end mounted on the top end of the vertical shaft and the other at the bottom end of the shaft. The resulting configuration is nearly circular in shape and looks something like an eggbeater. This device is currently under study by NASA as a means of supplying power to single-family dwellings. The NASA prototype is located atop a two-story building, about 50 ft above the ground. The two 14-ft diameter blades are made of balsa wood covered with fiberglass. The 15-ft vertical shaft is made of aluminum tubing.

The Darrieus device has an efficiency of about 35 percent and has a potentially low capital cost under mass production. A Darrieus rotor built by the National Research Council of Canada, using 15-foot diameter blades, produced 900 watts.

Researchers at Sandia Laboratory in Albuquerque, New Mexico, are combining the advantage of the self-starting Savonius rotor with the reliable action of the Darrieus rotor. [9] The Darrieus turbine blades are fabricated in three pieces since conventional blades are more difficult to build. Small S-rotor devices are installed at the top and bottom of the shaft to provide a self-starting feature for the device. The combination Savonius and Darrieus machine is shown in Fig. 14.6.

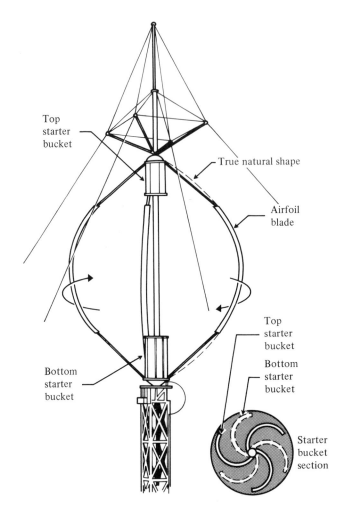

Fig. 14.6 *A combination Savonius and Darrieus machine. Cables stabilize the tower in the wind. The airfoil blade is fabricated from three straight pieces and varies from the natural airfoil shape slightly as shown.*

14.6 A WIND GENERATOR FOR EVERY HOME

A typical home in the U.S. requires a power source capable of supplying 10 kilowatts in order to supply the refrigerator, electric heater, and well pump, if they all have to be operated at the same time. Of course, one could use a 5-kilowatt wind drive and store the energy in batteries or pumped storage in a water tank. Excessive weight of the tower and rotor blades and high construction costs have deterred people from using wind machines.

An engineer at Princeton University, Thomas Sweeney, has designed and built a 7-kilowatt horizontal-axis windmill using a 25-foot sail-wing for a blade. This device is currently under license to Grumman Corporation of New York and is projected to sell for about $4000, exclusive of the tower.

Currently, for example, a person may purchase a Quirk wind generator, which is manufactured in Australia and sold in the U.S. The Quirk 110 provides two kilowatts in a 25-mph wind using a three-blade horizontal-axis rotor. The output is 110-volts dc and must be converted to ac for normal use. A total system including tower, blades, and storage batteries costs about $5000 to purchase and erect.

14.7 PROPOSED LARGE WIND-DRIVEN GENERATING SCHEMES

In many areas of the world the wind has a velocity, 20 feet above ground level, of 8 to 12 meters/second (18 to 27 mph). Several schemes incorporating large systems of wind machines have been advanced recently which would tap the energy in these winds and develop considerable amounts of power by means of a large number of interconnected wind generators connected together in a network grid or storage system.

Wind has a higher velocity as it flows up a ridge or a bald-faced mountain (as in the Alps). In the U.S. there are three regions that could be examined: (1) the Green and the White Mountains of New England; (2) the Koolau Range of Oahu, Hawaii; and (3) the coastal foothills of Alaska. In these areas, a system of wind machines could be built to capture the wind power; the average wind velocity would be approximately 8 to 10 meters/second.

An alternative way of obtaining wind speeds of 8 to 10 m/sec is to place the wind turbine on top of a very tall tower, say 800 feet high. Obviously such a tower would be difficult to build and certainly costly.

In either case—choosing an appropriate mountain range or constructing high towers—if we can assume wind velocities of 15 meters/second and a blade radius of 31 meters, we would produce, as we calculated in Eq. (14.8), a power of two megawatts. In order to equal the power of five 1000-megawatt power stations to be installed over the next several decades in New England, there would be required 2500 such wind machines on towers on the slopes of the New England mountains.

Professor William Heronemus has devised a system of offshore wind-power machines. [10] The offshore system proposes to place three rows of wind stations across the prevailing westerlies in the Gulf of Maine–Georges Bank area of the continental shelf. Each wind machine would have a capacity of 2 megawatts, and three wind machines would be provided at each station, mounted on floating platforms near the 100-meter depth contour. Wind stations would be clustered around an electrolyzer plant to yield hydrogen that could be stored in a collector. The total system would require 2000 such wind platforms and would provide a total of 12×10^6 kilowatts. The average wind speed at towers 100 feet high would be 20 miles/hour. A proposed 200-foot diameter wind machine with three sets of such units mounted on each platform is shown in Fig. 14.7. [10] The total project is estimated to cost 22×10^9, or 22 billion dollars, a large engineering project indeed! As a contrast, the Alaskan oil pipeline is expected to cost 4 billion dollars, and the Bay Area Rapid Transit System (BART) in California has cost over two billion dollars.

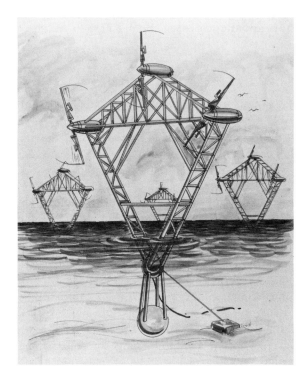

Fig. 14.7 *The Heronemus off-shore wind-machine concept. (Courtesy of W. Heronemus.)*

14.8 THE ECONOMICS OF WIND POWER

The economics of wind power utilized for generation of electric power is not a proven science since no actual systems have been constructed since the Smith–Putnam wind machine, beyond a few engineering prototypes.

Several studies have shown that large-scale wind generators could be constructed for about $400 to $600 per installed kilowatt compared with costs of $600/kilowatt for nuclear plants. [11, 12] However, these figures exclude the cost of provision of a storage system, such as compressed air or pumped storage.

As a simple example, consider a 2-megawatt wind turbine costing about $1 million to erect. This device, similar to the Smith–Putnam machine, would therefore cost $500/kilowatt. Of course, for wind machines, the operating expense is very low and the fuel is free.

Small wind machines useful for single-family dwellings provide 2 to 4 kilowatts at a cost of 5 to 10 thousand dollars. Therefore, these smaller devices cost the user about $2500 per kilowatt.

The National Aeronautics and Space Administration (NASA) and the Energy Research and Development Administration (ERDA) have engaged General Electric Company to build the biggest windmill in history. It will have two fiberglass blades that span 200 feet, be mounted on a 150-foot tower, and generate 1.5 megawatts in a 22-mph wind. The windmill will cost $7 million and be completed in 1978.

Wind power has approximately the same advantages and drawbacks as solar power (see Chapter 17) with regard to consistency, reliability, and the need for an auxiliary storage facility. As an economic commodity, wind power is a match for solar power.

14.9 ENERGY STORED IN THE OCEANS

Energy is stored in the tides, the currents, and the thermal gradients of the oceans. The tides and tidal currents have a power estimated at 3×10^{12} watts. The energy stored in the thermal layers of the ocean as energy received from the sun is vast, but is widely dispersed over the surface of all the oceans of the world. There is energy present in the wave motion of the surface of the oceans, also. The tapping of the energy of the ocean is a difficult and complex process, and the necessity of collecting the widely dispersed energy would make for high costs. The best opportunity for using the energy of the oceans lies, at present, in utilizing the tidal-energy sources of the world.

Tidal Energy

The energy of the tides was tapped as early as the 11th century, when tide mills were built along the Atlantic coast of France, Britain, and Spain. In London, in the 16th century, part of the water supply was pumped by 20-foot waterwheels installed under the arches of London Bridge. Tide mills were installed in Holland as early as 1200 and were brought to the U.S. in the 17th century. A tide-powered motor was built by the city of Santa Cruz, California, around 1898. [16] The power of these tide-driven devices ranged from 10 to 50 kilowatts and was largely used for grinding and milling.

Tides are generated by the gravitational and kinematic force interactions of the earth–moon–sun system. The gravitational force on the surface of the oceans depends upon the position of the moon and the sun in relation to the point of location as well as the distances to those bodies. The period of the tides depends upon the period of rotation of the moon around the earth and upon the orientation of the earth as it travels around the sun and rotates daily. The difference in the lengths of the lunar and solar days gives rise to spring (or maximum-amplitude) tides and minimum (or *neap*) tides. When the sun and moon are nearly in line with the earth, a spring tide results; and when they are 90° out of place with the earth, a neap tide results. Tidal forces are relatively small; and effects produced in bodies of water as large as the Great Lakes are virtually negligible.

In the open ocean, the tidal range R, or difference in amplitude between high and low tides, is only two feet. This range is amplified by the continental shelves to about 6-foot tides. In estuaries, or bays with a deep narrow basin that resonates with the wave motion flowing into it, the range of tides may grow to 30 feet. (One well-known example is the Bay of Fundy on the east coast of Canada.)

Tidal power is obtained from the oscillatory flow of water in the filling and emptying of partially enclosed coastal basins during the semidiurnal rise and fall of the oceanic tides. This energy may be partially converted into tidal electric power by enclosing such basins with dams to create a difference in water level between the ocean and the basin, and then using the water flow, while the basin is filling or emptying, to propel hydraulic turbines that drive electric generators.

Let us determine the maximum potential energy of a body of water with a tidal range R, held behind a dam in a basin of surface area A. The potential energy of a differential mass dm at a height y and of vertical thickness dy is

$$dE = yg\,dm, \tag{14.1}$$

where g is the gravitational constant. The mass of the water is

$$dm = \rho A\,dy, \tag{14.2}$$

where ρ is the density of water. Therefore,

$$dE = \rho Agy\,dy. \tag{14.3}$$

The total potential energy is then

$$E = \int_0^R \rho Agy\,dy, \tag{14.4}$$

where we will assume vertical sides for the basin. Then

$$E_{\max} = \tfrac{1}{2}\rho AgR^2. \tag{14.5}$$

This energy is available during both emptying and filling of the basin. The maximum average power of one tidal day of 24 hours and 48.8 minutes (8.92×10^4 seconds) is calculated for the period $T = 8.92 \times 10^4$ seconds as

$$P_{\max} = \frac{4E_{\max}}{T}, \tag{14.6}$$

since the tide empties and fills the basin twice a day. Then,

$$P_{\max} = \frac{2\rho AgR^2}{8.92 \times 10^4}, \tag{14.7}$$

where $\rho = 1.025 \times 10^3$ kg/m^3 and $g = 9.8$ m/s^2. Therefore,

$$P_{\max} = 0.225AR^2. \qquad (14.8)$$

Consider a basin 3 km by 20 km, where $R = 1.5$ meters. Then,

$$P_{\max} = 0.225(60 \times 10^6)(1.5)^2$$
$$= 3.04 \times 10^7 \text{ watts} = 30.4 \text{ megawatts}. \qquad (14.9)$$

The actual power output of a tidal electric-power plant is about 25 percent of the maximum; therefore,

$$P_{\text{actual}} = 0.056AR^2. \qquad (14.10)$$

Consider the bay of Rance in France with a basin area of 22×10^6 m^2 and a tidal range of 11.4 meters. We expect a power of 160 megawatts. This basin, which is being utilized for tidal power, has, in fact, an average power output of 160 megawatts and a maximum power capacity of 240 megawatts. Recall that a large nuclear-power plant such as those being built today would probably have a capacity of about 1000 megawatts.

The primary cost of a tidal power plant is the cost of building the dam (or barrage) across the opening of the bay. The cost of the dam is roughly proportional to the length, L, of the barriage. Therefore, desirable tidal-power sites have a relatively low L with a relatively large A. The ratio of L/A is often used to measure the desirability of a site, and sites with a ratio under 80 are considered possible sites. The ratio of L/P_{actual} is also a measure of the desirability of a site. The smaller these ratios, the more desirable the site.

Types of Tidal-Power Schemes

There are at least four basic types of operating schemes for utilizing tidal power, the chief differences lying in whether the tidal *ebb* is used as well as the flow, and whether one or two basins are used. We will consider these four basic combinations.

A single-basin, single-tide generation scheme uses a single basin, and it fills during the *rising* tides. When the water level is higher than the level of the sea, the potential energy is used to drive turbine generators. This system operates the turbine generators during about one-half of each tidal cycle. This scheme is not as desirable as the two-way flow scheme.

A single-basin, two-way-flow scheme uses both the rising and the falling tide to drive the power generators. Since the turbines must be capable of being driven in two directions, depending on the flow, they are relatively more expensive. The cycle of tides and the generation of electrical power is shown in Fig. 14.8 for a two-way flow scheme.

A double-basin, two-way-flow scheme uses a second basin to store water for generating electric power when peak needs are experienced. This scheme can

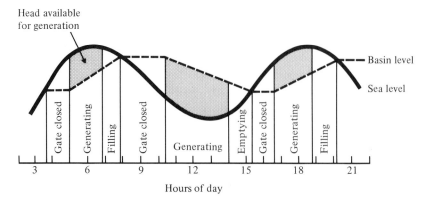

Fig. 14.8 *The cycle of the tides and the generation of electrical power using a two-way flow scheme. The dotted line is the level of the water in the basin, and the solid line is the level of the sea.*

also be used to generate power continuously as water is released from the high-level basin to a lower-level basin and then to the sea, through two sets of generators.

The two currently operating power sites are single-basin, two-way-flow generating plants.

Tidal Power Sites

There are several hundred potential tidal sites in the world, with varying measures of quality. In Table 14.1, seven useful sites are listed, of which several are currently being exploited. The Rance River project in France was completed in

Table 14.1
CHARACTERISTICS OF TIDAL POWER SITES

Site	Tidal range, R (meters)	Basin area A (km²)	P_{actual} (megawatts)	Barrage length, L (meters)	L/A (m/km²)	L/P_{actual} (m/MW)
Rance, France	11.4	22.0	160	725	33	4.5
Brest, France	6.4	92.0	211	3640	40	17.3
Chausey, France	12.4	610.0	5252	23500	39	4.5
Severn, Britain	11.5	50.0	370	3500	70	9.5
Passamaquoddy, U.S.A.	7.5	120.0	378	4270	36	11.3
Bay of Fundy, Cobequid, Canada	11	700	4743	8000	11.4	1.7
Kislogrub, U.S.S.R.	4	2	1.8	30.5	15	16.9

1966 at a cost of $100 million. The dam is 725 meters long. A sluice section is incorporated into the dam structure to improve power-station operation by speeding up the emptying and filling of the reservoir during the operating cycles. The gates are operated several times a day, and since they are subjected to head reversals, have both upstream and downstream sealing arrangements. Both incoming and outgoing tides are utilized by using a reversible axial-flow turbine. This turbine is essentially a horizontal Kaplan-type turbine with adjustable wicket gates and runner blades. The project has twenty-four 10,000-kW turbine generator sets, for a total installed capacity of 240 megawatts. The average actual power generated is estimated to be 160 megawatts, as shown in Table 16.1. The measures L/A and L/P_{actual} are both desirably low, and indicate the quality of the Rance River site. A photo of the Rance River power plant is shown in Fig. 14.9.

Tides with large ranges occur on the northeastern shores of Russia near Murmansk. A small 1200-kilowatt plant was installed in 1966 at Kislaya Bay, Kislogrub, utilizing a channel with a barrage of only 30.5 meters. The Russians used reinforced-concrete modular structures for the power plant. These structures were floated to the site and sunk to form the dam and house the turbines. The sea transport of floating modules considerably reduced the cost of construction of the dam by avoiding costly cofferdam construction. [17] The Russians are now considering a 6000-kilowatt station at Mezen Bay on Russia's Arctic coast as well as at Tugu Bay and Penginsk Bay.

The Passamaquoddy tidal project was started in the state of Maine in the 1930's by President Roosevelt, but was abandoned before much real progress was achieved. This site has a potential actual power of 378 megawatts, assuming a basin area of 120 km^2 in Cobscook Bay, Maine. The scheme called for two basins and two-way-flow generation. However, Congress failed to appropriate the financial resources in the 1930's, although it was seen to be an economical plant, compared to fossil-fuel power plants of that time.

The Passamaquoddy system was restudied in 1948, when it was recommended that a series of dams and sluice gates be used to fill Passamaquoddy Bay at high tide and empty it at low tide into Cobscook Bay, which empties into the Atlantic Ocean.

In 1961, at the request of President John F. Kennedy, the Department of the Interior restudied the project and recommended that the proposed plant's capacity be increased. Nevertheless, this project has not yet been commenced since it is expensive and is also opposed by the private New England power companies.

The provinces of Nova Scotia and New Brunswick, Canada, are currently considering utilization of the power in the Bay of Fundy. [18] This project, if funded, is expected to cost $2 to $3 billion dollars. In addition, an electrical-transmission line would have to be constructed, to connect to the Canadian and New England power grid.

Several engineers are advocating the construction of a large tidal-power project on the Severn estuary in England. [19, 20] A 4000-megawatt project is proposed, with an estimated cost of 3.2 billion dollars.

Fig. 14.9 *The Rance River Tidal Project in France. (a) The power plant is shown under construction; (b) the tidal dam with a road running along the top of the dam. The power plant appears on the left side of the dam. (Photos courtesy of the French Embassy.)*

14.10 ECONOMICS OF TIDAL-POWER PLANTS

Since tidal-power plants are located in a marine environment, the effects of corrosion must be compensated for by using appropriate metals and protective coatings. These factors increase the cost of the power plants.

The cost of constructing a tidal-power plant may be approximately $600/kilowatt of capacity. The Rance project was installed for $400/kilowatt in 1966, and might cost $600/kilowatt today. The proposed Severn project is estimated to cost $800/kilowatt of capacity, while the Bay of Fundy installation is estimated at $700/kilowatt. These capital costs are relatively higher than the $600/kilowatt for a nuclear plant or $500/kilowatt for some fossil-fuel plants.

The fact that we have only two operating tidal-power plants can be attributed to the high capital costs for construction and the intermittent nature of the power output. The capital costs may be reduced in the future by floating modular structures into place and thus eliminating the need for construction of cofferdams.

REFERENCES

1. H. WULFF, *Traditional Crafts of Persia*, M.I.T. Press, Cambridge, Mass. 1966.

2. V. W. TORREY, "Windmills in the history of technology," *Technology Review*, April 1975, pp. 8–10.

3. W. CLARK, "Interest in wind is picking up as fuels dwindle," *Smithsonian Magazine*, Nov. 1973, pp. 70–78.

4. P. H. THOMAS, *Electric Power from the Wind*, Federal Power Commission, Wash., D.C., 1945.

5. *Energy Facts*, Science Policy Research Division, Library of Congress, U.S. Government Printing Office, Wash., D.C., 1973.

6. S. S. PENNER AND L. ICERMAN, *Energy* (Vol. II), Addison-Wesley Publishing Co., Reading, Mass., 1974.

7. C. C. JOHNSON, *et. al.*, "Wind-power development and applications," *Power Engineering*, Oct. 1974, pp. 50–54.

8. M. KENWARD, "Energy file," *New Scientist*, 24 July 1975, p. 218.

9. R. STEPLER, "Eggbeater windmill," *Popular Science*, May 1975, pp. 74–75.

10. W. E. HERONEMUS, "Pollution-free energy from offshore winds," Proceedings of the 8th Annual Conference of the Marine Technology Society, Sept. 1972.

11. N. WADE, "Windmills: The resurrection of an ancient technology," *Science*, June 7, 1974, pp. 1055–58.

12. S. ENGSTROM, "Renewable energy resources: Wind energy from a Swedish viewpoint," *Ambio*, **4**, No. 2, 1975.

13. W. E. HERONEMUS, "The U.S. energy crisis: Some proposed gentle solutions," *Congressional Record*, No. 17, Part II, Feb. 9, 1972.

14. P. C. PUTNAM, *Power from the Wind*, Van Nostrand Co., New York, 1948.

15. J. M. SAVINO, "Wind power," *Astronautics and Aeronautics*, Nov. 1975, pp. 53–57.

16. P. N. GARAY, "The ebb and flow of tidal power," *Consulting Engineer*, Oct. 1974, pp. 66–70.

17. L. B. BERNSTEIN, "Russian tidal-power station is precast offsite, floated into place," *Civil Engineering*, April 1974, pp. 46–49.

18. "Tidal power may now make sense," *Business Week*, Nov. 9, 1974, pp. 115–116.

19. T. L. SHAW, "Tidal energy from the Severn estuary," *Nature*, June 21, 1974, pp. 730–32.

20. T. SHAW, "Tidal power and the environment," *New Scientist*, 23 October 1975, pp. 202–206.

21. E. W. GOLDING, *The Generation of Electricity by Wind Power*, Halsted Press, New York, 1976.

22. T. S. JAYADEV, "Windmills stage a comeback," *IEEE Spectrum*, November, 1976, pp. 45–49.

23. H. THIRRING, *Energy for Man: Windmills to Nuclear Power*, Indiana University Press, Bloomington, 1977.

24. M. F. MERRIAM, "Wind energy for human needs," *Technology Review*, January 1977, pp. 28–39.

EXERCISES

*14.1 The kinetic energy per unit mass of a windstream is proportional to the square of the wind velocity, V. The wind power is proportional to the rate at which its kinetic energy flows through a wind machine—that is, to the kinetic energy multiplied by V or, therefore, proportional to V^3. Determine the equation for the power of a horizontal-axis windmill with ρ = density of air and r = the radius of the rotating blades.

14.2 If wind-power systems are to be a significant part of our future power systems, we must develop wind-power systems that give long service life (over 20 years), operate unattended, and require low maintenance and capital costs. Existing wind-plant concepts need to be analyzed to define the lowest-cost configurations and components that can be used in most regions of the country, the most suitable applications, and the suitable windy regions of the nation. Locate a windy area in your state and determine the fluctuations in wind speeds over the year. Calculate the actual power available from a horizontal-axis windmill located at that site.

14.3 A 100-foot-tall windmill started operating in 1975 at NASA's station near Sandusky, Ohio. The 100-kilowatt device has two 125-foot-diameter blades [22] and is the largest wind-powered generator now in operation, and the second largest that has ever been built. The next machine in the proposed series of large machines will be a $\frac{1}{2}$-MW device followed by a $1\frac{1}{2}$-MW system. These developments are being pursued by the NASA Lewis Research Center of Cleveland, Ohio. Examine the feasibility of the extensive use of $\frac{1}{2}$-megawatt windmills to

---

*An asterisk indicates a relatively difficult or advanced exercise.

supply electricity. Contact NASA for further information regarding the devices under investigation.

14.4 Prepare an analysis of Professor Heronemus' scheme for offshore wind-power machines for the U.S. Energy Research and Development Administration. Include the impact on shipping lines, the environment, and coastal development in your analysis. Estimate how many machines you would deploy and what means you would employ to transport the energy to the user.

14.5 Consider the possibility of using several large wind machines to supply the electric power for your college or university. Determine the electric power used by your college, and estimate the number and size of machines that would be required.

14.6 Determine the price and availability of a Quirk or Grumman windmill-generator. Calculate the cost per kW to purchase and install the machine.

14.7 Calculate the actual power generated by a windmill experiencing winds of 8 meters/second. Complete these calculations for a machine with a rotor radius of 6, 10, and 14 meters.

14.8 The construction of a tidal-power station causes physical and oceanographic changes in the area surrounding the development. Determine and list at least five environmental effects of tidal-power stations.

14.9 The Mezen estuary in the U.S.S.R. has a tidal range of 10 meters and a basin area of 150 km^2. Calculate the actual power expected for this tidal-power project.

*****14.10** The Bay of Fundy–Passamaquoddy Bay area near the border between Maine, New Brunswick, and Nova Scotia is the single location in North America which has aroused serious interest in tidal-power plants. These bays experience tidal ranges up to 10 meters and could produce significant power. The plan for Passamaquoddy Bay uses a high pool, which is filled during the incoming tide while the locks of the low pool are cleared. The stored floodwaters are then released through the power house to produce electricity and to drain into the low pool and then back into the sea. This two-pool arrangement is more flexible than a single-pool system in that the "stored" water can be used to generate power at the most advantageous times. [16, 18] Investigate the economic and environmental factors of this project and estimate the power capacity of the project.

14.11 The environmental effects of a tidal-basin power project are largely uncharted due to limited experience with such projects. Some initial studies of the potential environmental effects of the proposed Severn Estuary project have been accomplished. [19, 20] Explore the potential effects of a tidal-power project and list the benefits and costs.

15

Geothermal Energy

15.1 INTRODUCTION

Over the life of the earth, thermal energy has been stored within its core. Of this energy stored within the earth, some is transmitted from the interior by means of conduction through the earth. The average rate of flow of heat to the surface has been found to be 0.063 watts per square meter. For the earth's surface of 510×10^{12} square meters, the total heat flow amounts to 32×10^{12} watts. Of this, only 1 percent of the total rate, or 0.3×10^{12} watts, is due to heat convection by hot springs and volcanoes.

Unfortunately, we cannot directly exploit this heat supply, but we can use local hot spots—subterranean reservoirs where the heat has been stored in the form of steam and hot water. Such reservoirs are the source of geothermal energy.

15.2 GEOTHERMAL RESERVOIRS

The earth's crust is four to five miles thick beneath the oceans and 20 to 35 miles thick under the continents. Below this crust lies the mantle, which extends down 1800 miles to the boundary of the earth's core. The upper part of the mantle, which is quite variable in composition and temperature, is believed to be the source of magma, that mixture of molten rock and gases that penetrates the crust and erupts at the surface of volcanoes. Magma can also remain within the crust, where it forms sources of heat.

Along the continental margins encircling the Pacific Ocean there is a belt of crustal deformation, evidenced by the squeezing and breaking of the rock strata. Volcanic eruptions and earthquakes are characteristic of this belt. A similar belt extends through southern Europe and across the middle of Asia, connecting with the Indonesian island chain.

Within the continental belts of recent volcanic action and deformation, heat flow is higher than the earth's average. The high-temperature geothermal areas are found in these belts of high heat flow.

Geothermal reservoirs consist of permeable and porous rock in which, by circulation of steam or hot water, a convection system can develop. The porosity and permeability need not be inherent in the type of rock itself, but can result from fracturing due to deformation. Ground water, which can percolate down to depths of several miles, is heated directly or indirectly by the underlying magmas, then expands, and ascends towards the surface.

In order for this water to form a heat reservoir, a cap or cover of some sort is required; otherwise the heat simply dissipates. Often this capping is provided by a layer of impervious rock overlying the permeable reservoir rock.

A geothermal reservoir can store enormous amounts of energy in the form of steam and hot water. Temperatures of the water or steam are as high as 700°C. Such reservoirs are most readily tapped within depths of 2000 to 8000 feet.

Another type of reservoir containing water at 100° to 150°C at relatively shallow depths is located in large sedimentary basins. Heat at this temperature is useful for space heating and for agricultural, mining, and other purposes.

Regions of high heat flow usually display natural hot springs, geysers, and steam vents such as in Yellowstone National Park. These features represent leakage from a geothermal reservoir.

A schematic diagram of a geothermal reservoir is shown in Fig. 15.1. Water entering the reservoir may come from ground water that seeps down from the surface, or from release of water from the molten rock (magma) of the earth's core. Heat energy from molten or hot intrusive rock heats the water in reservoirs. Cap rock traps the heat of the water until it is released by a well or a natural fracture.

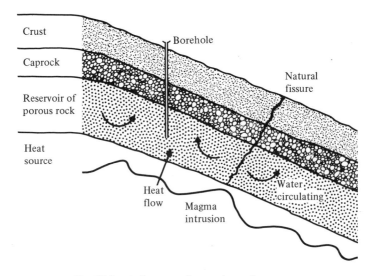

Fig. 15.1 *A diagram of a geothermal reservoir.*

15.3 USE OF GEOTHERMAL ENERGY

The Romans used geothermal springs to supply their mineral baths for thermal and medicinal value. An interesting example of an early mineral bath of the Roman era is located in Bath, England, and the old baths may still be visited. In the U.S. a fine example of a mineral hot springs is at Grand Junction, Colorado.

Hot steam or hot water escaping from a natural fissure at a temperature of from 50 to 100°C may be used to heat homes and industry. Hot water is used for heating all the dwellings and for industrial use in the City of Reykjavik, Iceland. The cities of Boise, Idaho, and Klamath Falls, Oregon, are partially heated by waters from nearby wells and hot springs.

The energy of hot-water sources has also been used in fish farming, pulp and paper processing and hydroponics. In Hungary the land area covered by green-houses heated by geothermal water was $1.5 \times 10^6 \, m^2$ in 1973. Any small industrial process that requires hot water or space heat can effectively use naturally escaping steam or heat. However, in order to effectively tap geothermal energy, wells have to be driven to the geothermal reservoir.

Roman documents 2100 years old tell of steam escaping at what is now Larderello, south of Florence, Italy. During the 19th century the natural steam was used for heating and limited mechanical power. In 1904, electricity was first produced at Larderello using the steam to drive a turbine and generator. Commercial operation at the Larderello plant began in 1912 when power was supplied to local customers, by using a 250-kilowatt generator. The current generating capacity at Larderello is 406 megawatts, and this steam source has served effectively for over 60 years.

The second commercial electric plant using geothermal energy was installed in the Wairakei area of New Zealand in 1958. The plant, located in the volcanic belt of the North Island, reached full power of 145 megawatts in 1964. The plant operates at an annual load factor of 0.9 and provides base-load electricity to the power network. [1] Further development of geothermal energy in New Zealand has been temporarily halted due to the recent discovery of natural gas.

Mexico began operation of a 75-megawatt plant in the Cerro Prieto region of northern Mexico in 1973.

In Japan, at Matsukawa, a geothermal-energy site has been developed and 30 megawatts of generating capacity is currently operating.

The Geysers Plant

The Geysers Power Plant in Sonoma County north of San Francisco, California, is the only geothermal plant active in the U.S. The Geysers is located on the steep slopes of Cobb Mountain, an extinct volcano. Steam is obtained from fumaroles and steam vents which emit steady steam vapor—not intermittent geysers, despite the name of the site. The first generating unit began operation in 1960 and provided 11 megawatts. Development of the site continued over the next 15 years, and the capacity of the site was 516 megawatts in 1975. The power

Fig. 15.2   *A portion of the Geysers geothermal power plant in Northern California. These two units went into operation in 1960 and 1963, respectively and have a combined capacity of 24 megawatts. (Courtesy of Pacific Gas and Electric Company.)*

is generated and distributed by Pacific Gas and Electric Company of California. A portion of the Geysers Power Plant is shown in Fig. 15.2.

Geothermal Electrical Generating Capacity

The capacity of the five predominant geothermal power-plant sites is summarized in Table 15.1. The world's total geothermal electric-power capacity under operation in 1975 is 1172 megawatts. Since this is only equal to the capacity of one large nuclear generating plant, the development of geothermal power can be seen to be in its infancy.

Growth in the use of geothermal power can be expected to provide ever increasing power for the world. For developing nations, hydroelectric and geothermal electric-power plants may be the most economical to develop over the next 30 years. [2] The comparatively small geothermal power stations will potentially

Table 15.1
ELECTRIC POWER CAPACITY OF
GEOTHERMAL PLANTS IN 1975

Larderello, Italy	406 megawatts
Wairakei, New Zealand	145 megawatts
The Geysers, California	516 megawatts
Matsukawa, Japan	30 megawatts
Cerro Prieto, Mexico	75 megawatts
World total	1172 megawatts

fit the initial needs of developing nations. Mexico recently installed a 75-megawatt plant at Cerro Prieto in the southern extension of the Imperial Valley. The U.S.S.R. is installing a small plant on the Kamchatka Peninsula, a region of recent volcanoes. Many countries in Central and South America are seeking suitable geothermal sites.

The U.S. in 1975 utilized some 1500 kilowatts (thermal) of geothermal power for generating electric power, heating, and other uses. [3] Given the actual constraints to exploration and utilization, one might expect, at best, a capacity of 7000 megawatts in 1985.

15.4   EXPANDED GEOTHERMAL ENERGY USE

The limited energy that has thus far been tapped from the earth's interior has come entirely from natural outcroppings of steam or hot water. Such sites occur in limited number in the world, and any widespread use of geothermal energy will require exploratory drilling deep into the earth to reach either steam, hot water, or heated rock.

These methods are as yet untried, but if they do prove to be economical and reliable, there is no question that geothermal energy offers a vast potential. For example, the Russians estimate that their geothermal potential is equal to their reserves of petroleum and coal combined. The basic technology for drilling to deep sources is available; the uncertainties lie in the geography and geology of locating suitable sites, and the economics of exploiting them.

According to one geothermal study, 20,000 to 30,000 megawatts of geothermal power could be produced in California's Imperial Valley after the necessary exploration and technological development have taken place. [4] In the case of the Imperial Valley, the product of the wells and plants would be electrical power, desalinated water and minerals.

15.5   TYPES OF GEOTHERMAL SOURCES

To obtain more power from geothermal sources, we will need to work with sources less readily usable than the convenient dry steam emerging from a source such as the Geysers. For our purposes, let us distinguish four types of sources: dry steam, wet steam, hot brines, and hot rock. The characteristics of these four types are summarized in Table 15.2. As we go across the list, from left to right, we note that each source is more abundant than the last, but harder to utilize. Using wet steam, for example, may require flashing and steam-separating units. With hot brine we need additional components.

Table 15.2
CHARACTERISTICS F FOUR TYPES OF GEOTHERMAL SOURCES

Type/Characteristic	Dry steam	Wet steam	Hot brine	Hot rock
Ease of utilization	Highest (greatest ease)	Somewhat difficult	Difficult	Greatest difficulty
Abundance	Lowest	Somewhat abundant	Abundant	Great abundance
Drilling depth	8000 ft (2440 m)	3000 ft (915 m)	4500 ft (1370 m)	10,000 ft (3050 m)
Pressure and temperature of steam or water	7 atmospheres, 200°C	2 atmospheres, 230°C	1 atmosphere, 240°C	400°C
Currently operating fields	Larderello, Italy; Geysers, U.S.A.; Matsukawa, Japan	Wairakei, New Zealand; Cerro Prieto, Mexico	None	None
Exploration sites		Kilauea Volcano, Hawaii	Imperial Valley, Calif.	Marysville, Montana; Vallez Caldera, New Mexico

The only currently operating systems are of the dry-steam and wet steam types. Research is proceeding to explore the necessary technology for the hot brine and the hot rock types.

Dry-Steam Geothermal Fields

The first dry-steam field to be developed was at Larderello, Italy, and was first utilized in 1904. The first geothermal power production in the U.S. was begun in 1960 at the Geysers field near San Francisco; and this field is expected to supply 900 megawatts by 1979.

The cost of a geothermal production well drilled to 8000 feet (2400 meters) is about $250,000. Production from such a well can reach 100 tons of steam per hour at a pressure of 10 atmospheres and a temperature over 200°C. [5] (Unfortunately, this type of geothermal source is the least common.) Little or no water remains with the superheated steam as it emerges, and therefore the steam is particularly easy to utilize.

A schematic diagram of a dry-steam electric power plant is shown in Fig. 15.3. The steam is separated from water particles in the separator and filtered to remove abrasive particles. The steam is at a relatively low pressure (typically 100 psi = 689 × 10^3 n/m^2 = 6.6. atmospheres) and temperature, 200°C. (Modern fossil-fuel power plants operate at much higher pressures and temperatures, typically 300 psi and 500°C, respectively.) Therefore, the turbines for geothermal plants have a different design. The steam from the turbine is condensed and the resulting water is cooled in cooling towers. Most of the water is ultimately evaporated to the atmosphere in these towers, conveying with it the waste heat from the plant. About 20 percent of the condensed water, containing trace chemicals such as boron and ammonia, which would pollute local streams if released, is reinjected into the ground through deep wells.

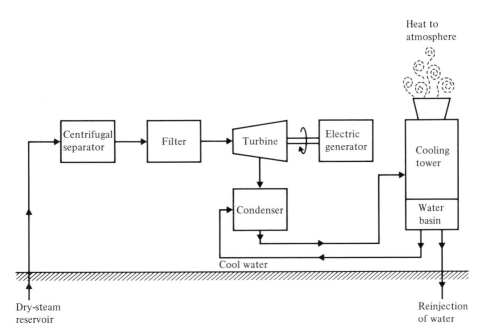

Fig. 15.3   *Schematic diagram of a dry-steam electric power plant, using a cooling tower and a reinjection well.*

Because of the lower pressures and temperatures at which they operate, the turbines of dry steam power plants are about one-third less efficient than those of conventional fossil-fuel plants; the dry steam plant has an efficiency of approximately 22 percent. Nonetheless, the dry-steam plants cost less to build and operate than do comparably sized fossil-fuel plants, and the electricity generated at the Geysers is less expensive for the utility than that from its other sources.

Wet-Steam Geothermal Fields

Geothermal wells commonly produce a mixture of steam and hot water, rather than steam alone. Wet-steam deposits are estimated to be 20 times more common than dry-steam deposits. Water-dominated sites are most commonly found in areas of volcanism, and in some cases the hot-water system has penetrated to the earth's surface in the form of hot springs or geysers.

The water contained in wet-steam deposits is often heavily contaminated by dissolved salts and minerals, which must be removed before the water enters the turbine, in order to avoid clogging and corrosion.

The first hot-water geothermal power plant was built in 1960 in Wairakei, New Zealand. Other plants were built in Cerror Prieto, Mexico, Kaweran, Japan, and Pauzhetska, U.S.S.R. Typically, the ratio of steam to water in the well output is 1:3, by weight.

A schematic diagram of a geothermal power plant using steam is shown in Fig. 15.4. A centrifugal separator is used to separate the steam and the water. Often, the water output of the separator is then allowed to flash at some suitable lower pressure, and the low-pressure steam is utilized while the water output is discarded. In New Zealand, where the salinity of the hot water is low, the waste water is simply discarded into a nearby river. When the water is relatively saline,

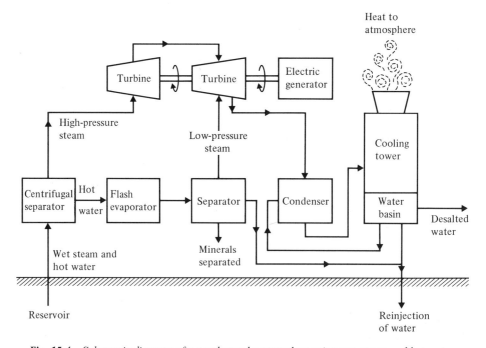

Fig. 15.4 *Schematic diagram of a geothermal power plant using wet steam and hot water.*

it is reinjected into the earth. As shown in Fig. 15.4, two turbines may be used, one each for high- and low-pressure steam, respectively. The output of the low-pressure turbine is condensed and then cooled. The Cerro Prieto plant uses an evaporative pond instead of a cooling tower. Surplus condensate may be purified as desalted water, if desired, as is the case in a design being studied by the government of Chile.

An average production well in a hot-water field is drilled to 3000 feet (915 m) and costs $150,000. Production of the well is about 400 tons of steam and water per hour at 230°C. Since the wells are relatively shallow, drilling costs are lower than in dry-steam fields. However, additional components, such as flash units and turbine sections, and the cost of disposal of waste water raises the total cost to about twice the cost of the dry-steam plant. [5] This cost is still attractive in comparison to that of fossil-fuel plants.

Hot-Brine Geothermal Fields

Many geothermal fields have reservoirs of water that has a salt content equal to or greater than that of sea water. For example, salt content of geothermal water in California's Imperial Valley generally runs from 30,000 to 300,000 mg per liter. By comparison, ocean water averages about 35,000 mg per liter. Indeed, a productive exploratory well in the Cesano area of Italy tapped hot-brine deposits containing about 35 percent dissolved salts.

The Imperial Valley area, east of San Diego, holds many suitable sites where the hot brine is at a temperature of 240°C while containing up to 25 percent dissolved salts. These dissolved salts are extremely corrosive; and once the steam has been removed, the hot brine becomes a thick syrup, almost impossible to handle. Therefore, research is working towards the removal of the salts from the hot brines by a distillation process. Then, since the water is desalted, it may later be used for irrigation purposes.

A schematic diagram of a power plant using a separate secondary fluid (such as isobutane or freon) is shown in Fig. 15.5. The corrosive brine solution is used to heat this secondary fluid in a series of heat exchangers and then reinjected into the reservoir in order to avoid land subsidence and ground- and surface-water degradation. Secondary fluids such as isobutane have lower boiling points than water and are effective high-pressure heat carriers. Prototype turbines are currently being constructed for an experimental plant in the Imperial Valley. In Russia a geothermal power plant using a secondary fluid is now operating on an experimental basis. This plant, located on the Kamchatka Peninsula, produces 440 kilowatts from 80°C water wells.

A research group at Lawrence Livermore Laboratory in California is working on the design of an impulse turbine that will be able to use the hot brine directly. [14] The hot brine is expanded through a nozzle and the water–steam combination drives the turbine wheel. This system, called a total flow system, may provide a 60 percent increase in efficiency over that of the separate fluid system of Fig. 15.5. The primary problem is to design a corrosion-resistant turbine wheel.

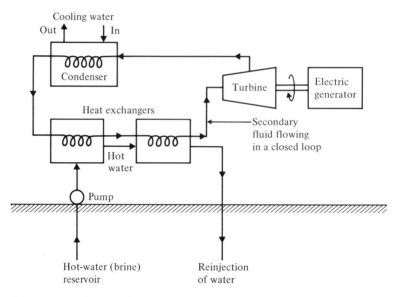

Fig. 15.5 *Schematic diagram of a power plant using a separate secondary fluid such as isobutane or freon. (A binary cycle system.)*

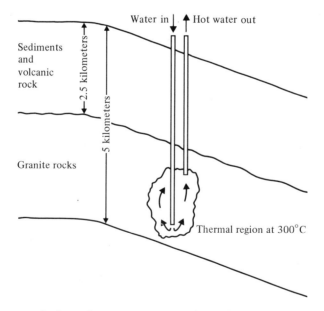

Fig. 15.6 *A proposed scheme for extracting energy from a hot rock site. The crack is not a three-dimensional cavity but rather lies in a vertical plane.*

Hot-Rock Geothermal Fields

Underground water systems do not come in contact with most of the near-surface deposits of geothermal heat. Thus, hot rock constitutes an important source of heat, but one that is difficult to exploit. These high-porosity rocks are called hot-rock deposits. Utilizing these deposits requires development of a technique for artificial fracture of the rock, the injection of water into the fractures, and subsequent recovery of steam for turbines.

The fracturing of the rock may be done by inducing thermal stresses with cold water or by chemical explosions. Some researchers feel that underground nuclear explosions are worth considering, but many environmental problems may preclude such use.

One experiment in New Mexico used water under pressure to open a crack. [6]

One proposed scheme for the extraction of energy from hot rocks is shown in Fig. 15.6. Water circulating down one well, through the crack in the hot rock, and up through the other well, would carry off heat that could be used to run a power turbine at the surface.

Experimental assessments are proceeding at a site near Helena, Montana, which may yield hot-rock energy within one or two kilometers of the surface.

15.6 ECONOMICS OF GEOTHERMAL POWER PLANTS

The economics of geothermal power plants are already quite attractive for the dry-steam and hot-water types of geothermal deposits. Both of these systems provide electrical energy to public utilities at a rate competitive with that of fossil-fueled or nuclear power plants. The hot-brine and hot-rock geothermal deposits are currently in the research stage, and it is only possible to estimate the actual costs of producing electric power from these sources.

The typical dry-steam or hot-water well operates at a rate of 5 megawatts. It normally takes about two drilling attempts to obtain one producing well. [5] If the U.S. wished to develop 1000 megawatts per year, we would need to drill 400 wells per year (at a cost of 95 million dollars yearly) for drilling, exploration, leasing, and rental costs. [5] The current rate of drilling is about 100 wells per year.

The leasing of land for the exploration of geothermal sites is necessary, but is developing slowly. The major geothermal resources of the U.S. are located in the western states, where 60 percent of the potentially desirable areas are federal lands and most have not yet been released for exploration. Federal leasing requirements are stricter for known geothermal resource areas than for other prospecting areas; and since many nonfederal prospecting areas are adjacent to federal lands, there is a reluctance to carry out exploration because adjacent areas will then be classified as known areas.

Nevertheless, recent leasing of federal lands for geothermal exploration brought forth many bids. Shell Oil bid $4.5 million for a tract near the Geysers area, while the city of Burbank bid $515,000 for rights to a tract near Mono Lake,

California. [7] The first areas released to bidding amounted to 5 million acres out of a potential total of 58 million acres of federal lands in western states.

Many of the available areas were not bid upon and only 69 tracts have been processed for actual leasing. [8] The exploration companies state that the rules for leasing are too strict, the leasing process is too bureaucratic, and geothermal drilling lacks many of the economic incentives that oil, gas, and coal enjoy.

One of the major obstacles to geothermal development in the United States is the fact that the federal tax laws do not refer to "geothermal resources" or provide explicit tax treatment for exploration and drilling expenses in connection with development of this energy source.

Geothermal energy is quite economical relative to other sources of energy. As long as a geothermal plant can be constructed for $200 to $250 per kilowatt and will provide electricity at a cost of 0.5 to 1.0 cents/kWh, then geothermal exploration will grow. The Geysers and Larderello meet these criteria easily and therefore they are continually expanding. It remains to be proved whether the fields in the Imperial Valley and the schemes for exploiting hot rock can be developed economically. However, as the price of other more conventional energy sources rises, geothermal plants will become more attractive to developers.

The Environmental Impact of Geothermal Power Plants

Geothermal power generation, like any thermal power generation method, is not without environmental impact. In many systems, towers must be erected to cool the surplus heat in the output steam. A plant's impact will vary widely with the quality of the steam or water that emerges from the condenser, and will be dependent upon whether subterranean pressures present obstacles in the way of returning the residues to the earth. If the waste water can be used for irrigation purposes, this is one offsetting advantage.

In the case of using hot dry rock, substantial quantities of surface water will be required for injection into the artificial fracture. In other respects, hot-rock systems seem to be of high environmental quality.

Steam coming from the earth may contain objectionable gaseous effluents. The Geysers power plant, for example, uses steam containing minute amounts of carbon dioxide, ammonia, methane, hydrogen sulfide, and other noncondensable gases.

The disposal of geothermal waste water from hot-brine plants will be a problem because of the mineral content, which can be as high as 25 percent. These minerals can cause problems initially by corroding, or solidifying, and blocking the pipes and turbines. Once through the turbine, the condensed water may be too full of contaminants to permit dumping into natural streams or lakes. Because the saline and siliceous solids precipitate out as the water temperature and pressure drop, reinjection into the ground also may be difficult. The solids may block the porosity of the underground rock. An entirely new water-treatment technology may have to be developed to handle large quantities of high-temperature, saline-rich water.

A recent environmental impact study was accomplished for the Wairakei power plant in New Zealand, which uses hot water and steam. [1] Water from a nearby river is used for cooling, and then discharged back into the river, which flows to the Tasman Sea, 320 kilometers away. The geothermal fluid is a 1:4 (by weight) mixture of steam and hot water. Substantially all the chemical effluents are dissolved in the waste water, which discharges into the river. However, the mineral content of the hot water and steam is low and therefore, after the waste water mixes with the river, even at the lowest river flow rate, the resulting concentration is still well below the 500-ppm* maximum recommended for potable water. [1] The Wairakei plant emits a thermal effluent in the waste water and there is some land subsidence.

The Wairakei plant discharges heat, water vapor, sulfur, H_2S, CO_2, arsenic, and mercury, at concentrations that have some limited environmental effects. Designed and built at a time when environmental sensibilities were less acute and geothermal technology was less developed, Wairakei produces an overall environmental impact that would be neither acceptable nor necessary in a new plant. Despite its imperfections, however, the Wairakei plant has been under development or in operation for more than 20 years without presenting any serious environmental problems for the local population. Reinjection of the hot-water waste, an as yet unproven procedure for liquid-dominated fields, would sharply reduce the plant's environmental impact.

The maintenance of the environmental quality of the air, water, and land surrounding a geothermal power plant will be a subject for research and development as new schemes for geothermal power advance.

REFERENCES

1. R. C. AXTMANN, "Environmental impact of a geothermal plant," *Science*, March 7, 1975, pp. 795–802.

2. J. H. KRIEGER, "Geothermal energy stirs worldwide action," *Chemical and Engineering News*, June 9, 1975, pp. 21–26.

3. *Energy Facts*, Science Policy Research Division, Library of Congress, U.S. Government Printing Office, Wash., D.C. 1973.

4. E. BURGESS, "Geologists say State officials must face up to major decisions regarding development of geothermal resources," *California Journal*, Feb., 1973, pp. 56–59.

5. G. R. ROBSON, "Geothermal electricity production," *Science*, April 19, 1974, pp. 371–375.

6. A. L. HAMMOND, "Dry geothermal wells: Promising experimental results," *Science*, Oct. 5, 1973, pp. 43–44.

7. E. C. GOTTSCHALK, JR., "Energy shortage inspires a big rush to develop geothermal power sources," *Wall Street Journal*, Feb. 21, 1974.

8. S. ATCHISON, "Geothermal power: Strangled by red tape," *Business Week*, Aug. 11, 1975, pp. 68–69.

---

* Parts per million.

9. J. BARNEA, "Geothermal power," *Scientific American*, Jan. 1972, pp. 70–77.

10. P. KRUGER AND C. OTLE (eds.), *Geothermal Energy*, Stanford University Press, Stanford, California, 1973.

11. ELLIS, A. J., "Geothermal systems and power development," *American Scientist*, Sept.-Oct. 1975, pp. 510–521.

12. E. DOUGLAS, "Arizona energy alternatives," *Christian Science Monitor*, Nov. 11, 1975, pg. 21.

13. D. E. WHITE AND D. L. WILLIAMS, "Assessment of geothermal resources of the U.S. 1975," U.S. Geological Survey Circular 726, 1975, Wash., D.C.

14. A. L. AUSTIN AND A. W. LUNDBERG, "Electric power generation from geothermal hot-water deposits," *Mechanical Engineering*, Dec. 1975, pp. 18–25.

15. S. L. MILORA AND J. W. TESTER, *Geothermal Energy as a Source of Electric Power*, The M.I.T. Press, Cambridge, Mass, 1976.

16. J. C. DENTON AND N. H. AFGAN, *Future Energy Production Systems*, Academic Press. New York, 1976, Vol. 2.

EXERCISES

15.1 Geothermal power stations are relatively inefficient in converting thermal energy to electric energy. [11] The limiting thermodynamic efficiency is approximately 30% because of the low average steam temperatures, but the actual efficiencies of conversion are approximately one-half this figure in steam-producing fields and about one-quarter in water-producing fields. There is an obvious incentive to make more efficient use of the energy from high-temperature geothermal fields.

Explore several schemes for increasing the efficiency of geothermal power plants. Consider various uses for the waste hot water.

15.2 Some scientists estimate that sources of geothermal steam exist 8000 feet below the surface in Arizona. [12] Examine the costs of tapping such a deep well, and determine the feasibility of such a project.

*15.3 The geopressured fluids of the U.S. Gulf Coast have a large geothermal potential. The energy deliverable at the wellhead in the onshore part of the Gulf Coast varies according to production plan, but may range from 31,000 to 115,000 megawatts for 30 years. [13] The actual recoverable resource depends upon the availability of water and economically feasible technologies. Nevertheless, if 20,000 megawatts was recoverable, then this geothermal resource could substitute for 20 nuclear power plants over the period 1985 to 2015. Explore the feasibility of developing this potential, and develop a plan for development over the next decade.

*15.4 The total-flow concept for the utilization of hot-water geothermal resources may be the most efficient. [14] Based on thermodynamic principles, a direct

---

* An asterisk indicates a relatively difficult or advanced exercise.

expansion from wellhead to sink condition has the potential for conversion of the greatest fraction of the available energy. Investigate the total-flow concept, and list the impediments to the development of the system.

15.5 Determine the geothermal potential of your state by contacting the state energy or utility commission. Obtain estimates of the potential and the cost of utilizing these resources.

16

Energy from Wastes

All human and industrial processes produce wastes, that is, normally unused and undesirable products of a specific process. The waste products of a home include paper, containers, tin cans, aluminum cans, and food scraps, as well as sewage. The waste products of industry and commerce include paper, wood, and metal scraps, as well as agricultural waste products. The U.S. produces approximately 0.9 tonne (one tonne = 1000 kilograms) of solid waste per person per year, or over 185 million tonnes in total. [1] This quantity has been growing at an annual rate of 5 percent per year until recently. Most of this waste is relegated to the land fills and garbage pits near our metropolitan areas.

Waste products of our industrialized society can be used to provide fuels for our electric power and steam plants. Furthermore, some valuable materials now classified as wastes can be recycled. The recycling of metals, such as scrap aluminum or steel, can help in conserving energy, since more energy is consumed in the original manufacture of these metals than is consumed in the recycling process.

16.1 RECYCLING OF WASTE PRODUCTS

The recycling of waste products such as metals, paper, and wood has recently become an active goal of much of American industry and of the home. Recycled newspaper can be used to make cardboard for containers, and scrap steel can be recycled and used in combination with new steel. It is less wasteful of energy to recycle aluminum than to produce new aluminum from bauxite ore.

Many materials from municipal refuse can be recycled: wood, ferrous metal, aluminum, copper, glass, tar, textiles, oil, paper, and plastics are examples.

16.2 COMBUSTION OF WASTES

It has been estimated that 50 to 60 percent of urban waste is combustible. Thus, the potential exists for burning some 90 million tonnes of refuse each year and using the heat to produce steam for heating, or for driving steam turbines to produce electricity. The heat content of refuse varies, but in general, two tonnes of refuse is equal to the heat content of one ton of coal. Therefore, the heat content of the total combustible refuse is 1.05×10^{18} joules, or about 1.3 percent of the total energy used in the U.S. in 1974. Of course, much of this waste is dispersed in small towns and farms. Nevertheless, in urban areas it may be practical to use waste as fuel for generating electricity or steam.

Table 16.1 lists the available organic wastes in the U.S. in 1971. [2] The net conversion of wastes to oil is based on 1.25 barrels per ton of dry organic waste. The estimate of gas potential is based on 5 cubic feet of methane produced from each pound of waste. While 880 million tons were produced in 1971, it is estimated that 1061 million tons of dry organic waste will be produced in 1980.

Table 16.1

AMOUNTS OF AVAILABLE DRY ORGANIC WASTES IN THE U.S.A.
IN 1971 IN MILLION TONS* (2×10^9 lbs)

Source	Wastes generated	Readily collectable
Manure	200	26.0
Urban refuse	129	71.0
Logging and wood manufacturing residues	55	5.0
Agricultural crops and food wastes	390	22.6
Industrial wastes	44	5.2
Municipal sewage solids	12	1.5
Miscellaneous	50	5.0
Total	880	136.3
Net oil potential (10^6 barrels)	1098	170
Net methane potential (10^9 cubic feet)	8.8	1.36

* One ton is 2000 lbs, which is equivalent to 907 kilograms, or 0.907 tonne.

Manure from farm animals is convertible to fuel, and it is estimated that wastes from animal feedlots having a thousand or more cattle, or large poultry or hog operations, would yield 26 million tons of readily collectible waste.

Because of increased concern of the public over the pollution of rivers, streams, lakes, and even the oceans, the usual methods for disposing of municipal sewage used in the past have become obsolete, and in some cases, illegal. The principal pollutant of this sewage is organic matter and its decomposition products. The availability of this organic material for conversion to fuel will depend somewhat on the legal requirements imposed on municipalities to treat sewage effluent before disposal or recycling. A conservative estimate for 1971 is that 1.5 million tons of organic solids are available from this source.

Large quantities of these available wastes are now concentrated at specific locations. For example, a single cattle feedlot with 100,000 head would produce about 410 tons of dry organic solids per day, or 150,000 tons per year. A city (and its nearby suburbs) with a population of 1 million would generate 1,750 tons of dry organic solids per day, or 640,000 tons per year. Of course at a feedlot (or in a city) of the above size, the amount of waste material to be handled would be considerably more than the amounts given, since the organic material is only part of the waste. There are several areas where a large city or several cities could collect their wastes in a common location.

Efforts to recover energy from solid waste can be classified in three distinct schemes: (1) direct heat recovery from special incinerators; (2) supplementary fueling of power plants with waste material; and (3) conversion of the waste to synthetic fuels.

16.3 INCINERATORS AS STEAM BOILERS

Several large incinerators have been built recently as add-on boilers for existing power plants, or as boilers for new electric generating plants or steam heating systems. The most notable of these are in Chicago and at Hempstead, New York. The Hempstead plant generates 68,000 kilograms of steam per hour both for in-plant use and for the desalinization of seawater. The Chicago Southwest plant currently has contracts for supplying 454,000 kilograms of steam per day to adjacent industries in the stockyard area.

Chicago also plans to sell the steam from a waste processing plant to Commonwealth Edison Company for its nearby generating station. The processing plant will process 250,000 tons annually and will cost about $14 million. [3] The city of Chicago expects to receive about $700,000 per year from Commonwealth Edison. In addition, it expects to receive $200,000 per year from the sale of separated metals.

An incinerator plant in Nashville, Tennessee, was completed in 1975 at a cost of $13 million to handle over half of the city's daily refuse of 700 tons and provide heat for the 32 buildings surrounding the facility. The steam produced is used for either heat or an air-conditioning plant. This facility should substantially reduce the need for sanitary landfill in Nashville and improve the air quality by burning the refuse in a central plant with all the necessary pollution-control equipment. For heating purposes, steam at 600 degrees goes directly from the boilers into noncondensing steam turbines and then into the distribution network. For cooling, steam generated in the boilers is piped to noncondensing turbines whose exhaust steam drives two condensing, steam-turbine-driven, air-conditioning refrigeration systems.

In order to burn refuse without a supplementary fuel, the noncombustible materials must be less than half of the waste matter and the material must be less than half water while the combustible material must be greater than 25 percent. [3] Therefore, the combustion requirements for self-burning refuse are: (1) water < 50 percent, (2) combustible material > 25 percent, and (3) non-combustible material < 50 percent. Thus, it is useful to remove the noncombustible material and water prior to incineration, and to recycle the useful metals obtained in the process of separation. The physical composition of typical municipal refuse is given in Table 16.2. [4]

Table 16.2
THE PHYSICAL COMPOSITION OF
TYPICAL U.S. MUNICIPAL WASTE

Category	Percent
Paper	50.7
Food waste	19.1
Yard waste	—
Metal	10.0
Glass	9.7
Wood	2.9
Textiles	2.6
Leather, rubber, and plastics	3.3
Miscellaneous	1.7
Total	100%

One of the first successful plants designed to use the energy contained in municipal refuse is the plant in Bern, Switzerland, constructed in 1954 with a capacity of 200 tonnes a day. Still in continuous operation, this plant produces steam, heated water, and electricity—for hospitals, schools, apartments, a railroad station, and a chocolate factory. Inspired by the success of the Bern plant, other cities soon accepted the idea—and today almost every major European population center has its refuse-energy plant. The refuse-to-energy concept next spread to Australia and Japan, where fuel costs are also high. In Canada, the Montreal 1080-tonne facility was built in 1969, and a 900-tonne facility was completed in Quebec City in 1974, providing energy to a nearby paper-pulp plant. [9]

In Paris, a principal use of the steam produced from several central incinerators has been space-heating of over 100,000 dwelling units. The city is underlaid with an elaborate steam-piping network to distribute the steam. Also, 200 million kilowatt-hours of electricity are generated and sold to the public utility.

16.4 WASTE MATTER AS A SUPPLEMENT TO FOSSIL FUEL

The principal problem affecting energy recovery directly from incinerators is the wide assortment of waste constituents and related moisture content. Variations up to 100 percent in heat content of waste can be expected; for this reason, incinerator plants have to provide for the use of supplementary fossil fuels to maintain steam pressure.

In other countries the burning of municipal waste for the generation of steam is also growing. In Germany, as of 1969, there were 13 refuse-burning installations, which handled the refuse of about 7.5 million inhabitants, or about 12.5 percent of total population. This amounted to 37 percent of the refuse in the large cities where the plants are located. By 1972, it is estimated that the refuse of 20 percent of the total population was being handled in refuse-burning installations. In the case of Munich, the plants consume *all* of the refuse generated within the city. These plants all showed a variable availability of heat present in the refuse, and used fossil fuels as supplemental fuels.

In St. Louis, Missouri, a system prototype has been operated, which uses solid waste as an auxiliary fuel for electric power generation. The system uses a waste-separation process to remove the noncombustibles.

Initial refuse preparation occurs at city-operated facilities where material to be processed is discharged from packer trucks to the floor of the raw-refuse receiving building. Waste is then shredded in a hammer mill and conveyed to a storage bin. The head end of the conveyor is equipped with a magnetic separator to remove ferrous metals for recycling. An air-classification system is used to remove most of the remaining noncombustibles.

The dry refuse is burned in two boilers powering turbo-generators with a rating of 135 megawatts each, and some 7500 tons are burned each day. The boilers run on 20 percent refuse and 80 percent coal. Union Electric Company invested $70 million to set up the trash centers and convert the boilers. However, the fuel saving alone will amount to $10 million a year (or more, as the cost of coal rises). Furthermore, the utility will be able to sell the recovered scrap metals to industry.

It is interesting to note that slightly less than 50 percent of the electrical energy generated in this country is the result of burning coal. This is equivalent to 8×10^{15} Btu (8.4×10^{18} J) of energy input. If we could convert half of the coal-fired burners so that they could burn refuse as 20 percent of the heat input into the boiler, then 40 percent of the total municipal solid waste in the U.S. could be converted to electrical energy.

It is particularly important to note that solid waste has a significantly lower sulfur content than coal, and therefore the pollution level would be reduced by these mixed-fuel systems.

The economics of using refuse as a supplementary fuel are not clear as yet. Union Electric of St. Louis will replace 6 percent of its coal with refuse. Coal costs $5 to $7 per ton while refuse may cost $3 per ton. Refuse has about half of the energy per unit weight of coal. To these figures must be added the costs of separation and of conversion of the boilers. The economic feasibility of the St. Louis system will be examined as it operates over the next several years.

16.5   SYNTHETIC FUELS FROM WASTES

There are three major methods for conversion of organic wastes to synthetic fuels: (1) hydrogenation; (2) pyrolysis; and (3) bioconversion. The first two have been advanced to the pilot-plant stage, while the third has been the subject of only minor research effort, but is a long term possibility. [5]

Hydrogenation is basically a deoxygenation (or chemical reduction) process. The principal chemical reaction is abstraction of oxygen from the cellulose present in municipal wastes, by carbon monoxide and steam. In the process, organic waste and as much as five percent of an alkaline catalyst such as sodium carbonate are placed in a reactor with carbon monoxide and steam at an initial pressure of 100 to 250 atmospheres, and heated at 240° to 380°C for as long as an hour. Under normal conditions, as much as 90 percent of the carbon content is converted to oil—about 2 barrels (0.32 m^3) of oil per ton of dry waste. Because some of the oil must be used to provide heat and carbon monoxide for the reaction, however, the net yield is about 1.25 barrels per ton of dry waste.

The product of this process is a heavy oil with an oxygen content of 10 percent and a sulfur content less than 0.4 percent. The oil has an energy of 15,000 Btu per pound (34.8 × 10^6 J/kg), similar to No. 6 fuel oil. Because of the need for pressurized reaction vessels, the hydrogenation process is relatively expensive.

A second method for converting wastes to synthetic fuel is *pyrolysis,* or destructive distillation. Pyrolysis is the thermal decomposition of matter in the absence of oxygen. The fuel output may be either gaseous or liquid, and several pilot projects are under development. While this process requires heating of the waste up to 500°C, it can be done at atmospheric pressure, thus keeping the cost of the reaction vessels lower than those used in hydrogenation. One pilot process produces 1 barrel of oil averaging 10,500 Btu per pound (24.4 × 10^6 J/kg) for each ton of refuse. A major disadvantage is that the process also produces gas and char, thus complicating the output collection process. The low-energy gas (500 Btu per cu ft) is recycled to provide the oxygen-free atmosphere for pyrolysis and, with part of the char, is burned to supply heat for the process.

A full-scale plant, processing 2000 tons of refuse per day from a city of 500,000, would cost about $12 million. It is believed that such a plant could produce heating oil worth more than the cost of operating the plant per unit of output. It is projected that one barrel of synthetic oil worth $6 or $7 would result for every ton of refuse costing $5.

One operating system that opened in 1974 in Baltimore, Maryland, consumes 1000 tons of refuse per day, half of the city's waste. The plant reduced the cost of waste disposal to half its former cost, since the output gas is used to generate steam for a nearby electric generating plant. [6] One study of the economics of pyrolysis plants showed that nine plants could recover enough energy from the 10,000 tons of garbage collected daily by the San Francisco Bay area counties to supply 10 percent of the region's electric energy. [6]

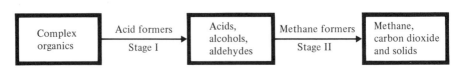

Fig. 16.1 *Anaerobic digestion occurs in two stages. Acid-forming bacteria are active in the first stage, and methane-forming bacteria in the second stage.*

A final method of converting refuse into fuel is a bacterial process, *anaerobic digestion*. This process is similar to those used in sewage-treatment facilities and yields methane gas as an output. In a typical process, the organic portion of the refuse is blended with nutrients in an aqueous slurry, heated to 72°C, and circulated through a digester for a week or more. The output is a low-energy gas at about 600 Btu/cu ft. The anaerobic digestion process is shown in Fig. 16.1. Several bioconversion schemes have been proposed using algae, phytoplankton, and other plants grown specifically for this use. However, there is limited practical experience with such schemes.

16.6 ORGANIZING FOR FUEL FROM WASTES

To a considerable extent, the problem involved in the use of solid wastes to produce fuels or energy directly is one of organization rather than of cost or technology. The wastes of a nation are widely dispersed, and have to be collected and disposed of by many local agencies or contractors. Many collection areas or companies are too small to be able to operate an incinerator or fuel converter of economic size to yield fuel gas, oil, or steam for a power plant or heating. Water-supply and liquid-waste treatment and disposal have long been handled on a regional basis in most areas, and the case for a similar solid-waste agency is said to be hardly less compelling.

The state of Connecticut was the first to not only adopt, but to start implementing, a statewide plan for solid waste management. The plan calls for 10 resource-recovery centers for the recycling of materials and the separation of organic wastes, as well as 45 coordinated waste-transfer stations. It is expected that 60 percent of the state's refuse will be recovered as material or energy. Also, air pollution from waste disposal will be reduced by 70 percent and land areas sacrificed to landfills will be reduced by 80 percent. The shredded fuel output by the ten resource-recovery centers will be sold to the electric utility. This planned system may well be a model for other regions of the nation.

REFERENCES

1. J. E. HEER, JR., AND D. J. HAGERTY, "Refuse turns resource," *IEEE Spectrum*, Sept. 74, pp. 83–87.

2. J. PAPAMARCOS, "Power from solid waste," *Power Engineering*, Sept. 1974, pp. 46–55.

3. L. L. ANDERSON, "Energy potential from organic wastes," Bureau of Mines Circular 8594, U.S. Government Printing Office, Washington, D.C., 1972.

4. "Experience in burning waste abounds throughout industry," *Power*, Feb. 1975, pp. 2–9.

5. T. H. MAUGH, "Fuel from wastes: A minor energy source," *Science*, Nov. 10, 1972, pp. 599–602.

6. "Converting garbage into energy," *Business Week*, March 30, 1974, pp. 45–46.

7. E. F. LINDSLEY, "Methane from waste," *Popular Science*, Dec. 1974, pp. 58–60.

8. D. P. BURKE, "Methanol," *Chemical Week*, Sept. 24, 1975, pp. 33–42.

9. W. K. MacAdam, "Megawatts from municipal waste," *IEEE Spectrum*, Nov. 1975, pp. 46–50.

10. A. L. Johnson, Jr., "Biomass energy," *Astronautics and Aeronautics*, Nov. 1975, pp. 64–70.

11. S. C. James, "The indispensable landfill," *Tech. Rev.*, Feb. 1977, pp. 39–47.

EXERCISES

16.1 Methane produced from the 172-acre sanitary landfill site at Palos Verdes near Los Angeles is already supplying 3500 homes. (New York City plans a similar program to serve up to 20,000 homes with gas from a 2800-acre landfill in Staten Island.) The methane process uses extraction pipes 6 inches in diameter, which have been sunk into the landfill to a depth of 100 feet. Contaminated methane gas generated by the decomposing garbage is drawn out by a vacuum and routed through pre-treatment tanks, where moisture, offensive odors, and some contaminants are removed. The gas then moves to vessels packed with claylike pellets, which draw off carbon dioxide and the remaining contaminants. Once cleaned, the gas is compressed and pumped directly into the lines of the Southern California Gas Company for transmission.

The landfill should continue to generate gas for 20 years after its scheduled close in 1980. Investigate this simple process of producing methane gas and determine the economics of such an approach.

16.2 Methanol, CH_3OH, may be produced from wastes. It is a good chemical feedstock and fuel and is convenient to use since it is in the liquid state. The city of Seattle has been planning to convert solid municipal waste into 31 million gallons per year of methanol. [8] Maine and California are interested in using forest product waste for the manufacture of methanol. Methanol has been used experimentally as an automotive fuel. [8] Investigate the economics and technical feasibility of generating and using methanol obtained from waste products.

16.3 Phoenix, Arizona, disposes of two million tons of refuse each day by dumping it into deep trenches and covering it with earth. It is estimated that this garbage then decomposes into 25 million cubic feet of methane over a period of time. Estimate this period of decomposition and determine whether it could be accelerated. What are the possible uses for the low-Btu gas that is produced? Determine what the size of a similar landfill is for your community.

16.4 The chief advantages of a refuse-fired boiler burning unprocessed waste are demonstrated reliability and a mature technology, coupled with good efficiency and low operating cost. Its main drawback is the high initial capital expenditure required, even though this is largely offset by low-cost operation. The shredded-refuse system offers an opportunity to supplement existing high-temperature, high-pressure utility boilers directly with air-classified trash. It entails a lower first cost for boiler modifications, but usually a higher total operating cost and shorter service life than a refuse boiler firing unprocessed refuse. Investigate the advantages and disadvantages of these two approaches.

16.5 It is possible to obtain a relatively high yield of methane from the decay of several nonfood plants such as water hyacinth, Sudan grass, and eucalyptus. [10] For example, an average annual eucalyptus yield in California could be 24 tons per acre. For cellulose crops such as eucalyptus, acid hydrolysis, followed by yeast fermentation of the saccharate to liquor, is the most highly developed process. The resultant products will be both a fuel and a chemical feedstock. As in the waste-management energy recovery systems, integration is necessary for economic viability. Calculate the number of acres of eucalyptus required to supply the methane needs of your city or state. State what efficiencies of conversion you assume.

16.6 a) Calculate the amount of energy contained in one ton of refuse that can be reclaimed for combustion.

b) A ton of refuse contains 40 kg (90 lb) of ferromagnetic material and almost 5 kg (11 lb) of aluminum. Calculate the intrinsic value of a ton of refuse, assuming all the energy and materials can be recovered.

16.7 It has been estimated that the energy cost of collecting and disposing of garbage is about 5×10^9 joules per metric ton. Calculate the overall energy cost of garbage collection and disposal in the U.S., and compare to the overall energy use in the U.S.

*16.8 Prepare an analysis and estimate of the upper limit on recoverable energy from crop residues, feedlot manure, and urban refuse. Compare this limit to the total energy use in the U.S.

---

*An asterisk indicates a relatively difficult or advanced exercise.

Solar Energy

17.1 INTRODUCTION

All our food and fuel (exclusive of nuclear energy) has been made possible by the sun through the photosynthetic combination of water and atmospheric carbon dioxide in growing plants. Solar energy is the basic energy support for life and underlies the wind, the climate, and fossil fuels. [1] The energy flow diagram of Fig. 3.7 portrays the flow of the sun's energy into the atmosphere and then ultimately into heat, wind, evaporation of water, and photosynthesis.

In this chapter we will discuss the direct use of the sun's energy as it impinges on the earth. Thus, we are interested here in the tapping of solar energy in the form of electromagnetic energy prior to its conversion to wind energy, fossil fuels, and plants. Solar energy in its converted form of wind and tidal energy is discussed in Chapter 14, and solar energy for growing plants is considered in Chapter 19.

17.2 SOLAR ENERGY REACHING THE EARTH

Immediately outside the earth's atmosphere the sun provides energy at the rate of 1,353 watts/m^2 normal to the sun. Since the area of the diametral plane of the earth is 1.27×10^{14} m^2, the solar input to the earth is 1.73×10^{17} watts.

The solar radiation received at the earth's surface is only a portion of the total outside the earth's atmosphere. A considerable portion is reflected back into outer space from the atmosphere and clouds. Some solar energy is scattered by water droplets in clouds and dust particles. In cloudy weather the radiation is reduced even further. Solar energy predominantly lies in the spectrum of light waves between 0.3 and 1.2 micrometers.

The average solar radiation received at the earth's surface is approximately 630 watts/m^2. This is an *average* figure; the actual radiation received at a specific location can be significantly higher or lower. Of course, this radiation is received only at a time when the point on the earth is normal to the sun's rays. Thus, no radiation will be received at night, and a reduced amount is received at times other than noon. The solar energy received also varies significantly with latitude, as well as with time of year. The solar energy received in a given location in the U.S. during a clear day in December is about one-half of that received in July.

The solar radiation at the earth's surface, averaged over the day and over the total year, has been recorded for the U.S., and is called the *average annual insolation.** The average annual insolation for several regions or cities in the U.S. is given in Table 17.1. The region consisting of the southern two-thirds of New Mexico and Arizona and the bordering desert regions of Nevada and California receive an average insolation of about 260 watts/m². This is about 40 percent more than the insolation of New York or the New England states.

AVERAGE ANNUAL INSOLATION IN
THE U.S.†

Location	Insolation (watts/m²)
Vermont	150
New York City	180
Washington, D.C.	200
Denver	220
San Francisco	230
Arizona–New Mexico	260

† The insolation is averaged over a 24-hour period.

17.3 WHY CONSIDER THE DIRECT USE OF SOLAR ENERGY?

The use of direct power from the sun's radiation has many advantages. Solar power is abundantly available, even in regions remote from the source of fossil fuels. It is essentially a nondepletable source of energy in comparison with fossil fuels or nuclear fission power, and it is cost-free in its original radiation form. Of course, there is a significant cost for the capital plant required for converting solar energy to other forms of energy. If solar energy is utilized locally, then the need for transporting the energy is avoided. Also, solar power can be used in small units, as for an individual building or home. Solar devices hold promise for the developing world as well as for the economically developed world. Since solar power burns no fuel, it causes no air or water pollution.

The average annual insolation for the continental U.S. is about 180 watts/m². This amounts to about 1600×10^{12} watts over the entire U.S. Assuming that one-half of the total time period available during a year was used for converting

* Insolation: the amount of solar radiation per unit of horizontal surface over a period of time.

solar power to useful energy, we have

$$E = 1600 \times 10^{12} \text{ watts} \times \frac{8760 \text{ hours}}{2}$$

$$= 7.00 \times 10^{18} \text{ watt-hours}$$

$$= 2.40 \times 10^{19} \text{ Btu.} \tag{17.1}$$

The U.S. consumed about 78×10^{15} Btu in 1975, so the solar energy available might be

$$\frac{2.4 \times 10^{19}}{78 \times 10^{15}} = 308 \text{ times that required by the nation.}$$

Alternatively, one could state that, if solar radiation could be collected over 1/308 of the surface of the U.S. with a load factor of one-half, then the energy consumption of the U.S. could be supplied. Of course, this calculation assumes that the collection efficiency is 100 percent. If we had collection and conversion efficiency of 10 percent, we would require $10/308 = 0.0325$ or 3.25 percent of the U.S. land surface.

A disadvantage of the utilization of solar radiation is its dispersal over the surface area of the earth. The average insolation in the U.S. is 180 watts/m² and the land area required to collect significant amounts of energy is large. For example, the land required to collect the equal of the output of a 1000-megawatt power plant would be

$$\text{Area} = \frac{1000 \times 10^6 \text{ watts}}{180 \text{ watts/m}^2} = 5.56 \times 10^6 \text{ m}^2$$

$$= 2.15 \text{ square miles}, \tag{17.2}$$

assuming a collector efficiency of 100 percent.

17.4 THE HISTORY OF THE USE OF SOLAR ENERGY

Solar radiation has been used by man since the beginnings of time for heating his domicile, for agriculture, and for personal comfort. Solar heating has been utilized in various forms since ancient times, when the sun's rays were focused for heating and cooling. In 1774 Joseph Priestly used lenses to concentrate solar rays to decompose mercuric oxide into mercury and oxygen. In 1872, in the desert of North Chile, a solar distillation unit covering 4750 square meters of land was built to provide fresh water from salt water. This plant was operated for 40 years, producing 6000 gallons of water per day.

At an exhibition in Paris in 1878, sunlight was focused onto a steam boiler that operated an engine which drove a printing press, as shown in Fig. 17.1. During the period 1901 to 1915, several solar collectors used with steam engines of several horsepower were constructed in California and Pennsylvania. In 1913, near Cairo, Egypt, F. Shuman and C. V. Boys (engineers from Philadelphia) built

Fig. 17.1 *A solar-energy-operated printing press at the Paris Exposition of 1878. (Courtesy of National Aeronautics and Space Administration.)*

a large solar engine of over 40 kilowatts. The long parabolic cylinders focused the solar radiation into a central pipe, as shown in Fig. 17.2.

Small solar-powered steam engines were built during the period 1930 to 1960, but it was difficult to market them in competition with engines running on inexpensive gasoline. Recently, however, with the increasing cost of fossil fuels

Fig. 17.2 *A solar-energy steam engine used to operate a steam-powered irrigation pump in Egypt in 1913. (Courtesy of NASA.)*

and the depletion of our resources, interest has been rekindled in the harnessing of solar energy for heating and cooling, the generation of electricity, and other purposes.

17.5 COLLECTORS FOR SOLAR RADIATION

While solar energy arriving at the land surface of the U.S. is about 500 times the present rate of consumption of energy, it is also both diffuse and intermittent. Gathering sunlight and providing it in useful form when needed at a competitive cost is the principal challenge to solar energy technology. The recent plan issued by the Energy Research and Development Administration calls for a significant contribution to the U.S. supply of energy by means of solar power. [2] Table 17.2 summarizes the ERDA projections for the future contribution of solar energy. These projections assume that solar energy will provide 1 percent and 6.7 percent of the total U.S. energy in 1985 and 2000, respectively. However, the projections call for solar energy providing 25 percent of the total energy consumption by 2020. Much work remains to be done to prove the economic feasibility of solar-energy units and develop technically sound applications.

Table 17.2
ESTIMATES OF THE FUTURE CONTRIBUTION OF SOLAR ENERGY

Solar technology	1985	2000	2020
Direct thermal applications (in units of 10^{15} Btu = 1 Q per year)			
Heating and cooling	0.15 Q	2.0 Q	15 Q
Agricultural applications	0.03	0.6	3
Industrial applications	0.02	0.4	2
Total	0.2 Q	3 Q	20 Q
Solar electric capacity (in units of 10^9 watts = 1 Gwe)			
Wind	1.0 Gwe	20 Gwe	60 Gwe
Photovoltaic	0.1	30	80
Solar thermal	0.05	20	70
Ocean thermal	0.1	10	40
Total	1.3 Gwe	80 Gwe	250 Gwe
Equivalent fuel energy	0.07 Q	5 Q	15 Q
Fuels from biomass	0.5 Q	3 Q	10 Q
Total solar energy	~1 Q	~11 Q	~45 Q
Projected U.S. energy demand	100 Q	150 Q	180 Q

The first stage in the process of utilizing solar energy is collecting by some means. Solar energy can be collected by three means: (1) thermal collectors, (2) photovoltaic collectors, and (3) biological collectors. In addition, the collectors can be either focusing or nonfocusing. Focusing (or concentrating) collectors can be lenses, mirrors, or selective films to concentrate the solar radiation. An example of a photovoltaic system is a collector consisting of solar electric cells, such as those currently used in space stations. Biological collection is facilitated by means of photosynthesis. We do not include ocean thermal gradients, hydroelectric power, or wind energy here (they are discussed in other chapters), although they are inherently a natural means of collecting the sun's radiant energy.

The thermal collector operates on the basis of absorbing the solar radiation at a surface and thus raising the temperature of the collector surface. The photovoltaic collector directly uses the photons in the solar beam.

There are about 630 watts/m^2 of solar radiation received in rays at the earth's surface at noon. The flux is higher where there are no clouds and lower with clouds. Solar radiation arrives at some angle to the earth's surface depending on latitude, time of day, and time of year. We designate the angle between a plane perpendicular to the radiation and the earth's surface as θ, and the intensity per unit of surface area on the ground is reduced by $\cos \theta$. For example, if the sun is 45° away from the zenith, the power per unit of surface area on the ground is reduced to

$$P = 630 \times \cos \theta = 445 \text{ watts/m}^2. \tag{17.3}$$

One way to overcome this loss of power is to place the collector at an angle so that it is perpendicular to the sun's rays. Some collectors are designed to track the movement of the sun across the sky so that the collector remains perpendicular to the incoming rays.

Nonfocusing collectors, often called *flat-plate collectors*, can be set at an angle so that they are nearly perpendicular to the sun's radiation for most of the day. Flat-plate collectors are generally less expensive than focusing collectors, and they use heat from diffuse solar radiation as well as from the direct radiation.

The surface of a flat-plate collector is blackened to absorb the radiation. The plate absorbing the radiation rises in temperature and transfers the heat to a fluid, usually air or water, flowing on the back side of the collector. The surface is usually heated to 100°C–300°C. One or more sheets of glass or plastic are placed over the absorbing surface in an airtight box. Sunlight with wavelengths of 0.2 to 2.0 $\times$ 10^{-6} meters passes through these transparent coverings, but the long-wavelength radiation (2 to 8 $\times$ 10^{-6} m) emitted by the heated absorber will not pass through the covering, and this remains within the box (this is often called the greenhouse effect). Flat-plate collector performance is given by the relation

$$q_a = S \cdot \overline{\tau \alpha} - U(T - T_a), \tag{17.4}$$

where q_a is the net absorbed flux density, S is the solar intensity measured in the plane of the collector, $\overline{\tau \alpha}$ is the effective product of transmittance of the coverplates by the absorptivity of the absorber plate, T is the mean temperature of the plate, T_a is the ambient temperature, and U is the overall coefficient of upward loss of the blackened plate to the ambient air. [3]

Flat-plate collectors can be improved by (1) increasing the number of coverings (glass plates), (2) reducing the glass reflectance, and (3) increasing the absorbance, α, of the absorber plate. Flat-plate collectors can achieve temperatures in the range of 50°C to 150°C.

With focusing collectors it is possible to obtain much higher temperatures, but they usually cost more, they usually need to be moved to track the sun, and they can only use the radiation that comes, unscattered by clouds, directly from the sun. Solar radiation can be focused by parabolic mirrors or lenses. Optical concentrators can bring the sun's energy into either an approximate point focus or a line focus. Conduction losses to the air can be eliminated by evacuating the region around the collector. However, radiation losses can be important. The radiation loss is represented by

$$P = Ae\sigma T^4, \tag{17.5}$$

where A is the area, e is the emissivity, σ is the Stefan–Boltzmann constant (5.67×10^{-8} W/m$^2 \cdot$k^4) and T is the kelvin temperature. The emissivity for a perfect blackbody is equal to one. Concentrators allow a small area, A, to be utilized, thus reducing the radiation losses. The ratio of concentrator area to the absorber area is usually in the range of 10:1 to 10,000:1. An example of a concentrating collector is shown in Fig. 17.3. This experimental collector is located at Odeillo, France, and has a concentration ratio equal to about 20,000 to 1.

Another concept for a focusing collector system uses a field of heliostats to focus the sun's rays on a tower that has an absorber located on the top. [3] The heliostats, or small mirrors, are located on the ground, and are used to track the sun and reflect it to the tower absorber. Several plans call for systems of 1500 to 3000 heliostats combined with 300- to 500-foot towers. A concentration ratio can be achieved of about 100 to 1 and an absorber temperature of 1500°C. A 100-megawatt power plant might require hundreds of acres devoted to thousands of heliostats. [3]

Concentration ratios of up to 20,000 may be achieved, with resulting high temperatures. The parabolic cylinders used in Meadi, Egypt (Fig. 17.2) achieved

Fig. 17.3   An experimental solar energy collector using a mirror system located at Odeillo, France, which achieves high temperatures. The absorber is located in the six-story building that appears in front of the mirror system. The parabolic mirror is about 30 meters high. (Courtesy of French Embassy Press and Information Division.)

a ratio of 10 to 1 and an absorber temperature of 200°C. With concentration ratios of 1000 to 1, absorber temperatures equal to 1200°C may be obtained.

Both the flat-plate collectors and the focusing collectors have problems because of their large size. They should be strong enough to withstand the most severe winds in their localities, and they should be fastened down in a safe position when a storm is approaching. They must be so arranged that they can be cleaned easily. Other factors, such as wind damage and the tenacity with which dust particles adhere to the surface of the collector, may mean success or failure in a solar-energy project.

17.6 SOLAR THERMAL ENERGY

A solar collector can be used to generate a reasonably high temperature, which can then be used to create steam to drive a turbine and ultimately generate electricity. A schematic diagram of a solar thermal electric generating system is shown in Fig. 17.4. In this system, a heat-transfer fluid (or gas) transfers the heat to a thermal storage tank until it is needed by the turbine.

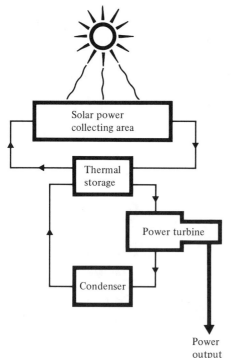

Fig. 17.4 *A schematic diagram of a solar thermal electric generating system.*

Aden and Marjorie Meinel have worked for several years at the University of Arizona on a system which could yield large power outputs. [10] They propose the use of a cylindrical mirror collector as shown in Fig. 17.5. The cylindrical mirror concentrates sunlight so that the focused beam enters an evacuated glass pipe through a transparent region. The remaining portion of the interior of the glass pipe is silvered to produce a high reflectivity. The sunlight entering this enclosure impinges either directly, or after a single reflection at the walls, on a steel pipe, which is coated on its exterior with a selective coating. The enclosure is evacuated so that no heat is lost by convection. The result is that the steel pipe gets very hot, and if no way of extracting the absorbed energy is available, the pipe will rise in temperature to the order of 600°C. The thin coating on the pipe surface should selectively absorb the sun's radiation while emitting a very low amount of radiation due to the high temperature of the fluid. A selective film looks like a window to the sun's spectrum while appearing as a mirror to the radiation from the heated fluid (in the range 2 to 10 × 10^{-6} m).

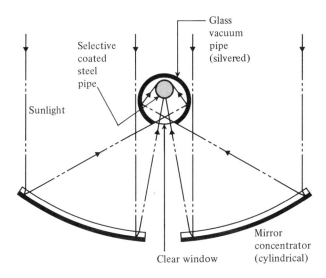

Fig. 17.5 *Cross-section diagram of a typical collector using a cylindrical rear-surfaced mirror as the concentrator.*

Let us take a sample calculation of the absorption and the radiation from such a system. The energy received in the pipe for a unit area of concentrator would be (where U is assumed to be equal to zero)

$$q_A = S\tau\alpha$$
$$= 600 \text{ W/m}^2 \times (0.9) \times (0.9)$$
$$= 486 \text{ watts/m}^2. \tag{17.6}$$

The radiated power (assuming 600°C for the absorber) would be

$$P = Ae\sigma T^4$$

$$= \left(\frac{1}{50}\right)(e)(5.67 \times 10^{-8})(873)^4$$

$$= 659e \text{ watts/m}^2, \tag{17.7}$$

assuming the concentration ratio is equal to 50. If $e = 0.05$, we have $P = 33$ watts/m², which is less than 10 percent of the incoming power. In general, we may represent the ratio of received to radiated power as

$$\frac{q_A}{P} = \frac{S\tau\alpha}{\left(\frac{1}{C}\right)eT^4}, \tag{17.8}$$

where c = concentration ratio. The key factor is $C\alpha/e$, which is the product of the concentration ratio and the solar absorptivity divided by the thermal emissivity. We wish to obtain highly selective films where α/e is large, and a high concentration ratio, C. Current values for α/e are approximately 10. A concentration ratio of 50 is achievable.

Referring to the example given in Eqs. (17.6) and (17.7) for $C = 50$ and $\alpha/e = 10$, we have $q_A = 486$ w/m² and $P = 59.3$ W/m². The net value of power obtained is then 427 W/m². In order to achieve a thermal power generation of 500 megawatts, approximately one million square meters of collector area, or an area of cylindrical mirrors of one km by one km, would be required. This arrangement might actually take up a land area of 2 kilometers on a side. This land area could be set aside in the Arizona desert and dedicated to the purposes of a solar power plant. Since the efficiency of a turbine-electric generating plant is only about one-third, a thermal power of 500 megawatts would yield only 167 megawatts of electric power. Of course, this thermal power is available only during the daylight hours, and provision must be made for storage of the thermal energy. [4] A solar collector power system operating during the daylight hours at an electric generating capacity of 1000 megawatts would require about 100 square kilometers (40 square miles) of land area. The cost of such a system could come to about $60/m² of parabolic collector. [5] Thus, a 1000-megawatt plant may cost $360 million, or $360/kw, exclusive of the turbines and storage system. The total cost of the system may be about $500/kW—very competitive with fossil-fuel or nuclear plants. Nevertheless, many consider these estimates of future costs to be excessively optimistic, and no actual plant has yet been built.

Two California utilities and the Federal government have committed $100 million to build the world's first full-scale solar power plant. [21] At Barstow, in the Mojave Desert, 3000 mirror arrays will focus the sun on a boiler atop a 400-ft tower. The output of the electric-power generator, utilizing the steam from the boiler, will be 10 megawatts. A molten-salt thermal storage will be used. Construction of the plant is scheduled for 1980–1981.

Energy Storage

Since solar energy is intermittently available, storage of the energy is desirable. A major cost of a solar thermal power plant is storage. Simple, short-term thermal-energy storage of 3 to 8 hours may comprise single tank units using the working fluid of the system as a storage medium. Besides providing storage, the system provides thermal inertia in the receiver loop carrying energy to the turbine generator, which is desirable to smooth out fluctuations in incident energy due to passage of clouds. However, since the turbine is more sensitive to inlet steam temperature, storage must be provided at the normal inlet temperature.

For thermal systems using steam, high-temperature storage requires an intermediate loop, as shown in Fig. 17.4. Desirable properties of a thermal storage medium are (1) low cost, (2) high heat capacity, (3) high-temperature capability, (4) noncorrosive properties, and (5) safety. If a water–steam system is used for the primary thermal loop, then practical heat-storage materials such as crushed rock are the candidates. Water and eutectic salts are attractive for temperature capabilities, but their costs are high. Several latent-heat phase-change materials have melting points near 116°F (47°C) and large heats of fusion, thus serving as good storage materials. Two examples are sodium thiosulfate pentahydrate and P-116 paraffin wax. [6]

Biological and Photochemical Conversion of Solar Energy

The solar energy incident upon the earth can be used for the direct biological process of photosynthesis. The growth of plants and trees to be harvested for an energy supply could be part of the agricultural system. Firewood remains as an important source of energy throughout the world.

The photosynthetic carbon-reduction cycle can be used as an energy source and as a material source. For example, sugar cane could be grown and then converted to alcohol to be used as a fuel. [7]

The Hevea rubber plant grown on plantations in Malaysia and Indonesia yields about 1 ton per acre and can be considered for a direct photosynthetic source of hydrocarbon for use in chemicals and materials in place of petroleum.

It is possible to consider the production of hydrogen by plants. In the normal process, the active hydrogen reduces carbon dioxide to make sugar; it is possible to limit the amount of carbon dioxide available, and thus force the plant to generate hydrogen. [7]

Perhaps water can be directly photodissociated by solar energy. [8] Water dissociation, in closed-cycle processes based on endothermic photochemical reactions, offers a potential source of hydrogen fuel. These proposed systems use transition-action complexes as catalysts in combination with the solar radiation. Systematic research into the photochemistry of transition-metal complexes, with the aim of designing suitable systems for solar energy conversion into hydrogen, may have long-range promise.

17.7 SOLAR HEATING AND COOLING OF BUILDINGS

About 25 percent of the energy now consumed in the U.S. is used to heat and cool buildings and to heat hot water. Furthermore, the use of energy for this purpose is growing rapidly, with air conditioning now being incorporated into 40 percent of all new building construction. Some studies have shown that solar energy could provide at least one-half of this energy consumption in the U.S. The potential savings through the use of solar energy for heating and cooling could amount to as much as 5 percent of our annual consumption of fossil fuels by the turn of the century.

Table 17.3
PERCENTAGE OF HEATING AND COOLING THAT
CAN BE READILY CONTRIBUTED BY SOLAR SYSTEMS

City	Percent of heating	Percent of cooling
Los Angeles	100%	98%
Atlanta	85%	63%
New York	84%	77%
Chicago	50%	67%

The amount of building heating and cooling that can readily be contributed by solar energy varies by location, as indicated in Table 17.3. In general, an average of 75 percent of the heating and cooling requirements of a city can be met by utilizing solar energy. The principal barrier to widespread use of solar heating and cooling systems is the relatively high cost of a system, with half of the cost attributable to the solar collector. The need for an auxiliary heating system is a financial deterrent to customers. Nevertheless, solar heating and cooling systems could save significant amounts of energy for the U.S.

Thermal energy is usually supplied to buildings at or below 100°C, a temperature readily achieved by heating a fluid on a solar flat-plate collector. A flat-plate collector can be built into the roof at an angle of 45° so that the collector can yield 600 W/m² during a six-hour day, or a total energy of

$$E = 6(3600 \text{ sec})600 \text{ W/m}^2$$
$$= 12.96 \times 10^6 \text{ joules/m}^2. \tag{17.9}$$

A typical residence of 1600 ft² might use 0.9×10^9 joule/day in December for space heating and water heating. Therefore, the house would require

$$\text{Area} = \frac{0.9 \times 10^9}{12.96 \times 10^6} = 69.4 \text{ m}^2$$

$$= 747 \text{ ft}^2. \tag{17.10}$$

Thus, one side of a 45°-angle roof, the south-facing side, mounted with flat-plate collectors, would provide sufficient energy for the daily requirements of the home on clear, sunny days.

For days when the sun does not shine, storage of energy will be required. Heat storage for house heating has been accomplished by means of water tanks, bins of rocks, or in hydrated chemicals such as sodium sulfate, whose solid–liquid phase change adds to the heat energy that can be stored. Assuming a 40°C temperature change, water will store 1.6×10^8 joules/m³. In order to store two days' energy consumption, one would require, for the typical residence,

$$\text{Volume} = 2(0.9 \times 10^9 \text{ joules})/1.6 \times 10^8 \text{ joules/m}^3$$
$$= 11.3 \text{ m}^3$$
$$= 399 \text{ ft}^3.$$

Thus, a water tank of 2.24 m (7.35 ft) on a side would be sufficient for two-day energy storage. This storage system might cost about $1000 to purchase and install.

The collectors might (optimistically) cost about $30/m², for a total of $2100. Thus, the cost of the total heating system for this typical residence would be about $3100. In addition, the usual heat-distribution system is required, as well as a small auxiliary heating system for occasions when solar energy is unavailable for longer than several days. It is estimated that a solar heating and hot-water system would cost $3000 to $4000 for a 1600-ft² home.

The schematic diagram of a domestic space-heating and hot-water-heating system is shown in Fig. 17.6. This system uses a 4000-gallon water tank (15 m³)

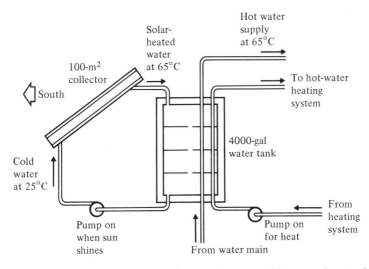

Fig. 17.6 *A schematic diagram of a system for space heating and hot water heating for a house of 1800 ft².*

to provide three days' heat storage. In this case, a collector temperature of 65°C is assumed. It is usually desirable to have a collector fluid temperature in the range 40°C to 70°C. Water can be heated to any temperature below 93°C in unpressurized storage. For forced-air space heating and with air as the heat fluid passing through the collector, rock beds are usually used for storage.

Solar air-conditioning systems are largely experimental at present. Absorption air-conditioning systems for solar applications usually use LiBr–H_2O as the coolant fluid. These systems require at least 80°C, and thus higher-temperature collectors are required than for heating only. [9] The lithium-bromide/water absorption system works best near 120°C. The solar powered cooling system will be able to provide most (perhaps 90 percent) of a building's needs, but some small augmentation may be required.

The solar-augmented heat pump presents an alternative for cooling and heating systems. Heat-pump technology is well developed, but it is not as yet extensively used.

Solar-heated commercial buildings are being built all over the world and particularly in the U.S. It is estimated that 50 to 80 solar-heated buildings now exist. For example, the George A. Towns Elementary School in Atlanta, Georgia, uses about 10,000 square feet of flat-plate solar collectors for sunlight needed to heat, air-condition, and supply hot water for, the school building. The school opened in 1976 and it is the fifth school in the nation to rely primarily on solar energy for space heating.

About 2000 m^2 of solar-energy collectors were recently put on top of a 56-story New York building. This experimental project, under the direction of the Massachusetts Institute of Technology, is investigating the use of solar energy to operate an air-conditioning (cooling) unit for the building (Citicorp Center).

The project at Citicorp Center is based on a technique of removing moisture from the incoming air at or near the outside temperature by passing the air through a spray of water-absorbing liquid. The diluted liquid can be regenerated by heating it to temperatures that can be achieved with solar-energy collectors.

Several solar homes have been built in the U.S. over the past two decades. Massachusetts Institute of Technology (M.I.T.) house IV, built in 1959, was the last in a series of experiments carried out by H. C. Hottel and his colleagues, and represented a cooperative effort of architects and engineers to develop a functional, energy-conserving home with a major part of the energy for space heating and water heating to be supplied from the flat-plate collector. The collector had an area of 60 m^2, two glass covers, and a flat, black-paint absorbing surface. The main storage tank held 5700 kg (1500 gallons) of water.

The Denver solar house built by Lof in 1958 uses air as the heat-transfer medium and a pebble-bed storage unit. Only 22 percent of the heating and hot-water loads were carried by solar energy for the period December to April.

An experimental house, Solar One, designed to obtain up to 80 percent of its heat and electricity from sunlight, was completed at the University of Delaware in 1973. The 1350-square-foot house uses 72 m^2 of solar collectors. Air is used to collect the heat from the flat-plate collectors, and the heat is stored by means of eutectic salts. As the heated air passes over the heated salts, the salt melts, absorbing heat. During the summer a heat pump is used as an air conditioner. This system is estimated to cost $3000, if produced in mass quantities.

A solar house was completed in Arizona in 1975, which provides 100 percent of its heating and 75 percent of its cooling requirements. This house uses combination copper-roof and solar-energy panels, as shown in Fig. 17.7.

Fig. 17.7 *The Decade 80 Solar house in Tucson, Arizona. The house uses a combination copper roof and solar energy panel. About 1800 sq ft of collectors are used to supply energy to a 3000-gallon water-storage tank. Two lithium-bromide water-absorption units are used for cooling. (Courtesy of the Copper Development Association.)*

A U.S. government building scheduled for Saginaw, Michigan, uses 743 m^2 of flat-plate collectors oriented at about 50° to take advantage of the sun. Water circulating through the collector absorbs the sun's energy and carries it to two 15,000-gallon (56,800-liter) tanks.

The actual economic savings of a solar system over a conventional heating and cooling system depends upon the local cost of fuel and the cost of the collector panels. One study of a Wisconsin system estimated a breakeven case for a fuel cost of $5/gigajoule ($5/10^6 Btu) and a collector cost of $60/m^2. As collector costs drop below $40/m^2, solar systems will receive more widespread use. Nevertheless, a solar system of 80 m^2 would still cost upwards of $4000, including a storage unit. Solar-energy systems will, however, pay off over a period of 10 to 20 years as fuel prices continue to rise.

While solar air-conditioning systems are not yet competitive, space-heating and hot-water-heating systems are cost-competitive in many regions now. Building-code revisions, easily obtained loans, and tax incentives for solar heating would accelerate the use of these systems. In order to achieve widespread use of solar-heating systems, unit price reduction and mass production of flat-plate collectors will be required. Perhaps this will be attained by 1985 or 1990.

17.8 PHOTOVOLTAIC SOLAR SYSTEMS

Solar cells that directly generate electricity by means of photovoltaic processes are the predominant source of power for space satellites. These cells convert sunlight directly into electricity without an intermediate thermodynamic cycle. A photovoltaic device requires a material in which mobile charge carriers can be generated by absorption of light, and a built-in potential barrier by which these charge carriers can be separated from the region generating them. Semiconductor materials with p–n junctions fulfill these requirements. Semiconductors manufactured from silicon (Si), gallium arsenide (GaAs), cadmium sulfide (CdS), and copper sulfide (Cu_2S) can be used for photovoltaic conversion. Existing silicon cells develop about 0.5 volt, so that large numbers of cells must be arranged in series in order to achieve high voltages. The output of the cells is a direct current, which is inverted to alternating current. [18, 20, 22]

Silicon solar cells are made from single crystals of the material, and production of these crystals is costly. The cadmium sulfide cell is a candidate for a low-cost cell within the near future. However, it converts only about 6 percent of the incident light to electricity, while silicon cells can achieve an efficiency of about 11 percent.

Gallium arsenide cells can achieve up to 18 percent efficiency and have been tested in several laboratories. Aluminum antimonide can also convert light to electricity, and may achieve efficiencies as high as 20 percent.

Designing the photovoltaic arrays directly into the building structures permits savings and offers esthetic advantages. In addition to the photovoltaic converters, photovoltaic power systems require a mounting structure for the converters, a power collection network, electrical regulation and control equipment, probably power-conditioning equipment to convert dc to ac, energy storage equipment, and possibly concentrators and cooling systems for the converters.

Silicon solar cells are currently produced in the U.S. at a rate of 500 m^2 per year for spacecraft at a price of \$10,000/m^2. [11] A total solar cell system costs about \$15,000 per kilowatt in 1977. As the volume of solar-cell production increases, it is estimated that this cost can be reduced by a factor of 30. Thus, by 1985, the cost may be \$500/kw, which is entirely competitive with other means of obtaining electricity.

A two-kilowatt rooftop solar panel costing $1000 and delivering an average of 10kWh per day would pay for itself in about six years, in comparison with commercially available electricity at a rate of 5¢/kWh.

The "Solar One" house at the University of Delaware uses cadmium-sulfide solar panels to generate electricity, which is stored in 110-volt, lead–acid batteries of 9 kWh capacity. This system is supplemented by power from the utility distribution network.

Since about 600 watts/m^2 is available in many parts of the U.S., we shall calculate the area required to provide a significant portion of the electric energy to the U.S. Assuming that the solar power is available over an average of five hours of a day, we can expect

$$E = 600 \text{ W/m}^2 \times 5 \text{ hr} \times 365 \times \varepsilon \times A = 1.1 \times 10\varepsilon A \text{ Wh} \qquad (17.12)$$

available per year for an area of solar cells equal to A and a cell efficiency of ε. The current electrical energy consumption in the U.S. is approximately 2100×10^9 kWh. When the efficiency of the cells is $\varepsilon = 0.10$, we have

$$E = 1.1 \times 10^2 A \text{ kWh}; \qquad (17.13)$$

that is, we require $A = 19,000$ sq km in order to provide the total electric energy supply.

In order to supply 1000 megawatts of power, when $\varepsilon = 0.1$, we would require

$$P = 10^9 \text{w} = 600 \text{ w/m}^2 \times 0.1 \times A_\text{p}, \qquad (17.14)$$

and therefore,

$$A_\text{p} = 1.67 \times 10^7 \text{ m}^2 = 16.7 \text{ sq km}. \qquad (17.15)$$

This system would require a square of solar panels 4 km on a side, and would yield only 1000 megawatts over five to eight hours of the day.

In order to achieve ready use of photovoltaic panels for the generation of electric power during the next decade, it will be necessary to reduce the cost by a factor of 100. Mass production of solar cells, system optimization, improved storage systems, and increased reliability and maintainability of solar panels will be required, in order to achieve wide acceptance of photovoltaic solar cells.

17.9 SATELLITE SOLAR POWER STATIONS

The concept of placing a large photovoltaic solar array in geosynchronous orbit 35,800 km from Earth and transmitting this power to Earth was first proposed by Glaser in 1968. [12] A satellite solar power station (SSPS) located in orbit at a distance of 35,800 km and parallel to the Earth's equatorial plane would be stationary with respect to a point on Earth that could serve as a receiving point for electric power transmitted from the SSPS. The satellite would generate electricity from sunlight, using either arrays of photovoltaic converters or a solar-thermal electric system to power microwave transmitters. These transmitters would beam microwave power to a line-of-sight receiving station on Earth, where special receiving antennas (rectennas) would convert it directly to dc power for distribution to consumers.

One concept of a SSPS uses concentrators on the satellite to reflect sunlight onto photovoltaic cells in two large symetrically-arranged low-mass arrays, which convert the sunlight to electricity that is fed to microwave generators within the transmitting antenna located between the two arrays, as shown in Fig. 17.8.

Fig. 17.8 *A satellite solar power station using concentrators and solar voltaic cells. The power is transmitted to earth by means of the microwave antenna located between the solar arrays. (Courtesy of Arthur D. Little, Inc.)*

A solar array can receive up to 15 times as much solar energy as would the same array on the ground, as shown in Table 17.4. SSPS systems are most efficiently operated between about 3000 and 15,000 megawatts. The microwave power would be transmitted at a frequency selected between 3 and 4 gigahertz.

Table 17.4
AVAILABILITY OF SOLAR ENERGY IN SYNCHRONOUS ORBIT

Availability factor	Average on earth	In synchronous orbit	Average ratio, ground/orbit
Solar energy density, kW/m^2	1.1	1.4	4/5
Hours of useful radiation/day	8	24	1/3
Percentage of clear skies	50	100	1/2
Cosine of angle of incidence	0.5	1	1/2
Product of all factors			1/15

The diameter of the antenna would be about 1 km and the microwave transmission efficiency would be about 77 percent. The diameter of the receiving antenna on the earth would be about 7 km for maximum power reception. The receiving antenna both collects and rectifies the microwave power, providing an output of dc power.

For an SSPS providing 5000 MW output on earth, two solar panels, each measuring 5.2 km by 4.33 km would be required. The overall length of the entire station would be 11.73 km. The panels would contain solar cells made of silicon or other semiconductor 0.05 mm thick. The total satellite mass would be 10 million kg, and the total cost would be about $10 billion, as summarized in Table 17.5.

Table 17.5
PARAMETERS OF A 5000-MW SSPS

Single solar panel size	5.2 km × 4.33 km
Total SSPS size	11.73 km × 4.33 km
Satellite mass	10×10^6 kg
Transmitting antenna diameter	1 km
Receiving antenna diameter	7 km
Total system cost	$10 billion
System cost per kW	$2000

The actual cost of an SSPS would depend primarily on four factors: the capital cost per kilowatt of the power plant for converting sunlight to electricity, which we designate S ($/kW); the specific mass of the power plant in kg per kW, M; the cost per kg of placing the power plant in synchronous orbit, $\mathcal{O}$ ($/kg), and the efficiency of converting electricity into microwave energy and receiving it on earth, ε. The capital cost of the SSPS per kW would then be represented as

$$\text{Cost/kW} = \frac{1}{\varepsilon}[S + M \cdot \mathcal{O}]. \qquad (17.16)$$

Let us assume an efficiency of 70 percent. Current values of M range from 10 to 30 kg/kW, but we will assume that $M = 2$ is achievable by the 1980's. The costs of the solar arrays could be $300/kW by the next decade. Substituting these figures, we obtain

$$\text{Cost/kW} = 1.43[300 + 2\mathcal{O}].$$

If we wish to achieve a cost/kW of $1000, we would require a transportation cost for placing the SSPS in orbit of $\mathcal{O} = \$200$/kg.

The actual construction of an SSPS is heavily dependent upon a relatively low cost of placing the satellite in orbit. One concept is to use a space manufacturing facility located on the moon and utilize the moon's materials for construction of the satellite. [4] This approach would significantly reduce the cost of placing the satellite in orbit; it could also reduce the number of space shuttle flights from the earth. The overall efficiency of an SSPS could be about 10 percent, as shown in Table 17.6.

Table 17.6
EFFICIENCY OF A SATELLITE SOLAR
POWER STATION

Unit	Efficiency
Solar cells	15%
Microwave transmitter	77%
Microwave receiving antenna	90%
Overall efficiency	10.4%

Another proposal is to use a large solar-thermal power generator in synchronous orbit. The efficiency of the solar-to-thermal-to-electricity conversion could be as much as twice that of the solar-cell system. Of course, this approach is equally dependent upon the economic construction and transportation of the concentrators, turbines, and microwave system.

The actual construction and utilization of an SSPS will depend upon the future technology of space flight to near-earth orbits. If new means of low-cost shuttles become readily available, we can expect some real interest in space power stations to grow.

17.10 SOLAR ENERGY UTILIZATION THROUGHOUT THE WORLD

The potential use of solar energy throughout the world is significant, particularly in the less developed nations near the equator.

Some 2.5 million solar water heaters were sold in Japan in 1973. Such heaters usually consist of a half-dozen black cylinders serving as a collector, and a storage unit housed in a box and mounted on the house roof.

In Australia, flat-plate collectors have been used to produce hot water at 60°C over the past 20 years. Existing installations are estimated to number 20,000.

A large part of Africa could profitably use solar energy for many purposes. For years, different types of solar devices (e.g., solar water heaters, stills, refrigerators, engines and batteries) have been used in countries such as Senegal, Mali, Upper Volta, Nigeria, and Mauritania. Solar cooking, even if used only one-half of the time, could result in an annual savings of 30 million tons of firewood in the Sahelian and Sudanese zones alone. [15] This use of solar energy could relieve the deforestation and erosion of the area. Widespread utilization of solar water heaters and solar distillation of water could also relieve the demand on firewood.

Solar-energy units capable of producing sufficient power to pump water for irrigation and watering lifestock would be very useful throughout the world. An integral solar-powered pump would be a welcome product in the near future.

Small, effective, solar cookers for domestic food cooking would also be useful throughout the less developed world. These products require the quality, economy, and mass production methods now commonly available for manufacturing transistor radios. Solar energy, if widely used, can be a magnificent boon to rich man and poor man alike.

REFERENCES

1. F. DANIELS, *Direct Use of the Sun's Energy*, Ballantine Books, 1974, New York.
2. "A national plan for energy research, development, and demonstration," U.S. Energy Research and Development Administration, Report No. 48, June 28, 1975.
3. H. C. HOTTEL, "Solar energy," *Chemical Engineering Progress*, July 1975, pp. 53–65.
4. F. VON HIPPEL AND R. H. WILLIAMS, "Solar technologies," *Bulletin of Atomic Scientists*, Nov. 1975, pp. 25–31.
5. R. L. GERVAIS AND P. B. BOS, "Solar thermal electric power," *Astronautics and Aeronautics*, Nov. 1975, pp. 38–45.
6. H. G. LORSCH, "Thermal energy storage for solar heating," *ASHRAE Journal*, Nov. 1975, pp. 47–52.
7. M. CALVIN, "Solar energy by photosynthesis," *Science*, April 19, 1974, pp. 375–381.
8. V. BALZANI, *et al.*, "Solar-energy conversion by water photodissociation," *Science*, Sept. 12, 1975, pp. 852–855.
9. J. A. DUFFIE AND W. A. BECKMAN, "Solar heating and cooling," *Science*, Jan. 16, 1976, pp. 143–149.
10. A. B. AND M. P. MEINEL, *Applied Solar Energy: An Introduction*, Addison-Wesley Publishing Co., 1976.
11. M. WOLF, "Photovoltaic power," *Astronautics and Aeronautics*, November 1975, pp. 28–32.
12. P. E. GLASER, "Power from the sun: Its future," *Science*, Nov. 22, 1968, pp. 857–861.
13. J. R. WILLIAMS, "Geosynchronous satellite solar power," *Astronautics and Aeronautics*, Nov. 1975, pp. 46–52.
14. G. K. O'NEILL, "Space colonies and energy supply to the earth," *Science*, Dec. 5, 1975, pp. 943–947.
15. A. MOUMOUNI, "Prospects of solar power," *Ambio*, vol. 2, No. 6, 1973, pp. 203–213.
16. B. COMMONER, "Energy," *New Yorker*, Feb. 9, 1976, pp. 72–73.
17. L. M. MURPHY AND A. C. SKINROOD, "Coming: Solar power plants", *Mechanical Engineering*, Nov. 1976, pp. 26–32.
18. J. DU BOW, "From photons to kilowatts—Can solar technology deliver?" *Electronics*, Nov. 11, 1976, pp. 86–90.
19. D. BEHRMAN, *Solar Energy: The Awakening Science*, Little, Brown and Co., Boston, 1976.
20. J. A. MERRIGAN, *Sunlight to Electricity—Prospects for Solar Energy Conversion by Photovoltaics*, MIT Press, Cambridge, Mass, 1975.

21. "A solar-power plant for California's desert," *Business Week*, Jan. 24, 1977, pp. 23–24.

22. C. E. BACKUS, ed., *Solar Cells*, IEEE Press, New York, 1976.

23. "An economic analysis of solar water and space heating," *ERDA report DSE-2322-1*, U.S. Gov't. Printing Office, Wash. D.C., Nov. 1976.

24. J. G. ASHBURY AND R. O. MUELLER, "Solar energy and electric utilities: Should they be interfaced?" *Science*, Feb. 4, 1977, pp. 445–450.

EXERCISES

*17.1 It is desired to achieve a ratio $C\alpha/e$ greater than 20 for solar collectors. If $C = 2$ then α/e must be greater than 10. Examine the α/e ratio achievable using interference stacks as the collector.

17.2 The city of Colorado Springs constructed a solar house using a hot-water storage system combined with a heat pump. A backup heating element is also used in the water tank. The additional cost of the solar heating system and hot-water heating system is about $6000 and the solar system supplies about 80% of the building's heating needs. Survey your neighborhood and friends and determine what price they are willing to pay for a solar heating system which provides 80% of the home's needs.

*17.3 If a solar heating, cooling, and hot-water-heating system is available at a price of $4000 and the local cost of electricity is 4¢/kWh, calculate how many years it takes to break even with such a system. Assume the solar system accounts for 100% of the house's requirements and that the house would otherwise require 10,000 kWh of electric energy for heating, cooling, and hot water. Repeat the calculation for the home using 15,000 kWh per year, where the charge is 5¢/kWh.

17.4 It is possible to build homes with flat roofs holding tanks of water in the space between the ceiling and the roof. Ceiling ponds are capable of maintaining comfortable temperatures throughout a normal year, in Phoenix, Arizona, without supplementary heating or cooling devices. These ponds collect and store winter solar heat during the day and release it into rooms at night. In summer they collect and store infiltrated and internally generated heat during the day, then radiate it to the night sky. Movable insulation panels over the ponds operate as a thermal valve directing heat flow to produce desired thermal effects. Assume that such a house has a water tank with a transparent side to the sky. Insulation panels can be moved over this tank at the appropriate times of the day. Explain when you would have the tank covered in the winter and in the summer.

17.5 Solar stills for the distillation of water can be built and operated relatively inexpensively. A production cost of $3.50 per 1000 gallons is typical. Determine the disadvantages of solar stills that desalt 50,000 gallons per day. Is the price per 1000 gallons excessive?

---

* An asterisk indicates a relatively difficult or advanced exercise.

17.6 Shallow water ponds can be used to collect solar energy and heat water to about 60°C. For example, a pond one km on a side could collect a significant amount of solar energy in locations with high insolation. The hot water would be used to heat a working fluid such as freon, which ultimately would drive a turbine. (a) Calculate the amount of energy such a plant might collect. (b) What are the disadvantages of such a system?

17.7 Calculate or otherwise determine the amount of energy used for a typical outdoor home pool to heat it to a comfortable temperature over the period June to September. Based on this estimate, calculate the size of a flat-plate collector required to heat the pool if it was placed on the south-facing sloped roof of the house. Use the solar insolation figure for your region if you can obtain it.

17.8 Contact a company which sells domestic solar hot-water heaters and determine the cost of such a heater. Calculate the cost of keeping 80 gallons of water hot in a typical home and calculate how many years it will take to break even relative to a (a) typical gas heater, (b) electric heater.

17.9 Determine the location of a solar home in your community or state and obtain some data on its performance. Is it economically attractive for a person to purchase?

17.10 If photovoltaic cells were available for your region with an efficiency of 20% and at an installed cost of $120/m² (including storage batteries), calculate the cost of such a system for a typical house in your state. How long would the system require to pay off if electricity cost 5¢/kWh?

17.11 List the advantages and disadvantages of a satellite solar power station. Estimate the total cost of a 1000-MW system.

*17.12 A recent study of the use of solar hot-water heaters has demonstrated that the solar heater should be supplemented with an electric or gas heater for optimum cost. [16]· Consider a heater with an electric heating element and a solar heater. Assume that the solar collector and associated plumbing cost is $100/m², the system is amortized through a loan at 9% interest over 10 years, and the cost of electricity is 2¢/kWh. The cost of the electric solar heater tank is $125 in addition, for the combined system. The cost of a storage tank for the total solar system is $400. The cost of a total electric heater alone is $120 and its expected lifetime is 10 years. Calculate the annual cost of heating water with 10,000 kWh per year of heat by (a) electricity alone; (b) electricity providing half of the heating and solar power providing one-half; (c) solar heating providing all of the heat.

17.13 The Boeing Company of Seattle, Washington, has proposed a power satellite concept which would be capable of beaming microwave energy from a satellite in synchronous orbit at a 10-gigawatt power level. The satellite would incorporate four solar collector arrays with a total surface area of over 20,000 acres, that would concentrate sunlight into a solar cavity. The resulting heat would drive a set of helium gas turbines, which would drive alternators. Once the electric power is generated, it would be routed to a microwave generator for transmission to earth.

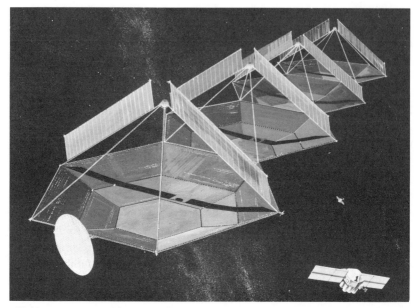

Fig. E17.13 *A conceptual drawing of the Powersat scheme. (Courtesy of Boeing Corporation.)*

A conceptual drawing of the satellite is shown in Fig. E17.13. Investigate the advantages and disadvantages of such a scheme. Contrast this scheme with a synchronous satellite using arrays of photovoltaic cells. It is estimated that one satellite would cost about $60 billion and the associated receiving station $13 billion. Calculate the cost per kW and compare with other schemes.

18

Hydrogen–An Energy Carrier for the 21st Century

In spite of the many advantages of electricity at the stage of fuel utilization, its generation and transmission creates many difficult problems. Electricity is difficult and costly to store for periods of peak demand and costly to transmit to the point of final use. Furthermore, the generation of electricity by the conversion of solar, wind, geothermal, nuclear, or thermal energy is a low-efficiency process. Electricity is a carrier of energy that has been converted from its original form in fossil fuels or the sun, for example, to electricity, which can be transmitted over distances to cities or towns where it is used.

Other means for storing and transporting energy could be developed over the next several decades. The qualities of a suitable energy carrier are summarized in Table 18.1. A suitable energy carrier is efficiently obtained from the original energy source, efficiently transported and distributed at a low cost, readily stored, readily used in industry and transportation, and is safe to handle.

In the U.S., we have made a large financial commitment to the generation, transmission, and distribution of electric power. Electric power is relatively safe to handle and can be generated and transmitted to the customers. However,

Table 18.1

CHARACTERISTICS OF A DESIRABLE ENERGY CARRIER

1. High efficiency of conversion of original energy source to the carrier fuel.
2. Availability of a method of transporting and distributing the carrier fuel with low attendant energy losses and at low cost per unit distance.
3. Availability of a method of storing the carrier fuel for relatively long periods of time and at low cost.
4. Availability of several methods of utilizing the carrier fuel for transportation and industry.
5. Safety in handling and storage.

electricity cannot be *stored* inexpensively. Pumped water storage is the most commonly used method at the present time. Furthermore, electric power cannot be readily used for transportation purposes without extensive efforts toward development of an electric auto system. Electricity, however, is used efficiently for electric streetcars and rapid-transit trains.

What is required for an energy carrier is a liquid or gaseous carrier, which can be obtained from abundant resources; this is especially important when it is desired to dissipate heat from nuclear reactors or other devices. This carrier should be capable of readily restoring the energy in the form of heat at the final consumption stage, without introducing significant environmental pollution. The desired process is shown in Fig. 18.1.

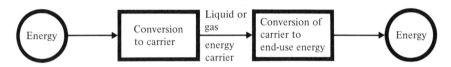

Fig. 18.1   *A schematic diagram of the desired energy carrier system. It is desired that a minimum of environmental pollution be generated at each conversion stage.*

18.1   HYDROGEN AS A CARRIER

In his 1847 novel, *The Mysterious Island*, Jules Verne envisioned the use of hydrogen as an energy carrier. "Water," says an engineer in the book. "Yes, my friends, I believe that ordinary water will one day be employed as fuel, that hydrogen and oxygen, which constitute it, used singly or together, will furnish an inexhaustible source of heat and light." Today Verne's dream is being taken seriously by many scientists and engineers. Hydrogen has the necessary properties and can fulfill the role of an energy carrier that can be derived from an abundant source, water. It can be substituted for petroleum and coal in almost all industrial processes which require a reducing agent, such as in steel manufacturing and other metallurgical operations. Further, hydrogen can easily be converted to a variety of fuel forms such as methanol, ammonia, and hydrazine. The use of hydrogen as a fuel would allow the industrial establishment to retain its present structure.

Hydrogen as an energy carrier can be favorably compared to electricity. In Table 18.2 we compare the projected characteristics of electricity and hydrogen which are expected to be true by the end of this century. Hydrogen is more efficiently converted to the carrier, transported readily and more inexpensively, stored less expensively, and readily used in industry and transportation, compared to electricity. Hydrogen does have some inherent safety hazards compared to electricity, which is used safely now.

The cost of transmission of an energy carrier becomes a very important characteristic as the new sources of energy become wind, solar, geothermal, and

Table 18.2

PROJECTED COMPARISON OF THE CHARACTERISTICS OF ELECTRICITY AND
HYDROGEN AT THE END OF THE 20TH CENTURY

Characteristic	Electricity	Hydrogen
1. Efficiency of conversion to the carrier	35 to 40%	35 to 50%
2. Transportation and distribution	More costly	Less costly
3. Storage	Costly	Low cost
4. Methods of using the carrier	Good in industry; poor in transportation	Good in industry and transportation
5. Safety	Proven safe	Some hazards

other renewable resources. These resources are not necessarily located near population centers; they require transportation of the energy over large distances. If we swing over to extensive use of nuclear power, it is expected that the plants will not be located near population centers, but rather in remote areas. Transmission cables are an environmental nuisance, their installation is costly, and the placing of cables underground is also costly. Economical storage of electricity by pumped storage is not feasible in most generating-plant locations. The present industrial and domestic economy is geared to combustible chemical fuels. Therefore, the use of a liquid or gaseous energy carrier would require a minimum alteration in all the existing end-use processes.

Hydrogen can be readily derived from the primary source by the decomposition of water. Hydrogen conversion and its end-use cycle is shown in Fig. 18.2. Energy is converted to hydrogen by the thermal or electrical decomposition of water, stored, and transported to its end use. Hydrogen is readily used as a combustible fuel, in combination with oxygen or air. The end product of the combustion is water, which recycles to the oceans and lakes. There is a small amount of nitrogen oxide (NO) that is formed from the air entering the combustion, but it can be controlled. Otherwise, the hydrogen flame is free of pollutants.

The idea of using hydrogen as an energy carrier and fuel is not new. In 1933 Rudolf A. Erren, a German inventor working in England, suggested the large-scale manufacture of hydrogen from offpeak electricity. He had done extensive work on modifying internal-combustion engines to run on hydrogen; and the main object of his suggestion was to eliminate automobile-exhaust pollution and to relieve pressure on the importation of oil. [2]

Transmission and Distribution of Hydrogen

Hydrogen can be transmitted and distributed in liquid or gaseous form. In gaseous form, hydrogen can be transmitted in much the same way that natural gas is today—by means of pipelines. The movement of a gas by pipeline is one of

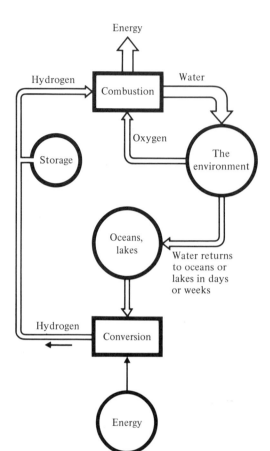

Fig. 18.2 *The hydrogen conversion and end-use cycle.*

the cheapest methods of energy transmission, and a gas delivery system is located underground where it is inconspicuous and safe. In the U.S. in 1970 there were 252,000 miles of trunk pipeline carrying a total of 22.4×10^{12} cubic feet of natural gas during the year. A similar pipeline system could be developed for hydrogen; or the existing system could be used, in part, for hydrogen. A typical pipe is 36 or 48 inches in diameter. Gas is pumped along the line by compressors spaced at 100-mile intervals. Pipelines operate at pressures of 600 psi (40 atmospheres) and are capable of power capacities of 10,000 megawatts, several times the capacity of an overhead electric-transmission line. A schematic diagram of a storage and transmission system for hydrogen is shown in Fig. 18.3.

The energy volume density of gaseous hydrogen is 325 Btu/ft^3, as shown in Table 18.3. By comparison, natural gas has a volume density of 1025 Btu/ft^3. Therefore, a pipe can handle the same energy per unit of time only if a flow rate for hydrogen is used that is three times that for natural gas. This can be achieved

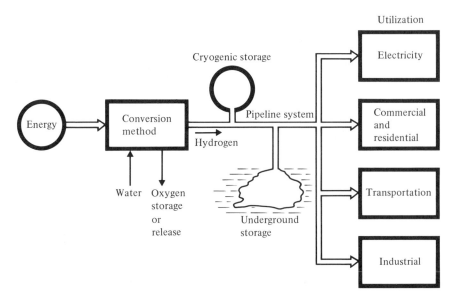

Fig. 18.3 *A storage and transmission system for hydrogen as an energy carrier.*

Table 18.3
CHARACTERISTICS OF ENERGY CARRIERS

Carrier characteristic	Hydrogen gas	Hydrogen liquid	Methanol	Gasoline	Methane
Energy volume density (Btu/ft³)	325	200,000	430,000	830,000	1,025
Energy per unit mass (Btu/lb)	58,000	58,000	9,200	19,400	23,000
Minimum ignition energy (mJ)	0.02	0.02	0.25	0.25	0.28
Ignition temperature (°K)	858	858	783	530	810
Flammability limits (% in air)	4.1–75	—	2.2–9.5	1.5–7.6	5.3–15

with a somewhat larger compressor. The estimated long-distance transportation costs, per 100 miles, to a local substation, including capital, operating, and maintenance costs, are given in Table 18.4. [1] Pipelines are clearly an attractive method of transportation of energy.

Hydrogen could be transported in liquid form, which it assumes at $-423°F$ ($-253°C$). This system would require continual cooling of the hydrogen. Liquid hydrogen is currently used for fuel in rockets for space ships. As can be noted

Table 18.4

ESTIMATED TRANSPORTATION COSTS PER
100 MILES OF THREE ENERGY CARRIERS

Transportation	Cost (Dollars/10^6 Btu)
Methane by pipeline	0.030
Hydrogen by pipeline	0.033
Electricity by high-voltage overhead transmission lines	0.21

in Table 18.3, hydrogen has a relatively high energy per unit mass, and thus is useful as a fuel for aircraft and rockets.

One serious problem associated with hydrogen is the potential embrittlement of metals and alloys used for its containment and transport, particularly pipelines. However, this effect can be decreased by the presence of relatively small concentrations of such impurities as oxygen, carbon dioxide, or water vapor. [3] Hydrogen pipelines have been in use in Texas and the Ruhr area of Germany for some time with no apparent degradation of the metals.

18.2 STORAGE OF HYDROGEN

The worldwide consumption of hydrogen in 1968 was 6.6 million tons, with 4.4 million tons consumed in the U.S. [3] Therefore, the U.S. has considerable experience with the storage, as well as the transportation, of hydrogen. With a power source such as the wind or solar energy, it is important to store the energy for use at times other than when the source is available. Hydrogen, as an energy carrier may be stored in the gaseous form in an underground space. Also, hydrogen may be stored in liquid form at cryogenic temperatures. Finally, hydrogen may be stored as a metal hydride. The cost of a storage facility for 10^9 Btu's of energy in the form of gaseous or liquid hydrogen might be $500,000 and $90,000, respectively.

Aquifers and underground rock formations are used for natural-gas storage and can be used for hydrogen, although the problems associated with the permeability of rocks are much more serious for hydrogen storage. Containers carrying up to 50,000 m^3 of liquefied hydrogen can be built without requiring any further basic research. Since gaseous hydrogen, however, requires much more storage space than natural gas, and liquid storage calls for costly cryogenic containers, alternatives are being investigated, including storage in the form of metal hydrides. While such metals as vanadium, niobium, and alloys such as the compound $LaNi_5$, when exposed to pressurized hydrogen, can store it efficiently, their cost is very high at present. Less expensive metals could make hydride storage feasible. New technology is required for storing small quantities of hydrogen; thus metal–air batteries, rechargeable with power from hydrogen, and titanium hydride storage in tanks are possibilities which have been considered.

The world's largest liquid-hydrogen storage tank at J. F. Kennedy Space Center has a capacity of 900,000 gallons equivalent to 37.7×10^9 Btu (11×10^6 kilowatt-hours). The energy content of this large tank is equal to 73 percent of the world's largest pumped-storage facility. [2]

18.3 HYDROGEN AS A FUEL

In Industry and the Home

H drogen can be readily used for space heating and, because it burns without n xious exhaust products, can be used in unvented appliances without hazard. Hence it is possible to conceive of a home heating furnace operating without a flue, thereby saving the cost of a chimney and adding as much as 30 percent to the efficiency of a gas-fired home heating system. More radical changes are possible, moreover, because without the need for a flue the concept of central heating itself is no longer necessary. Each room can have its heat supplied by unflued peripheral heating devices operating on hydrogen independently of one another. Indeed, the vented water vapor would provide beneficial humidification.

Another important possibility is the use of hydrogen for catalytic combustion. Therefore, flameless heating is possible for use in domestic heating and cooling facilities operating at temperatures as low as 100°C.

Hydrogen may be used in the chemical reduction of iron oxide ore to metallic iron, thus substituting for carbon or natural gas in the currently used process.

Ammonia, which serves as a nitrogen carrier in fertilizers, is presently made by the Haber-Bosch process, in which nitrogen and hydrogen react in the presence of catalysts. In the U.S. alone, 20 million tons of ammonia are produced each year; and 80 percent of the cost of manufacturing the ammonia is the cost of the hydrogen. [3] Therefore, as the demand for ammonia continues to grow, we will require an increasing source of relatively inexpensive hydrogen.

A number of additional uses for hydrogen are being explored. An Aphodial burner, for example, exploits the heat of combustion of hydrogen by burning the hydrogen with oxygen in the presence of water vapor rather than air. This device could be used in steam power plants.

Fuel cells, like batteries, transform the energy of a chemical reaction directly into electrical energy. One fuel-cell reaction is that of hydrogen with oxygen to produce water and electricity with an efficiency of 93 percent. Since fuel cells are efficient and virtually pollution-free, development for commercial applications should progress as hydrogen becomes inexpensive and readily available.

Hydrogen as an Automobile Fuel

One of the largest uses for hydrogen could be as a fuel for road vehicles. Practical experience has shown that hydrogen can be used as an auto fuel while requiring little alteration to the internal combustion engine on which so much development has been spent. The environmental pollution from hydrogen would be much less than from gasoline. Hydrogen can be used as a fuel for road transport in the liquid or gaseous form or as a metal hydride. Liquid or cryogenic hydrogen may be the best candidate. There are problems regarding the weight of the tank required to contain either the compressed gas or the liquid hydrogen. Nevertheless, the belief that cryogenic hydrogen can be handled safely and adequately is based on extensive industrial and research experience.

The lower ignition energy of hydrogen, compared to gasoline (see Table 18.3) may yield the advantage of requiring a less sophisticated ignition system for the modified auto engine. On the other hand, the lower ignition energy makes hydrogen less safe to handle since it can be ignited even by a static electricity discharge. The high diffusivity of hydrogen is considered a definite safety advantage since the consequent rapid dispersion of hydrogen into the atmosphere will greatly reduce the possibility of a combustible mixture.

France's Renault Company is working on the development of a car using a hydrogen fuel cell. Vehicles might also be powered by metal hydrides, which are solid compounds that decompose to give off hydrogen gas when they are heated. Less than 600 pounds of magnesium hydride, for example, contains the energy equivalent of 20 gallons of gasoline. Unfortunately, this weight of the hydride (600 lbs) is four times the weight of the 20 gallons of gasoline.

Another possibility is a combination of a cryogenic and solid hydride system. A cryogenic container stores liquid hydrogen, while the metal hydride is an additional source stored in the normal fuel-tank area. [5] Billings Research Corporation has built a prototype auto operating on hydrogen, and is using hydrogen at a fuel operating cost of 2.5 cents/mile, somewhat less than the fuel cost of gasoline. The Billings auto uses the cryogenic tank jointly with an iron–titanium hydride, which is heated by the coolant return from the engine.

The burning of hydrogen as fuel in an auto eliminates all pollutants except the troublesome oxides of nitrogen. At the high combustion temperature, some atmospheric nitrogen and oxygen combine to yield NO_x out of the exhaust, and NO_x acts as a catalyst to produce smog from burned hydrocarbons (produced by other vehicles).

Hydrogen is an excellent fuel for gas-turbine engines, and has been proposed as a fuel for supersonic jet transports. For this kind of use, fuel storage and tankage as liquid hydrogen are practical. Although the large volume required may make its use less attractive for subsonic aircraft, the very considerable saving in weight over an equivalent fuel load of kerosene gives hydrogen a distinct advantage. The range of aircraft would be increased two to three times for the same weight of fuel.

Methanol, a synthetic compound of hydrogen (CH_3OH) might also be used as a fuel for road vehicles. Methanol could be synthesized from hydrogen and then used as the fuel for autos and other road vehicles. Methanol, as may be noted in Table 18.3, has an energy volume density about 52 percent that of gasoline and much greater than that of liquid hydrogen and gaseous hydrogen. A methanol-powered car would have a clean exhaust characteristic. Methanol can be synthesized directly from carbon dioxide and hydrogen. The carbon dioxide could come from virtually unlimited supplies of limestone. Moreover, the use of methanol in autos with internal combustion engines may result in lower performance and poorer fuel economy than is obtained with gasoline. [6]

18.4 THE SAFETY OF USING HYDROGEN

The safe handling of hydrogen presents a challenge to technology to work out the methods necessary to utilize hydrogen for commercial, residential, and transportation uses. Many people remember May 6, 1937, when as the German airship *Hindenburg* was landing in Lakehurst, New Jersey, a random spark (or perhaps a saboteur) ignited a leaking hydrogen tank and the dirigible burst into a hovering ball of fire. The blaze killed 36 of the 97 passengers on board and set back the use of hydrogen for many decades. In this case, hydrogen was used as a buoyant, not as a fuel. Hydrogen has a relatively low ignition temperature, and a spark can ignite a leaking tank or gas bag such as that of the *Hindenburg*.

However, in open air and well-ventilated places, leaks or spills diffuse so rapidly (hydrogen being the lightest of all elements) that the risks of ignition or spreading flames are actually less than those for gasoline. As a consequence, hydrogen explosions are a rarity. For over a century before natural gas was widely distributed, cities and industries depended on so-called "town gas," manufactured from coal or coke, which contained up to 50 percent hydrogen, without danger to homes or industries. In general, hydrogen is hardly more hazardous than gasoline or even natural gas, although, having different characteristics, it requires different treatment. The basic requirements are adequate ventilation, leak prevention, and elimination of stray ignition sources. Today, liquid nitrogen can be stored and shipped around the country. What will be required will be the development of adequate safety methods for hydrogen during the next decade.

18.5 PRODUCTION OF HYDROGEN

Two processes are regarded as competitors for the method used to decompose water to obtain hydrogen. The first method, electrolysis, uses electric power generated by wind, geothermal, solar, or nuclear energy to decompose the water into hydrogen and oxygen. The second method uses high-temperature heat to directly decompose water, and the process is called *thermochemical decomposition*.

Electrolysis requires the step of electric production prior to the production of the hydrogen, and up to two thirds of the original energy is lost. Nevertheless, the process is presently utilized and could be expanded and used with the remote solar or wind generators, for example. Electrolysis of water involves the passage of an electric current through water by means of inert electrodes. As water is virtually a nonconductor, an electrolytic such as KOH is added. Equivalent amounts of oxygen and hydrogen are liberated at the anode and cathode, respectively. The gases may then be collected from the electrolytic cells. The chemical process is:

$$H_2O \rightarrow H_2 + \tfrac{1}{2}O_2 + 68.3 \text{ kilocalories/mole.} \qquad (18.1)$$

The process yields two gases and absorbs 68.3 kilocalories/mole of energy at standard conditions. The amount of energy required to decompose the water decreases with increasing temperature. Large electrolysis plants operate at 100 megawatts and an efficiency of 60 to 90 percent. It is possible to obtain even higher efficiencies; new electrolytes will be examined towards this end. The material used for the porous separators in the electrolytic cell prevents mixing of the two gases; and new materials may be examined for their beneficial advantages.

Electrolytic cells have a life of up to 25 years and need relatively low maintenance. Therefore, the electrolytic process is particularly adaptable to use at remote sites, such as wind-generator sites, for the generation of hydrogen. Electric power capacities of up to 1000 megawatts are envisioned for use with electrolytic plants. Each 1000-megawatt facility should be capable of producing about 150,000 tons per year of hydrogen. [4] Currently practical electrolytic plants produce about 2000 tons per year, so new developments will be required in order to increase the operating size of the plant.

An alternative to electrolysis is the closed-cycle thermochemical methods of directly dissociating water into hydrogen and oxygen. This process will be more efficient overall than the electrolytic process, since it does not require that the original energy be used to generate electricity (at an efficiency of 35 percent) and then the electricity be used to generate hydrogen. The thermochemical methods use the original energy, in the form of heat, to directly separate the hydrogen and oxygen. Using an undetermined chemical compound X, we examine the chemical reaction

$$H_2O + X \rightarrow XO + H_2, \qquad (18.2)$$

$$\text{Heat} + XO \rightarrow X + \tfrac{1}{2}O_2. \qquad (18.3)$$

This reaction takes place at some high temperature, and X is yet to be determined. At least ten research centers are exploring suitable thermochemical processes.

An example of one thermochemical process is based on the reactions of iron chlorides,

(1) $6FeCl_2 + 8H_2O \rightarrow 2Fe_3O_4 + 12HCl + 2H_2,$ (650°C)

(2) $2Fe_3O_4 + 3Cl_2 + 12HCl \rightarrow 6FeCl_3 + 6H_2O + O_2,$ (200°C) $\qquad (18.4)$

(3) $6FeCl_3 \rightarrow 6FeCl_2 + 3Cl_2,$ (420°C)

where each reaction takes place at the temperature shown.

The efficiency of a thermochemical reaction can approach the theoretical Carnot efficiency. Since conventional steam electric-generation plants operate

not on a Carnot cycle but on the less efficient Rankine cycle, it is conceivable
that a thermochemical cycle could more closely approach the efficiency of the
limiting Carnot cycle than conventional cycles do. The theoretical efficiency is:

$$\eta = \frac{\Delta H}{\Delta G} \frac{T_2 - T_1}{T_2}, \tag{18.5}$$

where ΔH and ΔG are, respectively, the heat and work for the dissociation of
water at 25°C and one atmosphere. For water, we have

$$\eta = \frac{1.2(T_2 - T_1)}{T_2}; \tag{18.6}$$

and therefore the maximum thermal efficiency is 20 percent higher than the
Carnot efficiency of a heat engine.

The major technical problems of the closed-cycle systems are corrosion and
materials compatibility, reaction kinetics, and the development of high-efficiency
heat exchangers. For example, consider the following cycle, designated as the
Mark I cycle:

$$\left. \begin{array}{l} CaBr_2 + 2H_2O \xrightarrow{730°C} Ca(OH)_2 + 2HBr, \\ Hg + 2HBr \xrightarrow{250°C} HgBr_2 + H_2, \\ HgBr_2 + Ca(OH)_2 \xrightarrow{200°C} CaBr_2 + HgO + H_2O, \\ HgO \xrightarrow{600°C} Hg + \tfrac{1}{2}O_2. \end{array} \right\} \tag{18.7}$$

The maximum temperature required is only 730°C, which is lower than the temperature of a nuclear-reactor coolant. However, the use of mercury and corrosive
HBr presents a significant potential hazard and control problem.

Commercial procedures for the thermochemical decomposition of water are
not yet available. However, even a new lower-efficiency process may still have an
economic advantage over a conventional process if the capital costs of the plant
are lower. Such reduced capital costs are a possibility, since a plant involving
chemical processes only may well be less expensive than one employing electrical
generation and electrolytic equipment. Because of the large scale of required
hydrogen production, even a small relative cost reduction could bring enormous
dollar savings.

18.6   THE ECONOMICS OF HYDROGEN PRODUCTION

The costs of producing, storing and transporting hydrogen at the levels of production envisioned for the 21st century are difficult to assess, since the large scale of
the system will yield favorable cost reductions over present methods. The production of hydrogen by means of the electrolytic process is currently about \$1.03/
10^6 Btu, including the cost of the fuel. The cost of transporting hydrogen via a

pipeline can be estimated to be \$0.001/kilowatt-hour (\$0.29/10^6 Btu) per 1609 kilometers of pipeline. Considering average lengths of pipeline to customer, we might expect a transmission and distribution cost of \$0.95/10^6 Btu. [1, 7] This compares with the transmission and distribution cost of electricity, which is \$2.85/10^6 Btu for a transmission distance of 200 miles (322 km).

The cost of hydrogen is dependent upon the cost of the energy used to decompose the water. Winsche, *et al.*, analyzed the cost of delivering hydrogen to residences for space heating, air conditioning, water heating, and for the automobile, for seven different possible initial energy sources. In the case of using a nuclear power plant to generate electricity to yield hydrogen, they conclude that the total annual cost is \$1081 for supplying one household with 165×10^6 Btu's per year via an underground distribution system. The total system consumes 810×10^6 Btu's as original energy and, therefore, has an overall efficiency of 20 percent. This system uses a fuel cell to serve as an energy storage device for the automobile. The cost to the customer is then \$1081/165×10^6 Btu, or \$6.55/10^6 Btu. The data for the seven systems, including a reference electric system, show that the delivery of hydrogen energy is economically competitive with conventional energy carrier systems.

The cost of producing, transmitting, and distributing hydrogen generated by means of using ocean thermal gradients or wind power could be approximately \$2.30/10^6 Btu. [8]

Even if the hydrogen must be converted to a more easily stored fuel for automotive use, such as methanol, ammonia, or even propane, some form of hydrogen energy distribution system is an attractive alternative which should be examined in more detail.

REFERENCES

1. W. E. WINSCHE, K. C. HOFFMAN, AND F. J. SALZANO, "Hydrogen: Its future role in the nation's energy economy," *Science*, 29 June 1973, pp. 1325–1332.

2. D. P. GREGORY, "The hydrogen economy," *Scientific American*, January 1973, pp. 13–21.

3. C. E. BAMBERGER AND J. BRAUNSTEIN, "Hydrogen: A versatile element," *American Scientist*, July-August, 1975, pp. 438–447.

4. J. K. DAWSON, "Prospects for hydrogen as an energy resource," *Nature*, June 21, 1974, pp. 724–726.

5. S. WALTERS, "Hydrogen age rolls forward," *Mechanical Engineering*, March 1974, pp. 40–41.

6. N. VALERY, "The best substitute for petrol may be petrol," *New Scientist*, 24 Jan. 1974, pp. 203–205.

7. T. H. MAUGH, "Hydrogen: Synthetic fuel of the future," *Science*, Nov. 24, 1972, pp. 849–852.

8. A. LAVI, C. ZENER, "Plumbing the ocean depths: A new source of power," *IEEE Spectrum*, Oct. 1973, pp. 22–27.

EXERCISES

18.1 Compare electricity and hydrogen as energy carriers in light of the desired characteristics of a carrier as listed in Table 18.1, for the present time and the expected comparison in 1985.

18.2 Investigate the safety and cost of distributing gaseous or liquid hydrogen. Prepare a position paper for your state legislature on whether your state should permit a full-scale system using hydrogen as an energy carrier.

18.3 In what form can hydrogen be used as a fuel for an automobile? Would you recommend that major auto companies design some autos to use hydrogen power? State your reasons for your answers. Contact the auto manufacturers and determine what their position is on the use of hydrogen as an auto fuel.

18.4 A recent movie entitled "Hindenburg" graphically portrays the disastrous burning of the dirigible *Hindenburg*. Discuss the impact of that film and other knowledge of the dirigible tragedy on the future use of hydrogen in industry and for transportation fuel.

*__18.5__ What is the energy-per-unit-weight advantage of liquid hydrogen over gasoline? For what type of transport is this advantage particularly useful?

---

*An asterisk indicates a relatively difficult or advanced exercise.

Alternative Conversion and Storage Systems

Alternative energy conversion and storage systems may emerge by the end of this century. While it is difficult to predict at this time what may emerge several decades from now, one can envision several new possible approaches.

19.1 BIOMASS FUEL SYSTEMS

One source of fuel is organic biological materials such as agricultural crops, grown specifically to serve as a feedstock, to be converted to methane, hydrogen, electricity, or another energy carrier.

Photosynthesis is a mechanism for converting solar energy into the form of plant material. Photosynthesis and plant growth can be used as a step in a solar-energy converter. Once converted into organic plant material, the solar energy can be converted to heat directly, or to an economically transportable fuel (solid, liquid, or gaseous) by chemical, thermochemical, or biological processes. Thus, photosynthesis serves as a means of collecting solar energy and converting it to a useful fuel material. The efficiency of conversion from solar energy to plant material is about 3 to 5 percent, with an upper limit of about 11 percent attainable. [1]

The productivity of a crop varies with geographical location. The productivity of several crops is given in Table 19.1. Many crops are candidates for use as fuels in energy-conversion systems. [12]

Wood as a Fuel

Wood was used as a fuel by people for centuries prior to the introduction of coal and oil within the past 200 years. The possibility of cultivating and harvesting large forests and converting the energy to electricity or another storage carrier

Table 19.1
PLANT PRODUCTIVITY

Location	Plant	Annual production (Metric tons/acre)*
California	Algae	20–30
Germany	Reedswamp	18.6
Mississippi	Water hyacinth	4.5–13.4
England	Coniferous forest	13.8
West Indies	Tropical forest	23.9
Minnesota	Maize	9.7
Hawaii	Sugar cane	30.4
Georgia	Sycamore	11–16

* One acre = 0.405 hectare.

is attracting interest. Forests may yield 10 to 20 metric tons/acre per year, and the period over which these average rates are obtained ranges from 20 to 50 years.

One concept incorporates a plan for large tree farms, covering areas from 50 to 200 square miles, where trees are grown and harvested by means of automated methods. At the center of the tree plantation would be a wood-burning electrical power plant. The wood would be chipped in the field, then crushed to the consistency of powder before being burned.

Forests of sycamore trees may yield up to 16 metric tons/acre per year. This is possible with new methods such as by harvesting sycamore "shoots" only two to four years old. Almost everything from trunks to twigs is used. Only the foliage, which contains most of the plant nutrients, is returned to the soil. Once harvested, the perennial sycamore sends out a new crop of sprouts, which are again ready for harvesting in two or three years. It is not necessary to replant, and little fertilizer is required.

Several advantages accompany the use of wood as a fuel. These include the following:

a) Trees improve the oxygen–carbon dioxide balance in the atmosphere;
b) They reduce runoff of soils and improve stream and river clarity;
c) They are renewable;
d) They can be grown and harvested with virtually no depletion of soil nutrients; and
e) The wood ash from power plants would provide a valuable fertilizer.

A tree farm of about 350 square miles would be required to produce enough wood to fuel a 400-megawatt electric generating plant. [2] It is estimated that electricity produced on such a farm would cost about $3/million Btu, which is about 1.5 times the cost of the energy of petroleum in 1977 in the U.S. The capital costs of developing an automated tree farm and associated power plant may amount to $1000 per kilowatt, which is competitive with a nuclear power plant's construction cost.

However, when one examines the tree-farm concept, several advantages are noted:

1. The fuel energy is stored in the trees in readily usable form;
2. The fuel contains less than 0.1 percent sulfur; and
3. The long-range ecological implications are minimal.

Also, it is important to note that wood burning boilers that generate steam for electric power plants are commercially available.

Algae and Plants

Algae, tropical grasses, water hyacinth, and other plants also may serve as fuel for energy plants. It has been estimated that giant floating kelp could yield methane or liquid natural gas at a cost of $2.50 per million Btu. [3] Another possibility is to use other aquatic plants such as pond algae and other seaweeds. If such plants were grown in controlled circumstances, yields up to several hundred tons per acre per year are possible.

The waste heat from the electric power plant could be cycled into the algae pond to maintain the growth process. A typical algae farm might involve an acre of land excavated to a depth of one and a half feet and flooded. Algae would be introduced, the pond would be covered with plastic, carbon dioxide would be injected, and a temperature of 100 degrees Fahrenheit would be maintained by using the power-plant cooling water. As the algae are extracted from the pond, the water used for growth would be returned, together with a small amount of the algae as "seed" to restart the process. The closed-loop cycle would be nearly self-perpetuating and economical, and could be installed virtually anywhere in the world.

19.2   MAGNETOHYDRODYNAMICS

Magnetohydrodynamics (MHD) is the study of the flow of a conducting fluid in a magnetic field. MHD using coal as a fuel was briefly discussed in Chapter 6. The use of MHD electric-power plants is possible by the end of this century if several technical problems can be solved. The process relies on the passage of a hot, ionized gas under high pressure through a magnetic coil. The source of the plasma gas can be petroleum, coal, or natural gas. The Russians are studying natural gas units, while the U.S. is exploring coal-fired units. The fuel is subjected to heat at a temperature of 2500°C; it ionizes and is capable of carrying an electric current. With such a high temperature, a theoretical efficiency of 80 percent could be achieved. In practice, an efficiency of 60 percent is expected, which is almost twice that normally achieved in a fossil-fuel boiler plant. [11]

The ionized gas consists of positive and negative ions and electrons, but is neutral overall and is called a *plasma*. There is a force acting on each particle, represented by

$$F = q\mathbf{v} \times \mathbf{B}, \tag{19.1}$$

where q is the charge, $\mathbf{v}$ is the velocity of the particle, and $\mathbf{B}$ is the magnetic field. The direction of the force depends upon the sign of the particle; collector plates are arranged to collect the charges, yielding a current. A seed material, such as cesium or potassium, is used to enhance the ionization process. This seed material is collected before the gases are exhausted. At the same time, air pollutants such as particulates and sulfur oxides are removed. The seed material reacts with the sulfur, improving its removal as a compound. It is possible to use gas turbines in line after the MHD generator in order to utilize all the energy. A schematic of an electric power plant using MHD and a steam-turbine cycle, called the open-cycle system, is shown in Fig. 19.1.

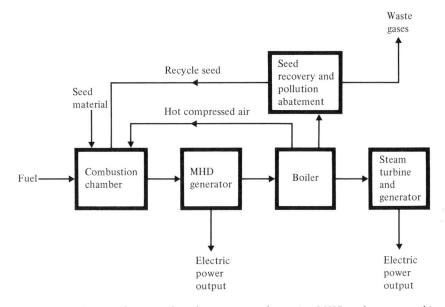

Fig. 19.1 *A schematic diagram of an electric power plant using MHD and a steam turbine.*

Technical problems include corrosion of the walls of the MHD device due to erosion caused by the rapidly moving ions. Pilot plants have running times of only a few hours at present. The world's largest MHD plant is the Soviet U-25 designed for an ultimate capacity of 25 megawatts. Thus far, U-25 has been operated only a few hours, at four megawatts or less. Also, engineers are planning to run the MHD generator using a superconducting magnet.

An alternative MHD generator, called a closed-cycle system, recirculates the hot gases within the MHD unit and does not use an additional turbine generator.

This system operates at about 1650°C, and thus avoids a need to develop new high-temperature materials to withstand the hot gases characteristic of the open-cycle system.

The desired advantages of the MHD system are primarily the increase in system efficiency by a factor of two and the ready control of pollutants. This approach to generating electricity may be realized in practice by the year 2000.

19.3 FUEL CELLS

The fuel cell was first discovered by Sir William Grove in 1839, but the first practical cell was not demonstrated until 120 years later. [4] While fuel cells are used in space vehicles for electric power, they have not been widely used due to high costs.

A single fuel cell is a single set of two electrodes joined by an electrolyte which conducts charge between the electrodes and an external circuit. One electrode is the fuel electrode (or anode) and the other is called the oxident or air electrode (cathode). In a typical fuel cell, hydrogen is fed to the anode where it is catalytically converted to hydrogen ions, releasing electrons to an external circuit. At the cathode, these electrons reduce oxygen to ions which then migrate through the electrolyte and combine with hydrogen ions to form water. This process is effectively the reverse of electrolysis. Fuel cells can use a wide variety of catalysts and fuels. The schematic diagram of a fuel cell is shown in Fig. 19.2. The process at the anode is:

$$H_2 \rightarrow 2H^+ + 2e^-. \tag{19.2}$$

The electrons pass through the external circuit while the H^+ ions move to the cathode. The reaction at the cathode is:

$$2H^+ + 2e^- + \tfrac{1}{2}O_2 \rightarrow H_2O. \tag{19.3}$$

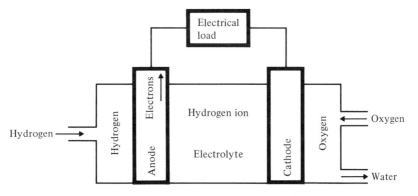

Fig. 19.2 *A schematic diagram of a hydrogen/oxygen fuel cell. Hydrogen is fed to the anode where it forms hydrogen ions, H^+, by giving up electrons that travel through the external circuit. At the cathode oxygen combines with the hydrogen ions and electrons to form water.*

The overall reaction is:

$$H_2 + \tfrac{1}{2}O_2 \rightarrow H_2O - Q, \qquad (19.4)$$

where H_2 and O_2 are input gases. The output is water (liquid) at $300°K$, and the heat energy, Q, released is 68.3 kilocalorie per mole.

Fuel cells can theoretically approach an efficiency of 100 percent in converting chemical energy to electrical energy. The efficiency of a fuel cell may be defined as:

$$\eta = \frac{\text{Useful work}}{\text{Heat released}} = \frac{\text{Gibbs free-energy change}}{Q}$$

$$= \frac{\Delta F}{Q}. \qquad (19.5)$$

For a hydrogen/oxygen cell, we have

$$\eta = \frac{\Delta F}{Q} \equiv \frac{56.62 \text{ kcal/mole}}{68.30 \text{ kcal/mole}} = 0.83. \qquad (19.6)$$

For a carbon monoxide gas and oxygen cell, the overall reaction at $300°C$ is:

$$CO + \tfrac{1}{2}O_2 \rightarrow CO_2 - Q, \qquad (19.7)$$

and carbon dioxide is taken off. The efficiency is:

$$\eta = \frac{\Delta F}{Q} = \frac{61.41}{67.64} = 0.91. \qquad (19.8)$$

In actual fuel cells, energy losses occur during the process, and overall losses may approach 25 percent during the operation. Thus the overall efficiency of a hydrogen/oxygen fuel cell is usually about 60 percent. This is a significant improvement over the efficiency of a thermal–fossil-fuel system with an efficiency of 35 to 40 percent.

Each single fuel cell has a potential difference of about one volt so many cells are placed in series to yield a larger voltage and power output. Using 1000 cells would yield a 1000-volt array with a power output of 10 megawatts if each cell had a cross-sectional area of five square meters.

Fuel cells are particularly appropriate for use as local energy sources since they do not require transmission lines for power transportation. Fuel cells can be located in neighborhoods or districts of cities and used to supply peak power at the necessary periods in the day. Fuel cells retain their higher efficiency over the power range of 10 kilowatts to 100 megawatts, and thus are more useful, particularly at the lower power levels, than low-efficiency gas-turbine systems or diesel electric generators for peak electric supply.

Fuel cells can be placed quickly anywhere without disruption to the environment and may be used to supply new growing areas of cities. Also, if hydrogen is generated by windmills, or solar energy, it can be transported to cities and used in the fuel cells. Also, hydrogen could be generated by offpeak electric power from hydroelectric or other types of plants.

The only waste products from fuel cells are water and heat, thus resulting in minimal environmental impact. Furthermore, the heat could be used for heating homes and water.

It is possible to use natural gas as the fuel, rather than hydrogen, but that possibility is counterbalanced by the growing shortage of natural gas in the U.S. This possibility would be very real in nations such as Algeria or in the Middle East, where natural-gas fuel cells operating at 50 percent efficiency are a significant improvement over thermal-boiler electric generators, which operate at an efficiency of 35 percent.

The economics of fuel cells remain to be developed to an attractive level. The present capital cost of fuel cells is about $500 per kilowatt, and they usually have an operating lifetime of several years. It will be necessary to decrease the capital, the fuel, and other operating costs of fuel cells and significantly increase their lifetime before they become useful for our electric utility systems. Perhaps that will occur by the end of this century, particularly if hydrogen becomes readily available as a fuel.

19.4   ENERGY STORAGE SYSTEMS

In earlier chapters we considered the pumped-storage concept and the generation of hydrogen to store energy converted from other sources. Energy storage involves the collection and retention of readily available energy for later use. Energy available at periods of low demand should be stored for times of peak demand. Often the peak power demand of an electric distribution system can be twice the minimum level of power demand. Thus, storage of hydroelectric, wind-energy, or solar-energy will be necessary in the last decades of this century.

In Chapter 11 we considered pumped storage for use with hydroelectric schemes. In Chapter 18 we discussed the use of hydrogen as a storage material for energy harnessed by windmills, tidal plants, or solar devices. In the following sections, we shall consider the usefulness of batteries, compressed air, and flywheels as storage devices. While the only currently economic means of storing significant amounts of energy is the use of pumped storage, we can expect new storage schemes to emerge as economically feasible approaches by 1990.

Storage Batteries

Storage batteries may reach economical use for the storage of large amounts of energy within the next 20 years. Batteries have the advantage of minimal siting problems and can be placed near to the anticipated load, thus reducing transmission costs. Efforts are needed to produce a battery with a specific energy of over 220 watt-hours per kilogram (100 watt-hrs/pound) and a specific power of over 55 watts per kilogram (25 watts/pound). In addition, a lifetime of four years or more and a life of 1000 cycles is desired. [4, 10] Furthermore, the efficiency of storage and discharge should be relatively high—normally more than 70 percent in total. Unfortunately, no presently available storage batteries meet these performance standards; and it may require years of research and development to obtain such batteries.

The common lead–acid battery used for automobiles is not designed for electric utility use, and would have a short life under rigid charging cycle requirements. The lead–acid battery can store only 22 to 33 watt-hours/kilogram (10 to 15 watt-hours/pound), and is capable of only 300 recharging cycles. This battery, originally devised in 1859, goes through the following chemical reaction:

$$PbO_2 + Pb + 2H_2SO_4 \underset{\overset{\longleftarrow}{\text{Charge}}}{\overset{\overset{\longrightarrow}{\text{Discharge}}}{\rightleftarrows}} 2PbSO_4 + 2H_2O. \qquad (19.9)$$

A storage battery designed for commercial use should have a high energy density, durability, and low production cost. Extensive work on candidates for commercially usable batteries is progressing steadily.

Nickel–iron and nickel–cadmium batteries have been studied, but they involve high costs. Silver–zinc batteries have an energy density range of 60 to 180 watt-hour/kilogram and a relatively limited number of cycles. However, the cost of silver makes silver–zinc and silver–cadmium batteries uneconomical.

Zinc–air storage batteries have an energy-density range of 130 to 200 watt-hour/kilogram, and are potentially economical to manufacture in quantity.

Organic-electrolyte storage batteries use lithium or sodium as one electrode. Lithium–sulfur batteries operate at 370°C, as one example. The cells will consist of positive electrodes of iron sulfide (either FeS or FeS_2), negative electrodes of a solid lithium–aluminum alloy, and an electrolyte of a molten salt (a mixture of lithium chloride and potassium chloride). Because the salt has a melting point of 352°C, the cells must be operated at temperatures of about 375°C. Test cells have yielded 150 watt-hours/kilogram. Ultimately these cells may have a lifetime of over five years and yield 1500 charge–discharge cycles.

Another battery that can operate at higher temperatures is the sodium–sulfur battery, which operates at 300°C. The electrolyte is a solid ceramic material called beta alumina. This battery may store 200 to 400 watt-hours/kilogram. The reactants of the battery are a liquid-sodium center core and a conducting sulfur electrode. The device is capable of a power density of 220 watts per kilogram, and it is hoped that it can be produced for about $2 per pound, or about $20 per kilowatt-hour of stored energy. Problems of corrosion of the containment vessel remain to be solved, as well as the development of a commercial version of this battery.

It is expected to be several decades before storage batteries are commercially available for energy storage, for use in electric vehicles, and for peak-power utilization by electric utilities.

Compressed-Air Storage

A compressed-air storage system uses an air compressor to place air in an underground cavity or a storage tank. When the stored energy is required, the compressed air is used to drive a turbine and an electric generator. During offpeak hours, energy is used to drive the pump and store air in the cavern or tank. Air can be stored in salt domes, mine caverns, depleted gas fields, abandoned mines, or an aquifer.

The overall efficiency of a compressed-air storage system is estimated to be 70 percent. The schematic of a basic compressed-air energy-storage system is shown in Fig. 19.3. During offpeak hours, electricity is fed to the motor which drives the compressor, thus sending air to the storage vessel. At times of peak demand, the air is released to the turbine generator to yield electric power out to the consumer.

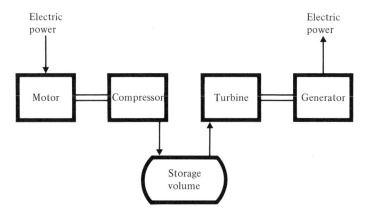

Fig. 19.3   *The basic schematic diagram of a constant-volume compressed-air energy-storage system.*

The use of underground, porous, water-filled structures, called aquifers, is being investigated for compressed-air storage. As air is pumped in, the water is displaced, and the reverse occurs in use of the air through the turbine.

The compressed-air storage concept has not yet been put into practice. In calculating the economics of such systems, the cost of developing the cavern and drilling the connecting well would have to be accounted for. Nevertheless, the compressed-air system may have an efficiency of about 75 percent compared to 65 percent for a pumped-storage system. Also, if an existing mine or aquifer is used, the capital costs of the air system may be significantly lower than that of a pumped-storage system.

Inertial-Energy Storage

The flywheel, which is one of the oldest of human inventions, is an inertial storage device. The flywheel offers the potential for storing energy on a large scale to handle the peak loads of utilities and to power electric vehicles. The principle of a flywheel is that a spinning wheel stores mechanical energy according to the equation

$$E = \tfrac{1}{2}I\omega^2, \tag{19.10}$$

where I is the moment of inertia and ω the angular velocity in radians/second. Flywheels are used to turn potter's wheels and in automobiles to smooth out the fluctuations of the engine speed.

One cannot store an unlimited amount of energy in a flywheel, since the energy storage is ultimately limited by the strength of the flywheel material. In Table 19.2 the characteristics of several energy-storage devices are listed. For a maraging* steel flywheel, a reasonably high energy density is achieved compared to a lead–acid battery. A flywheel vehicle was used in Switzerland from 1952 to 1969 which used a conventional 3300-pound steel wheel to store 6.6 watt-hour/kg. An electric motor was used to store energy in the flywheel during bus stops. Between stops, the flywheel turned a generator which fed the drive motor of the bus. [5]

Table 19.2
ENERGY-STORAGE DEVICES

Phenomena	Energy density (Watt-hr/kg)	Power density (Watts/kg)	Discharge cycle Life (Cycles)
Air compression	7.7	10^4	10^7
Lead–acid battery	17.8	80	500
Nickel-cadmium battery	30.6	80	2000
4340 steel flywheel	33.2	10^4	10^5
Maraging steel flywheel	55.0	10^4	10^5

The moment of inertia of a flywheel can be written as

$$I = cMR^2, \tag{19.11}$$

where M is the total mass and R is a linear dimension such as the radius of the disk. For example, $c = \tfrac{1}{2}$ for a solid disk and $c = 1$ for a constant-stress tapered disk. The maximum stress force on a flywheel disk is at the center. Relating the tensile strength of the disk to the maximum angular velocity, we have

$$T = \frac{\text{Force}}{\text{Area}} = \tfrac{1}{2}\rho\omega_{\text{max}}^2 R^2, \tag{19.12}$$

where $\rho = $ mass density. Solving for ω_{max} we have

$$\omega_{\text{max}} = \left(\frac{2T}{\rho R^2}\right)^{1/2}. \tag{19.13}$$

---

* Maraging is a heat treatment for steel that yields a high-strength, tough steel.

Therefore the maximum energy stored in a given flywheel is

$$E_{max} = \tfrac{1}{2}I\omega_{max}^2$$

$$= \tfrac{1}{2}(cMR^2)\left(\frac{2T}{\rho R^2}\right)$$

$$= \frac{cMT}{\rho}. \tag{19.14}$$

Since $M = \rho V$ we have

$$E_{max} = cVT. \tag{19.15}$$

Thus, the maximum energy stored depends upon the volume of the wheel and the tensile strength, so high-strength, low-density materials are necessary for a flywheel. Table 19.3 lists the strength and density characteristics of several materials. [6] Newly available materials, such as PRD-49 and fused silica, provide the potential for flywheels with large energy-storage density. Practical usable flywheels may have actual maximum energy-storage characteristics about one-half of those listed in Table 19.3.

Table 19.3
MAXIMUM ENERGY STORAGE OF A FLYWHEEL

Material	Tensile strength $(10^6 \ kgs/m^2)$	Density (gms/cm^3)	Maximum energy storage $(Watt\text{-}hrs/kg)$
Maraging steel	281	8.0	55
E glass	337	2.5	190
PRD-49	359	1.5	350
Fused silica	1406	2.1	870

The fiber materials such as E-glass, PRD-49′ or fused silica offer the characteristics desired in an energy-storage flywheel. E-glass is commercially available, and PRD is used for automobile tires and aerospace applications. Fused-silica fibers are still to be made available commercially, but they offer the greatest potential for a flywheel material.

Since the strength of fiber composites derives from the strength of the individual fibers, they will develop maximum strength in tension when all the fibers lie parallel to one another, lined up in the direction of the applied tensile force. Thus the idea arises of using a rotor which consists of many spokes. Each spoke is a bar or rod fabricated of composite fiber material with the direction of the fibers being parallel to the length of the rod. It is also possible to construct a flywheel by assembling several rings concentrically. [6] The rotor consists of a group of concentric fiberglass-reinforced epoxy rings supported by flexible spacers. The reinforcing fibers in the rings run continuously in the circumferential direction. By limiting the radial thickness of each ring, the radial stress is held within the comparatively low strength of the epoxy matrix material, allowing full use to be made of the strength of the fiberglass reinforcement without failure in radial tension.

Fiber flywheels have the additional advantage that if they fail, they merely shred or turn to powder, in contrast to the failure of steel flywheels, which causes pieces of steel to fly off at high speeds.

A flywheel device for storing energy for an electric utility might be 15 feet in diameter and weigh from 100 to 200 tons. Each unit could store 10 megawatt-hours of energy at a rotation speed of 3500 revolutions/minute. Such a system might be able to yield 3 megawatts of power at an efficiency of more than 90 percent, at a capital cost of $110/kW. [6]

Flywheels could supply power to consumers at peak periods of demand with negligible environmental impact. Flywheel storage systems can be made to store any amount of energy by locating flywheels at dispersed locations near consumer neighborhoods.

Flywheels can also be used to power electric vehicles. A flywheel of PRD-49 weighing 600 kilograms (270 lbs) could store about 100 kWh of energy. This would be sufficient to provide a vehicle with a range of 200 miles at 55 miles per hour, with full power to accelerate on a hill.

Flywheels constructed of fiber materials are not currently available; it may be another decade before they are proven commercially and economically.

19.5   TOTAL ENERGY SYSTEMS

Within the next 10 years, total energy systems may be developed and implemented on a sufficiently wide basis to become an important part of the U.S. energy system. *Total energy* or *integrated utility systems* are combined processing plants that generate electricity; use residual and recycled energy for heating, air conditioning, and hot water; treat water; process solid wastes, and treat liquid wastes. These systems are often called cogeneration systems.

Normally about 65 percent of the fuel energy content is wasted in the generation of electricity. In an integrated system, better than half of this "waste" energy can be recovered and used for space heating, air conditioning, domestic hot water, water and liquid waste treatment, etc. Utilizing thermal energy in this manner, as a substitute for electrical energy, results in major reductions in fuel requirements and attendant reductions in combustion products and thermal pollution.

The total or integrated system is located near the ultimate consumer in order to minimize the distribution costs and attendant energy losses.

The total system is an integrated modular system providing the five necessary utility services for community development: electricity; environmental conditioning; solid-waste processing; liquid-waste processing; domestic water.

A total energy system addresses the following objectives:

a) conserve natural resources;
b) reduce energy consumption;
c) minimize environmental impact.

In addition, it should:

d) be installed on a schedule consistent with development or redevelopment of a community;
e) eliminate impact of local restrictions on waste treatment which act as a deterrent to the construction of housing for low- and moderate-income families;
f) provide a transportable system for emergency operations (flood relief, earthquakes, etc.);
g) reduce total cost to the nation.

It has been estimated that by 1980, 15 quads per year of waste heat will be discharged to the environment. A total energy system attempts to utilize this heat energy for heating and other purposes.

One possible schematic of a total energy scheme is shown in Fig. 19.4. The products of this system are: (1) electric power, (2) heating steam and domestic hot water, (3) cooling water for air conditioning, and (4) desalted drinking water.

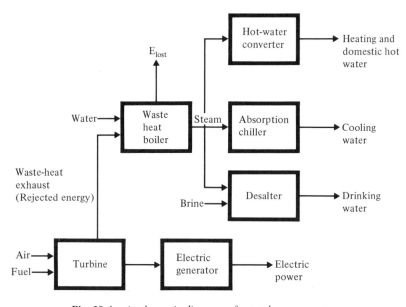

Fig. 19.4 *A schematic diagram of a total energy system.*

In a standard electric-power generating system, only 35 percent of the energy is converted to electric energy, and 10 percent of that energy is lost over transmission lines. If locally placed total systems can use the waste heat (normally unused thermal energy), then the net loss of heat is reduced significantly. The energy rejected, E_R, in a power plant is:

$$E_R = (1 - \eta)E_{in}, \tag{19.16}$$

where E_{in} = the input energy to the plant and η = efficiency of the turbine generator. Typically, $\eta = 0.33$ or

$$E_R = 0.67E_{in}. \tag{19.17}$$

However, if we can use one-half of the rejected energy from the turbine for other uses, as shown in Fig. 19.4, the ultimate loss of energy, E_{lost}, is then:

$$E_{lost} = \tfrac{1}{2}E_R$$
$$= \tfrac{1}{2}(1 - \eta)E_{in}. \tag{19.18}$$

When $\eta = 0.33$ and we define the overall system efficiency as E_{system}, we have

$$E_{system} = \frac{E_{in} - E_{lost}}{E_{in}}$$
$$= 1 - \tfrac{1}{2}(1 - \eta)$$
$$= 1 - 0.335$$
$$= 0.665. \tag{19.19}$$

An overall efficiency of 67 percent, compared to a turbine efficiency of only 33 percent, is a significant improvement indeed and worth pursuing! Depending upon the designer's ability to use the rejected heat, one can expect a system efficiency in the range of 60 to 80 percent. Thus, with a total energy system, one can achieve a savings of energy and thus reduce operating costs and lessen environmental pollution. [7, 14]

The rejected heat from a total energy system can also be used agriculturally, for soil heating, for environmental control of animal shelters and greenhouses, and for aquaculture. In combination with sewage disposal, waste heat can be used to provide an enhanced environment for agricultural crops.

Thus, it would appear that the disposal of sewage waste-water effluent on land areas and dissipation of waste heat from power plants could be complementary to each other in solving two environmental pollution problems—namely, nutrient pollution and thermal pollution of our surface waters. At the same time, the reuse of two wastes—sewage waste and water effluent, and power plant thermal wastes—would be accomplished.

While total energy systems are several years from wide use, they hold significant promise for an increase in total system efficiency. Therefore, total energy systems may become commonly used for neighborhoods or by industries or colleges within this century.

REFERENCES

1. T. R. SCHNEIDER, "Efficiency of photosynthesis as a solar energy converter," *Energy Conversion*, Vol. 13, 1973, pp. 77–85.

2. G. C. SZEGO AND C. C. KEMP, "Energy forests and fuel plantations," *Chemtech*, May 1973, pp. 275–285.

3. "Fuel processes must consider energy costs," *Chemical and Engineering News*, June 2, 1975, pp. 18–19.

4. A. L. ROBINSON, "Energy storage: Using electricity more efficiently," *Science*, May 17, 1974, pp. 785–787.

5. A. L. ROBINSON, "Energy storage: Developing advance technologies," *Science*, May 24, 1974, pp. 884–887.

6. R. F. Post and S. F. Post, "Flywheels," *Scientific American*, Dec. 1973, pp. 17–23.

7. R. MACKAY, "Generating power at high efficiency," *Power*, June 1975, pp. 87–89.

8. "Total energy," Report of the Educational Facilities Laboratories, New York, 1970.

9. H. C. HOTEL AND J. B. HOWARD, *New Energy Technology*, M.I.T. Press, Cambridge, Mass, 1971, Chapter 5.

10. A. L. ROBINSON, "Advanced storage batteries," *Science*, May 7, 1976, pp. 541–543.

11. S. W. ANGRIST, *Direct Energy Conversion*, 3rd Ed., Addison-Wesley, Reading, Mass, 1976.

12. W. G. POLLARD, "The long-range prospects for solar-derived fuels," *American Scientist*, Oct. 1976, pp. 509–513.

13. J. P. HARTNETT, *Alternative Energy Sources*, Academic Press New York, 1976.

14. G. W. NEAL, "Advantages of combined power/process generating plants," *Power Engineering*, February, 1977, pp. 56–59.

EXERCISES

19.1 Which agricultural crop has a high productivity yield in your state? Is this crop a good candidate for a fuel in an energy conversion system? Sketch a system for use of the fuel and outline the design of the system. Determine what size farm would be necessary to yield a crop sufficient for conversion to electric power for your college or city.

19.2 Two research laboratories that are aggressively pursuing MHD power are the AVCO Company near Boston, Mass., and Stanford Univ., Stanford, Calif. Obtain information from these laboratories regarding their progress on MHD.

*19.3 Obstacles to the development of commercial MHD generators are (1) the need for high-efficiency seed recovery and (2) corrosion of the MHD vessel. [9] Examine these factors and determine whether it is feasible to overcome these obstacles during the next decade. What steps of research would you recommend?

19.4 Since fuel cells have efficiencies of 60 to 70%, why are they not widely used? What steps can you suggest that would lead to the wide use of fuel cells for a local energy source?

19.5 Extensive research on rechargeable batteries is underway at Argonne National Laboratory in Illinois and at Ford Motor Company Laboratories near Detroit, Michigan. Obtain information from these laboratories regarding their

---

* An asterisk indicates a relatively difficult or advanced exercise.

progress to date. At what period do you estimate that batteries will be readily available for storage of electric energy by facilities to meet peak power demands?

19.6 Determine whether there are suitable sites for an underground compressed-air storage system in your state. Examine the economic feasibility of using the most suitable site. Determine the size of the air compressor necessary for the site and a suitable turbine–generator system for generating 50 kilowatts.

19.7 Flywheel storage devices with an energy storage capacity of over 100 watt-hrs/kg are not commercially available at present. Investigate the research progress on flywheels constructed of fiber materials.

19.8 Calculate the size of the flywheel system using PRD-49 that could store 500 kilowatt-hours and yield 50 kilowatts of peak-demand power for a neighborhood in your city.

****19.9** Consider a spinning flywheel rod and a small portion of the rod lying between x and $x + \Delta x$, from the center of the rod (the pivot). Using Newton's Law and integral calculus, determine the force $F(x)$ as a function of x. Show that the maximum force occurs at the center (pivot point) of the rod.

19.10 A total energy system can be designed for a neighborhood. Determine the electric power requirements of your neighborhood or college dormitory, and design the outline of a total energy system. Determine what fuel you would use and what overall efficiency you expect to achieve. Sketch a diagram of the system.

19.11 A total energy system is particularly applicable to a large college with the need for heating, cooling, electric power, and other energy uses. [8] Determine the overall energy efficiency of one of the following two high schools using a total energy system: (1) O'Rafferty High School, Lansing, Michigan; (2) McAllen High School, McAllen, Texas. Sketch the schematic of the energy system of one of these schools.

20

Conservation of Energy

20.1 INTRODUCTION

It is important not only to increase the supply of energy to meet the demands of the nations of the world, but equally important to reduce the demand for fossil fuels. The conservation of fossil fuels, and all forms of energy, can be achieved by an integrated, well-planned approach.

The conservation of energy can reduce the energy needs of the U.S. by a significant percentage. Estimates of ten to forty percent reduction in our overall energy consumption have been proposed as realistically achievable over the next decade. There are essentially four methods for energy conservation, which are: (1) elimination of waste, (2) shifting to less energy-intensive processes, (3) reduction of energy-consuming activities, and (4) improving efficiency of energy-consuming activities.

Waste can be eliminated by simple measures like repairing hot water leaks and broken windows and turning off unused lights. A shift to less energy-intensive processes may be illustrated by the use of mass transportation or the recycling of materials. A reduction in energy-consuming activities essentially implies a change of lifestyle to one less dependent upon automobiles, jet airplanes, and air-conditioned homes. The improvement of efficiency can be achieved in heating and cooling systems and industrial processes, for example.

One national energy-conservation goal is provided in the 1975 plan of the Energy Research and Development Administration (ERDA). This plan, summarized in Table 20.1, calls for an annual energy saving of about 10 to 30 $\times$ 10^{15} Btu in 1985. Recall that the projected consumption in 1985 is 100 $\times$ 10^{15} Btu. (See Chapter 4.) This is a significant conservation goal and yet it has been criticized by some as being too low. With a reasonable effort, energy conservation might account for a ten-percent savings in the total projected energy consumed in 1985.

Table 20.1

THE ERDA CONSERVATION PLAN OF 1975

Energy-saving source	Annual energy savings (10^{15} btu)	
	1985	2000
Conversion of waste materials	2.5	4.5–9
Electric conversion efficiency improvement	0	4.5
Electric power transmission and distribution	0	4.5
Electric transportation	0	4.5
Transportation efficiency improvement	2.5–6	4.5–9
Industrial energy efficiency	2.5–6	4.5–9
Conservation in buildings and consumer products	2.5–6	4.5–9
Total	10–27	31.5–49.5

The transportation sector consumes about 25 percent of the total energy consumed in the U.S., while industry consumes about 41 percent. Residential consumption of energy is about 20 percent, while the commercial sector uses 14 percent. It is important to develop energy-conserving schemes for each of these sectors. Space heating and cooling accounts for about 60 percent of the energy use in the residential and commercial sectors, or about 20 percent of the total U.S. energy consumption.

Thus, there are essentially three primary areas of immediate concern for energy conservation: (1) space heating and cooling (20 percent of total); (2) urban and intercity passenger travel and freight transport (25 percent of total energy use); and (3) industrial processes (40 percent of total).

For the period 1935 to 1965 the GNP generated per unit of energy consumed increased from 50 in 1935 to 67 in 1965. By 1972 it had dropped back to 62 as a result of additional requirements for environmental cleanup and protection activities.

Several energy-conservation schemes are presented in Table 20.2. Many of these schemes are estimated to yield an energy savings of up to 5 percent with only moderate effort. Of course, if a more concerted effort were made to improve the effectiveness of these schemes, the savings could be increased. Nevertheless, it would take several years to implement any of these schemes to the point of economic feasibility.

The most obvious way to reduce demand is to reduce waste. This approach includes the overcoming of loss due to (1) design malfunctions, (2) poor state of maintenance and repair, (3) use of unnecessary lights and heating, (4) hot-water leaks, and similar examples of waste.

Table 20.2

ENERGY CONSERVATION SCHEMES

	Potential near-term savings as percent of total for that item
I. *Residential*	
a) Insulation (homes)	5%
b) Air conditioning	1%
c) Lighting, appliances	5%
d) Solar heating	5%
II. *Industry*	
a) Process steam	5–10%
b) Recycling of materials	3%
c) Total energy scheme	5–10%
III. *Transportation*	
a) Mass transit for 10% of urban passenger-miles	2–5%
b) Smaller, more efficient autos	2–5%
c) Shift some freight and passengers to trains	1–2%
IV. *Commercial*	
a) Air conditioning	2%
b) Solar heating	5%
c) Building design	3%

A shift to less energy-intensive products and processes includes such actions as a shift to walking or bicycling (instead of driving an automobile), car-pooling, and the increased use of mass transit.

The U.S. would reduce energy-consuming activities and thus the demand for energy. Possibilities could include restricting shops to a five-day week, reducing outdoor advertising, and eliminating night-lighted sports activities.

Energy-conservation schemes individually contribute small savings. It is only the combination of all these savings that adds up to a worthwhile total savings. For example, shifting 50 percent of urban passenger miles to mass transit may save only 5 percent of the total energy consumed by urban transportation, and thus only 0.5 percent of the total U.S. energy consumption. However, if a large number of energy schemes are implemented, perhaps an overall energy savings may be achieved amounting to 5 percent of the total U.S. consumption within the next few years. The challenge to energy-conservation planners is to convince the consumer that many small energy savings can add up to something important for the nation in the aggregate. In addition, the long-term savings are particularly attractive.

20.2   ENERGY AND STANDARD OF LIVING

As we noted in Chapters 1 and 2, a relationship can be inferred between energy use and gross national product. Thus, one might be concerned that a conservation program necessarily implies a reduced standard of living. However, the GNP is an aggregate measure of standard of living and does not represent differences in cultural, economic, and physical patterns of life between nations. [19]

In Sweden, for example, a significant fraction of all energy consumed in thermal power plants is utilized as low-temperature process or space heating, rather than simply wasted in the cooling water or tower cooling as in the U.S. In Sweden and other European nations, an effective railroad system and lighter automobiles provide effective transportation at lower rates of energy consumption than in the U.S. [17] Mass transit is more highly utilized in Europe than in the U.S.

Overall, Sweden requires about one-half as much energy per dollar of GNP as the U.S., while maintaining the same overall GNP per capita. West Germany, which also has a high GNP per capita (about 15 percent less than the U.S.), also uses about one-half as much energy per dollar of GNP as does the U.S. These European nations use more mass transit, more efficient industrial processes, and less energy-intensive buildings, with little evidence of a reduced standard of living.

20.3 EFFICIENCY OF ENERGY UTILIZATION

Traditionally, the First Law of Thermodynamics has been used to express efficiency, as discussed in Chapter 3. First Law efficiency is defined as

$$\eta = \frac{\text{Energy output}}{\text{Energy input}}. \tag{20.1}$$

First Law efficiencies include the effects of the uses and losses of energy in a system.

The Second Law of Thermodynamics gives a measure of efficiency based on comparing the energy theoretically required to perform a task with that actually used. Second Law efficiency is defined as [1, 2]:

$$\varepsilon = \frac{\text{Theoretical minimum energy required}}{\text{Energy actually consumed}} = \frac{B_{min}}{B_{act}}, \tag{20.2}$$

where B = available work. Therefore, the amount of work needed to heat a home is small compared with the amount of work that would be extracted from combustion of fuels. By contrast, these fuels could be used to generate high-temperature differences in modern plants and would make possible the more efficient use of the available work.

Improving heat-transfer properties from flame to boiler in a power plant, which is an improvement in First Law efficiency, also will increase the temperature difference and therefore improve Second Law efficiency.

The Second Law efficiency of many processes is given in Table 20.3. The First Law efficiency of space heating, using a gas furnace, is approximately 0.70, while the Second Law efficiency is about 0.06. This difference of a factor of ten arises because the gas, which contains a high potential to do work, is being used to develop a low quality of work, that is, to develop a small temperature difference.

Table 20.3
SECOND LAW EFFICIENCY

Use	Relative thermodynamic quality	Percent of U.S. consumption (1968)	Second-law efficiency
Space heating	Lowest	18	0.06
Water heating	Low	4	0.03
Air conditioning	Lowest	2.5	0.05
Industrial Uses			
a) Process steam	Low	17	0.25
b) Direct heat	High	11	0.3
c) Electric drive	High	8	0.3
Transportation			
a) Automobile	High	13	0.1
b) Truck	High	5	0.1

The same gas, if used in a direct-heat industrial process, can be utilized at a Second Law efficiency of 0.3, since it will be used directly to develop a high temperature.

The minimum available work required, B_{min}, varies with the task. For a task that involves direct work, for example, turning a shaft, we have B_{min} = work. For the transfer of thermal energy Q to a reservoir (for example, a room being heated) at temperature T, we have

$$B_{min} = Q\left(1 - \frac{T_0}{T}\right), \tag{20.3}$$

where T_0 = ambient environment temperature (in kelvins). Therefore, if T = 1500 k in the case of a boiler process,

$$B_{min} = Q\left(1 - \frac{300}{1500}\right) = Q(0.8). \tag{20.4}$$

If the energy is used to heat a house with hot air or hot water at a temperature of 320 k and the ambient temperature is 0°C, we have

$$B_{min} = Q\left(1 - \frac{273}{320}\right) = 0.147Q. \tag{20.5}$$

Also, $Q = \eta B_{act}$ for a furnace that delivers energy to the desired space. Therefore, if $\eta = 0.60$ (a 60-percent efficient space heater) we have

$$\varepsilon = \frac{B_{min}}{B_{act}} = \frac{0.15Q}{(Q/0.6)} = 0.09. \tag{20.6}$$

In general we may write,

$$\varepsilon = \frac{B_{min}}{B_{act}} = \frac{Q(1 - T_0/T)}{(Q/\eta)}$$

$$= \eta(1 - T_0/T). \tag{20.7}$$

Therefore, the Second Law efficiency, E, is related to the First Law efficiency η. They are approximately equal when T is very large. However, in the case of low-temperature uses of available work such as in space heating, T_0/T is not negligible and ε can be significantly lower than η.

The shortcoming of the usual First Law definition of energy efficiency is most apparent for tasks in which fossil fuels are used to produce low-temperature heat. Since fossil fuels burn at very high flame temperatures (up to 2200°C), the available work produced by fossil fuels is largely wasted when it is used for hot-water heating, space heating, and industrial steam production, since these are relatively low-temperature processes. [21]

The First Law energy efficiency of several devices are given in Table 20.4. It is interesting to note that many devices, such as the fluorescent or incandescent lamp, are low-efficiency devices. Increasing the efficiency of lighting devices, motors, and space heaters are examples of energy conservation. For example, the efficiency, η, of a large (15-kW) electric motor is about 90 percent, while the efficiency of a smaller motor may be only 50 percent. In terms of energy efficiency, it is desirable to use the higher-efficiency motors.

Table 20.4
FIRST LAW ENERGY EFFICIENCIES OF
SEVERAL PROCESSES

Device	Efficiency (Percent)
Electric generator	98
Electric motor	90
Steam boiler	88
Home gas furnace	85
Home oil furnace	65
Steam-electric power plant	40
Diesel engine	38
Automobile engine	25
Fluorescent lamp	20
Photovoltaic cell	10
Incandescent lamp	5

Energy Efficiency in Industry

Industry uses about 41 percent of the total energy consumed in the U.S. and has always been concerned with efficient use of this energy. However, with increased energy prices and the declining availability of a secure supply of fossil fuels, industry has intensified its efforts toward energy conservation.

The use of energy in industrial processes is illustrated schematically in Fig. 20.1. An industry attempts to minimize its costs for energy, labor, and materials in order to economically produce goods or services. It is often possible to recycle waste materials and energy, as shown in the diagram.

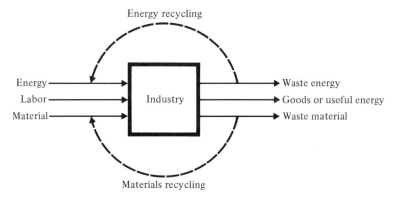

Fig 20.1   *The use of energy in industrial processes.*

The primary use of energy in the industrial sector is for direct heat, process steam, and electric drive, as shown in Fig. 20.2. As industries strive to conserve energy, they focus their efforts on more efficient use of direct heat, improved electrolytic processes, and well maintained and insulated uses of process steam. On examining their systems, many firms have found that up to one-half of their steam lines are not insulated, and many leaks in the pipes are not sealed.

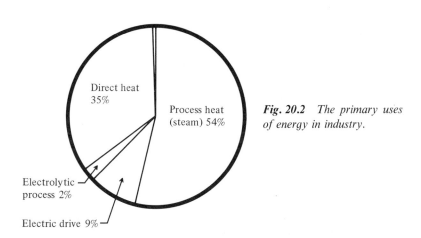

Fig. 20.2   *The primary uses of energy in industry.*

It has been estimated that seven industries—chemicals, iron and steel, farming and food processing, petroleum refining, paper, cement, and aluminum—account for about two-thirds of all the energy used by U.S. industry. The percent of industrial energy consumption by several industries is given in Table 20.5.

The energy intensity of selected industrial processes is given in Table 20.6. Intensity of energy use per dollar of value of the product indicates that aluminum,

Table 20.5
INDUSTRIAL ENERGY CONSUMPTION

Industry or product	% of industrial energy consumption	% of total U.S. energy consumption
Iron and steel	13.6	5.6
Petroleum refining	11.3	4.6
Paper and paperboard	5.2	2.1
Petrochemical feedstock	4.9	2.0
Aluminum	2.8	1.1
Cement	2.1	0.9
Ammonia	2.0	0.8
Ferrous foundries	2.0	0.8

Table 20.6
ENERGY INTENSITY OF SELECTED INDUSTRIAL
PROCESSES

Product	Energy intensity $(10^6 \text{ joules/dollar, U.S. 1975})$
Primary aluminum	409
Plastics	229
Manmade fibers	214
Airlines	202
Paper mills	187
Fertilizers	183
Glass	109
Fabricated metal products	97
Motor vehicles	71
Wearing apparel	53
Computers	38
Banking	19
U.S. Average	84

plastics, and manmade fibers, while they are are convenient modern materials, nevertheless require a high energy consumption to produce.

Until recently, U.S. industry spent only about 5 percent of its total expenditures on energy. With energy a low-cost item, energy-intensive processes and automation were often substituted for labor-intensive processes. As the price of energy increases, many firms are instituting energy-management programs, which focus on (1) the monitoring and control of steam processes, (2) waste heat recovery, (3) fuel substitutions, (4) added insulation, and (5) replacement and maintenance of equipment.

The Second Law efficiencies of the steelmaking and aluminum industries are 21 percent and 13 percent, respectively. One way that efficiency could be increased would be to use the steam heat for the generation of electricity as well as for the industrial process. A need also exists for instrumentation for these processes, to indicate the energy-consumption efficiency.

Conservation in Industry

One means of improving the use of process steam in industry is to use a two-step boiler scheme. At high temperature a vaporized carrier, such as potassium vapor at 800°C, would be used to drive a turbine and an electric generator. It would emerge from the turbine still hot enough to make steam (water) at the customary 540°C. This steam would be used for the regular process-steam needs in industry. Waste of steam would thus be cut in half.

The making of aluminum is an energy-intensive process. About 17 million metric tons (tonnes) of aluminum were produced in the world in 1975, and the figure is expected to double by 1985. Production of aluminum in the U.S. amounted to 7 million tonnes in 1975. U.S. aluminum consumption is 65 lbs (29 kg) per person compared to about 25 lbs (11 kg) per person in Britain and France. The aluminum industry's energy costs account for 25 percent of production costs. In 1970, for example, U.S. aluminum production required 64 billion kilowatt-hours of electricity, nearly 4 percent of all electric energy consumed that year.

The Bayer–Hall process for producing aluminum from bauxite ore required 12 kWh per pound of aluminum in 1939; this has been reduced to an average of 7 kWh today. A new process, under study at a pilot plant, uses chlorine, combined with the aluminum oxide recovered from bauxite, to make an aluminum chloride powder which is then dissolved in an electrolyte. The chlorine process is estimated to use 4.5 kWh per pound of aluminum produced.

The increasing substitution of aluminum for wood in our buildings has resulted in a more energy-intensive economy. The more efficient production of aluminum is a priority item for research and development during the next decade.

It requires 112 million Btu's per ton (130×10^9 joules per tonne) to process copper from a sulfide ore containing 0.7 percent copper. [3] It requires 26 million Btu's per ton (30×10^9 joules/tonne) to process iron ore to produce steel slabs by means of the open-hearth method.

The total energy consumed for the production of iron and steel in the U.S. was 3.7×10^{15} Btu in 1973. The total energy consumed to produce aluminum and copper was 1.2×10^{15} Btu and 0.24×10^{15} Btu, respectively. As extraction of metal ores becomes more difficult and more refining is necessary, more energy is lost. It will be necessary in the future to improve the energy efficiency of the refining processes in order to overcome the increasing energy costs of refining lower-grade ores.

It is estimated that the world's steel industry preempts about ten percent of the world's total use of energy. Therefore, the recycling of steel becomes an important method of conserving energy for the future.

Computer control of electricity usage and boiler operation is a useful approach to energy conservation. A user pays an electric utility bill based on consumption and peak demand. Some savings in energy could be realized by using computer control systems to selectively shut off fans, compressors, chillers, and other devices for short periods of time. Computers can also be used in multiboiler installations to optimize combinations of fuels, boilers, and turbines; and the ratio of generated to purchased power can be determined and controlled by a computer.

20.4 TRANSPORTATION

Approximately 25 percent of the energy consumed in the U.S. is used for transportation. This subject was discussed in Chapter 9; we will only note here some important aspects of potential energy conservation in this field.

It has been estimated that about 30 percent of the energy devoted to transportation could be saved through the use of more energy-efficient transport modes. The energy intensity of an automobile is about twice that of a bus. Thus, a shift of passengers from autos to urban buses or mass transit would be an important energy-conservation measure. The energy intensity of shipping freight by truck (in energy per tonne-km) is about four times that of railroads. Again, if long-distance freight shipments could be shifted from trucks to railroads, a significant savings could be achieved.

One conservation scheme suggests that people should be encouraged to walk or bicycle short distances in urban areas. Walking and bicycling require only 270×10^3 joules/passenger-km and 60×10^3 joules/passenger-km, respectively, compared to 450×10^3 joules/passenger-km for mass transit. In urban areas bicycles provide door-to-door convenience and shorter trip times than automobiles for short trips (less than 4 km).

The U.S. has taken the step of reducing the highway speed limit to 55 mph. In a recent study it was shown that the percent increase in gasoline consumption caused by an increase in speed from 50 to 60 mph was 11 percent. Thus, it is of value to maintain a reasonable speed limit in order to conserve petroleum.

20.5 CONSERVATION OF ELECTRICITY

The conservation of energy in the use of electricity holds significant potential. The U.S. government has called for a reduction of 50 percent in the power required by television sets. The voluntary appliance-efficiency program for electric appliances calls for a reduction of power requirements of 42 percent for refrigerators, 22 percent for room air conditioners, and 20 percent for electric ranges. By 1980, it is hoped to achieve an overall reduction for appliances equal to 20 percent compared to 1975.

One possibility is to read electric meters by remote control and use the same remote system to turn on and off low-priority, high-drain appliances such as water heaters and air conditioners. With the use of radio control, utilities could turn off such low-priority devices at peak demand periods, thus reducing peak-capacity requirements.

Other electric energy savings techniques are to use new modern methods in place of earlier approaches. As solid-state low-power devices replace earlier higher-power devices, a substantial savings in energy can be achieved. For example, fiber optics can replace copper cable requirements, and thus reduce the energy required to manufacture cables as well as the losses on the cables.

The energy efficacies of electric lights varies quite widely, as can be seen in Table 20.7. In this case we express the efficacy as the ratio of light output in lumens per watt input. Higher-wattage incandescent lamps are more efficacious than lower-wattage ones. For example, one 100-watt lamp produces more light than two 60-watt lamps. [4]

Table 20.7
EFFICACY OF ELECTRIC LAMPS

Type	Efficacy (Lumens/watt)
Low-pressure sodium	180
High-pressure sodium	130
Metal halide	100
Fluorescent	80
Mercury vapor	56
Incandescent	23

Fluorescent lamps offer about four times the efficacy of incandescent sources; however, incandescent lamps are still largely used in homes because of their convenience.

Mercury lamps were popular for street lighting in the early 1960's. Sodium lamps are widely used today for street lighting and can be distinguished by their yellow light. Low-pressure sodium lamps have a high efficacy and typically operate at 180 watts. Many cities are converting street lighting to sodium lamps in order to save energy and energy costs.

Lighting consumed about 20 percent of U.S. electric energy or about 5 percent of all U.S. energy, in 1975. About two-thirds of the lighting energy is used by incandescent lamps and one-third by fluorescent and vapor lamps. It would be useful to switch this ratio to two-thirds fluorescent and vapor lamps.

The use of electric heat pumps in place of electric space heating would be an important conservation measure. A heat pump is a device that transfers heat from a cooler to a warmer place. Thus the pump transfers heat from outside to inside a building. The Second Law efficiency of a heat pump is

$$\varepsilon = (Q/W)[1 - (T_0/T)], \tag{20.8}$$

where Q is the heat delivered to inside and W is the work (electric energy) required to achieve this. The outside ambient temperature is T_0 and the inside temperature is T.

The coefficient of performance (C.O.P.) of a heat pump is:

$$\text{C.O.P.} = \frac{Q}{W} = \frac{\varepsilon}{[1 - (T_0/T)]}. \tag{20.9}$$

For example, let $\varepsilon = 0.3$ and $T/T_0 = 1.1$; we then have

$$\text{C.O.P.} = \frac{0.3}{[1 - 0.91]} = 3.3. \tag{20.10}$$

However, if the ratio of outside to inside temperature is too high, the C.O.P. drops below 1.0. Let us consider a heat pump delivering indoor heat at 52°C when the outdoor temperature is $-3°C$. Then we have

$$\text{C.O.P.} = \frac{0.3}{1 - \left(\dfrac{270}{325}\right)} = 1.8. \tag{20.11}$$

As the C.O.P. approaches 1.0, the heat pump essentially operates as an electric resistance heater.

Electric Appliances and Air Conditioners

Electrical appliances account for 8 percent of the total use of energy in the U.S. The major appliance users of electrical energy are shown in Table 20.8. Appliances can be designed to reduce energy losses as much as 50 percent with some attendant increase in their price. However, much of the increased price would be recovered in reduced electricity expenditures.

Refrigerators account for about 20 percent of residential electricity used in the U.S. today, or about 7 percent of the total electricity consumption. The

Table 20.8
ENERGY REQUIREMENTS OF ELECTRIC APPLIANCES

Appliance	Average power (Watts)	Average annual energy consumption (kWh)
Water heater	2500	5,000
Refrigerator		
(14 cu ft frostless)	215	1,800
Freezer (15 cu ft)	341	1,200
Range (with oven)	12000	1,200
Dishwasher	1200	360
Color television	200	440
Air conditioners:		
Room	860	800
Central	1300	2,000

increase in the number of refrigerators and in refrigerator energy requirements have together caused total electric energy used by refrigerators to increase from 18 billion kWh in 1950 to 60 billion kWh in 1965. [5] It is estimated that electric energy for refrigerators may be 190 billion kWh by 1980. Current frostless refrigerators require about 14 watts per cubic foot. In order to reduce the power and energy requirements, it is possible to (1) increase insulation, (2) improve motor efficiency, (3) enlarge the evaporator surface, and (4) improve condenser design. It is estimated that the energy requirement could be reduced from 1800 kWh per year to less than 1000 kWh per year. [5]

A typical modern room air conditioner removes 5000 Btu/hour and uses 860 watts. This unit is said to have an energy-efficiency ratio (EER) as follows:

$$\text{EER} = \frac{5,000 \text{ Btu/hr}}{860 \text{ watts}} = 5.8 \text{ Btu/watt-hr.} \tag{20.12}$$

The EER of common air conditioners varies from 5 to 12.

The average operating time of an air conditioner is 900 hours per year and all such units consumed about 40 billion kWh during 1970 in the U.S. Room air conditioners use about 800 kWh per year each, and there were about 55 million installed in 1975. Thus, the total room air-conditioning load was 44 billion kWh. The connected power requirement was about 48 gigawatts. If the assumed EER were raised from 6 to 10 Btu/watt-hr, the annual consumption would be:

$$E = 500 \text{ watts} \times 900 \text{ hr} \times 55 \times 10^6 \text{ units}$$
$$= 25 \times 10^9 \text{ kWh.} \tag{20.13}$$

Thus, energy consumption for air conditioners could have been reduced in 1975 by 19 billion kWh (43 percent reduction) if all room air conditioners were required to have an EER equal to or larger than 10. Besides reducing the energy requirement, the power capacity required would be reduced to 28 gigawatts, thus relieving the peak power requirement.

Again, it is possible to use radio signals to turn off air conditioners for a short period of time, for example, for 5 minutes once or twice per hour, and relieve the load on the power utility. The homeowner does not find this interruption a problem, but if 5 minutes twice per hour is distributed equally over all the users, then the average power load can be reduced by one-sixth.

20.6 ENERGY CONSERVATION THROUGH BUILDING DESIGN

The design of commercial buildings can have an important effect on energy savings over the next decade. The recently adopted ASHRAE* Standard 90-75, if adopted by all 50 states, could cut energy consumed in new buildings by 25 percent. The standard establishes levels of thermal efficiency for (1) the building

---

* American Society of Heating, Refrigerating, and Air-Conditioning Engineers.

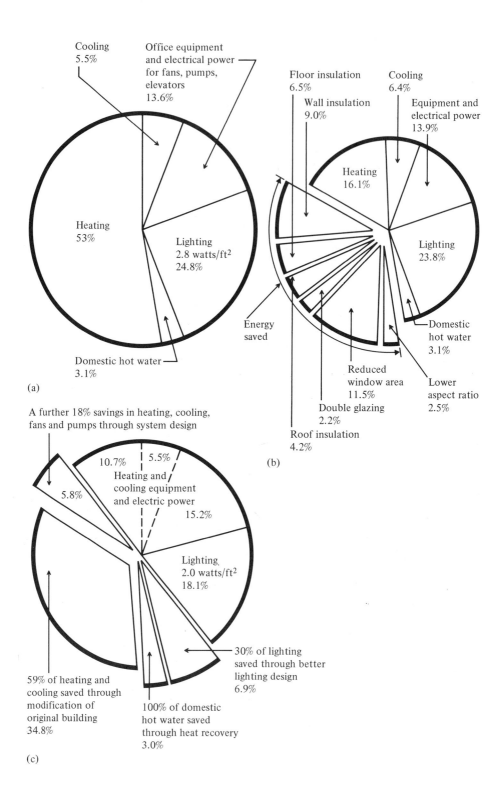

Cooling
5.5%

Office equipment
and electrical power
for fans, pumps,
elevators
13.6%

Heating
53%

Lighting
2.8 watts/ft²
24.8%

Domestic hot water
3.1%

(a)

Floor insulation
6.5%

Wall insulation
9.0%

Cooling
6.4%

Equipment and
electrical power
13.9%

Heating
16.1%

Lighting
23.8%

Energy
saved

Domestic
hot water
3.1%

Reduced
window area
11.5%

Lower
aspect ratio
2.5%

Double glazing
2.2%

Roof insulation
4.2%

(b)

A further 18% savings in heating, cooling,
fans and pumps through system design

10.7%   5.5%

Heating and
cooling equipment
and electric power
15.2%

5.8%

Lighting
2.0 watts/ft²
18.1%

30% of lighting
saved through better
lighting design
6.9%

59% of heating and
cooling saved through
modification of
original building
34.8%

100% of domestic
hot water saved
through heat recovery
3.0%

(c)

382

envelope; (2) heating, cooling, and ventilating systems and components; (3) domestic hot water systems; (4) use of energy for illumination; and (5) electrical distribution systems.

In the past, many large commercial buildings were built with little attention to energy use. The World Trade Center in New York City, for example, uses enough electricity for a city of 100,000 people.

A recently completed design study and construction project illustrates the potential energy savings in building design. A Federal office building was completed in Manchester, New Hampshire, in 1976. The building was built with several energy-saving measures such as a heavily insulated north wall with no windows, double-glazed windows, a low window-to-wall ratio, increased floor, roof, and wall insulation, and south-facing windows for winter solar heat. The Manchester building cost about 10 percent more to construct but will consume about one-half as much energy as a typical building. [6] All the energy savings are summarized in Fig. 20.3.

Heating of Homes

In a typical home in northeastern U.S., 50 percent of the heating load is created by infiltration of outside air and 45 percent by heat loss through ceilings, walls, and doors. Doubling the insulation protection of such a home would reduce the energy consumed for heating by about 22 percent.

The most significant environmental parameter for energy consumption in a house is, of course, the average daily outside temperature. The next most important parameter is wind speed. The effects of wind speed on energy consumption in a house are only now being studied, but it is becoming clear that wind dominates the dynamics of the exchange of air between inside and outside, known as "air infiltration." One-third or more of the total heat loss from the interior of the house is associated with this convective exchange with outside air. [1, 12]

Setting back the thermostat saves a significant amount of energy. For example, setting back the thermostat from 72 to 68°F (22 to 20°C) in Minneapolis saves 14 percent on heating energy use, when the thermostat is set to 60° at night. If the thermostat is set to 68°F in the day and 55°F at night in Minneapolis, the savings can be increased to 23 percent.

Fig. 20.3 *(a) Energy consumed in a typical 126,000-ft² office building in Manchester, N.H. is 13 × 10⁹ Btu/year (14 × 10¹² joules/year). (b) Improved design consisting of increasing wall, floor, and roof insulation reducing window area from 50 to 10%; adding double glazing; and changing the length/width of the building to 1/1 yielded a savings of 34%. (c) Improved lighting and heating, ventilating, and air conditioning systems cut another 16% off the annual energy consumption of the typical building.*

Better insulation of buildings is one method of conserving energy. Estimates are that up to one-fourth of all energy for space heating could be saved if insulation was improved. [18] The basic heat-flow equation is:

$$\frac{dQ}{dt} = \frac{kA\,\Delta T}{d}, \tag{20.14}$$

where dQ/dT is the heat flowing per unit time, k is the thermal conductivity, A is the area of the surface, d is the thickness of the surface, and ΔT is the temperature difference. Various values of k are given in Table 20.9. In order to convert the power requirement, we may take 0.293 watts as equal to 1 Btu/hr.

Table 20.9

THERMAL CONDUCTIVITY OF SELECTED MATERIALS

Material	Conductivity, k (Btu/hr)*	Conductivity, k (Calories/second)
Air	0.165	0.05×10^{-3}
Rock wool	0.26	0.08×10^{-3}
Glass wool	0.29	0.09×10^{-3}
Asbestos	0.55	0.17×10^{-3}
Wood	0.7	0.22×10^{-3}
Cinder block	2.5	0.78×10^{-3}
Brick	5	1.6×10^{-3}
Concrete	7	2.2×10^{-3}
Glass	6	1.4×10^{-3}
Aluminum	1450	0.48

* Btu/hr for one ft² of area, a thickness of one inch, and $\Delta T = 1$ degree Fahrenheit. Cal/sec for one cm² area, a thickness of one cm, and $\Delta T = 1$ degree centigrade. 1 Btu/hr = 0.293 watt.

Consider the heat loss of a typical home in New York, as represented in Fig. 20.4. [7] The annual heat loss is a function of ceiling and wall insulation, as well as the use of storm windows (or dual glazing). The upper point A may represent most of the housing built prior to 1970. If storm windows and insulation are added, so that the home operates at point B, the heat loss is reduced from 134 to 78×10^6 Btu/year. This 42 percent energy savings is applicable throughout the northeastern U.S.

The savings for a model home in Minneapolis would be a reduction from 209×10^6 Btu to 119×10^6 Btu for a change from point A to point B. This results in a savings of 90×10^6 Btu/year or a 43 percent reduction.

About 7.5×10^{15} Btu were consumed in the U.S. in 1975 for residential space heating. Assuming that an average of 25 percent could be saved in space heating energy if homes were insulated and storm windows were added, a total savings of 1.9×10^{15} Btu could be achieved. This savings amounts to 2.5 percent of the total energy consumed in the U.S.

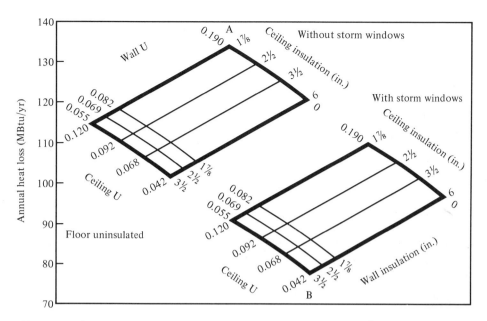

Fig. 20.4 *The annual heat loss from a typical home in New York in 10^6 Btu per year (U = Btu/ft^2 $hr°F$).*

Table 20.10

ENERGY BALANCE IN A STANDARD HOUSE AND IN A TARGET HOUSE
(POWER IN BTU/HR)

	Standard house		Target house	
	Mild, clear day ($\Delta T = 30°F$)	Cold, partly cloudy day ($\Delta T = 70°F$)	Mild, clear day ($\Delta T = 30°F$)	Cold, partly cloudy day ($\Delta T = 70°F$)
Energy losses				
walls and roof	7,200	16,800	2,800	6,500
floor	1,100	1,100	400	400
windows	6,600	15,400	1,800	4,200
incoming air (heating)	8,000	18,700	1,600	3,700
incoming air (humidification)	3,800	3,800	800	800
Total	26,700	55,800	7,400	15,600
Energy gains				
solar through windows	5,200	2,600	4,000	4,200
3 people	1,000	1,000	1,000	1,000
electric load	5,700	5,700	2,400	2,400
gas furnace load to balance	14,800	46,500	—	8,000
Total	26,700	55,800	7,400	15,600

Consider a two-story home with dimensions 25 ft wide × 30 ft long × 20 ft high (7.6 m × 9.1 m × 6.0 m). We will consider two different days: (a) mild and clear with $\Delta T = 30°F$ and (b) cold and partly cloudy with $\Delta T = 70°F$. The energy balance for the standard home and the target (or optimum) house is given in Table 21.10 for a mild, clear day and a cold, partly cloudy day. The gas-furnace load of the target home is reduced significantly by insulation and reduced infiltration of outside air.

20.7 RECYCLING

Materials and waste energy from industrial and commercial processes may be recycled (as shown in Fig. 20.1) in order to conserve resources. The typical energy content of materials and manufactured products is given in Table 20.11. It is definitely worthwhile to recycle steel, copper, aluminum, and magnesium when feasible.

Table 20.11
ENERGY CONTENT OF MATERIALS AND MANUFACTURED PRODUCTS

	Energy (Megajoules/kg)	Cost of energy/value of product
Metals		
Steel	40	0.3
Aluminum	100	0.4
Copper	30	0.05
Magnesium	90	0.1
Products		
Glass bottles	40	0.3
Plastic	10	0.04
Paper	25	0.3
Inorganic chemicals	12	0.2

The recycling of used metals reduces the need for additional raw materials and energy use. At present, only one-fourth of the scrap which the steel industry recycles comes from junkyards, dumps, and recycling centers. The rest of the scrap comes from the steel mill itself or from machine shops. Transportation costs, depletion allowances for raw materials, and the reluctance of steelmakers are some of the deterrents to a more complete recycling of steel. [20]

The energy cost in kWh/ton is given for several materials in Table 20.12 for the raw and recycled material. [8] It is possible to save about 50 percent of the energy for steel and paperboard. However, it is necessary to separate the paper or steel from waste products, which can often be a costly process. As we are aware, even aluminum cans are separated only by hand.

Table 20.12
RECYCLING OF MATERIALS

Material	Point of impact	Energy cost (kWh/ton)		Percent saved	Ease of recycling
		Virgin material	Recycled material		
Steel	Molten steel	13,680	6,636	52	Present technology limited by impurities, separation, and collection.
Aluminum	Molten aluminum	65,780	2,400	96	Present technology limited by separation and collection.
Plastics	Molten polymer	13,238	586	96	Extremely difficult; no satisfactory technology known.
Paperboard	Pulp	1,923	970	50	Present technology for separated waste; no satisfactory technology for mixed waste.
Glass	Transportation	2,287	2,287	0	Difficult and costly.

In principle, plastics could be recycled and would achieve as much as 75 percent energy savings. This requires that each type of plastic be separated from the rest, an entirely impractical goal in the foreseeable future because the densities and other physical properties of various polymers overlap too much. It has been calculated that replacement of refillable bottles saves about 1.4 kWh/filling for soft-drink, throwaway bottles, and 0.7 kWh/filling for soft-drink cans. If the 40 billion glass containers that were used in 1975 had been reusable, the savings would have been about 42 million kWh.

About 95 percent of the beverage cans now used in the U.S. are steel and the other 5 percent are aluminum. For the aluminum can we would save about 1.7 kWh/can, or a total of 7.2 billion kWh per year, by recycling all cans. Recycling all steel cans could save 34 billion kWh per year.

If we recycled every automobile by shredding and segregating, we could save about 11,000 kWh/car, or about 30 percent of the energy required for production of a new car using new materials. The net savings, if all scrapped automobiles in the US were recycled, could be about 100 billion kWh/year.

Recycling waste paper, if it is economically achievable, could save about 50 billion kWh/year. However, this is difficult to achieve due to the difficulty of separation of the paper from waste.

The potential energy savings from recycling of materials is summarized in Table 20.13. If all these savings were achievable by means of recycling, about

Table 20.13

ENERGY SAVINGS FROM RECYCLING MATERIALS IN THE U.S.

Material	kWh saved/ton of recycled material	Estimated total savings if all material was recycled (kWh)
Paper	4,210	50×10^9
Steel for autos	7,000	100×10^9
Aluminum beverage cans	49,000	7.2×10^9
Steel beverage cans	7,000	34×10^9
Total		191.2×10^9
Total electrical energy use in the U.S. in 1975		$1,900 \times 10^9$

10 percent of the electrical energy consumption in the U.S. could be conserved each year. A strategy which encourages the recycling of steel, aluminum, and other materials would not only be conserving of energy but also conserving of raw materials. Recycling will become more a part of our economy in the decades ahead.

20.8 TOTAL ENERGY SYSTEMS

In Chapter 19 we defined total energy systems (often called cogeneration systems) as combined processing plants that generate electricity; use residual and recycled energy for heating, air conditioning, and hot water; treat water; process solid wastes, and treat liquid wastes.

Regional heating with hot water from a central electric power plant used to heat buildings has been pioneered in Sweden. A district heating system does away with individual boilers in each building but instead connects each building into a network of hot-water pipes. The electric utility authorities in Sweden are responsible for supplying heat as well as electricity. [17]

As a result of using power-plant heat for space heating, the final efficiency of power plants in Sweden is 53 percent. This figure can be compared with an average electric power-plant efficiency of 33 percent in the U.S. Examples of cities using power-plant waste heat for district heating are Helsinki, Stockholm, Munich, Hamburg, Leningrad, and New York.

It is also feasible to store large amounts of byproduct heat for several months in underground water wells. More than three-fourths of the stored energy would be recoverable after 90 days. [9] If such a storage system could be developed, it would allow the use of heat energy from an electric power plant developed during periods other than winter.

It has been estimated that 75 percent of the Soviet Union's space-heating requirement is met by using hot-water at about 150°C from central power plants and heating plants.

The fuel input to electric power plants in the U.S. was about 16×10^{15} Btu in 1975. With an efficiency of one-third, about 10.7 quads are lost as waste heat. If total energy systems could be installed for one-half of the plants and the net efficiency of those plants raised to two-thirds by using the waste heat for hot-water heating and other total energy system uses, the energy saved would be 2.7 quads $(2.8 \times 10^{18}$ joules). Since the total energy consumption in 1975 was 78 quads, the net energy savings would be 3.5 percent. The conservation of energy using total energy schemes can result in significant savings in fuel consumption.

20.9 ECONOMICS AND CONSERVATION

The economic efficiency of various processes is measured by comparing the total cost of using energy. Expenditures for energy require between 2×10^5 and 10×10^5 Btu per dollar. The same dollars spent on conservation measures require only about 6 to 8×10^4 Btu per dollar. Thus, with a factor of ten in energy cost per dollar, it pays to save energy.

Some energy-conservation strategies, such as insulation in refrigerators and water heaters, pay back the initial incremental investments in months, while others such as industrial techniques, building insulation, and more efficient heating/ cooling systems require longer periods. It is usually less expensive for the user to invest in higher efficiency than for the local electric utility to invest in extra power capacity.

Strategies of energy-conservation substitute capital, materials, labor, know-how, and management skills for energy. This trend is the reverse of the earlier trend to substitute energy-intensive automated processes for labor.

It can be shown that conservation policies save more energy than new energy sources can produce, per dollar invested. Thus, it pays to invest in conservation schemes. An insulation program that would result in energy savings in the home and commercial buildings could replace the need for capital investment in oil refineries or nuclear power plants.

Since 1972, energy prices have risen dramatically. For airlines the cost of fuel is about 20 percent of the total cost of operating. For other industries the cost is about 5 percent of the total operating cost. Thus, rising energy costs are our incentive to energy conservation.

The ratio of the cost of labor to energy has increased from 1936 to 1973 by a factor of ten. However, the ratio of labor costs to capital costs has been constant since 1950, thus deterring a more rapid switch to a more energy-intensive production system. [10] As wages increase relative to other costs, energy use increases through the process of mechanization. However, in recent years, because of high interest rates and the cost of capital, the shift to a more intense energy economy has been impeded. The energy intensity of various industries, as contrasted to labor intensity, is summarized in Table 20.14.

One method of encouraging energy conservation is to institute an energy tax. This tax would be applied to all nonrenewable energy sources at the point

Table 20.14

ENERGY INTENSITY OF SELECTED
INDUSTRIES PER EMPLOYED PERSON

Industry	Energy intensity per employed person (10^{15} Btu per employee)
Chemical	14.0
Automobile	8.0
Paper	7.0
Construction	5.5
Food	3.0
Electrical	3.0
Wood	2.0

of extraction. Such a tax, to be effective, might have to double the price of gasoline or home heating oil. This approach attempts to restrict energy use by the utilization of the price mechanism.

Another approach is to institute a tax credit for individuals and industries that adopt approved conservation methods. For example, homeowners and commercial establishments could be allowed a tax credit for 20 percent of the cost of installing insulation. Perhaps a 50 percent credit could be provided for those homes and businesses which install solar heating and cooling equipment. Congress has proposed that a tax credit up to $500 be enacted for such conservation measures. Another Congressional proposal calls for a $10 billion subsidized-interest loan-guarantee program for conservation measures.

Another means of conserving energy is to institute a rationing scheme. The term rationing represents an essentially political process, rather than economic, in which allocation is determined by regulations enacted by a legislative body. An energy-rationing scheme would require a large bureaucratic structure to administer. One example of a rationing scheme would allow customers some percentage of their actual consumption in previous years. Another would allocate each driver a fixed number of gallons of gasoline per month. Rationing schemes can be envisioned which are set up on the basis of coupons for amounts of energy, for example, a set number of joules per month to be used by a person in any manner.

Industries, commercial establishments, and many homeowners measure their investment in a new venture by rate of return on capital, R, so that

$$R = \frac{P}{C}, \tag{20.15}$$

where P is the profit and C is the capital invested. An additional investment in energy-conserving alternatives and equipment ΔC produces a reduction of operating expenses, which in turn produces an increased profit ΔP. In the case of a homeowner, the reduced expense results in an incremental amount of available money which we can continue to call ΔP. A person or company will be inclined to make the investment if $R \leq \Delta P/\Delta C$ where, for simplicity, we can include the cost of capital in ΔC. An industry may seek a rate of return, R, equal to 10 percent while a homeowner may be willing to settle for R equal to 5 percent or less. With a high cost of capital (high interest rates) and low availability of capital, many conservation schemes fail this test. Thus, in order to encourage energy-conservation schemes, the cost of capital and the availability of capital must be improved.

The investments required for conservation measures are the result of millions of small investments by homeowners and small business owners. [11] Many of the benefits of conservation accrue to society as a whole and not to the individual homeowner or business operator. Reduced energy use will result in cleaner air, lower energy prices, less traffic congestion, and diminished threat to national security. Thus, the homeowner or business operator might be willing to settle for a rate of return, R, equal to zero, if an added tax-credit incentive were available.

One possibility of overcoming the shortage of capital for conservation measures would be for utilities to carry out the measure, such as installing home insulation, and charge for the investment on the customer's power bill. This scheme is particularly attractive if it can be shown to the homeowner or commercial building owner that the monthly cost of paying off the insulation is equal to (or less than) the savings in cost of the energy. Then the person's utility bill is the same, but he has the insulation in the home and is saving energy.

20.10 A PLAN FOR CONSERVATION

A modest energy-conservation plan for the U.S. could achieve significant savings of energy resources. One possible plan is listed in Table 20.15. This plan envisions using wastes to yield recyclable materials such as aluminum, steel, and paper, and to burn a portion of the remaining waste for heat generation. The efficiency of electric-energy conversion and transmission would be increased. The waste heat from electric power plants would be used in one-fourth of electric plants to provide district heating. The efficiency of appliances, lighting, and air conditioners would be improved and efficiency labelling would be instituted. All new 1985 automobiles would achieve at least an average of 27 miles/gallon. Industrial energy efficiency would be significantly improved and conservation in building heating and cooling would be developed. Finally, a revision of peak-load pricing schemes for electric utilities would be instituted so as to achieve reduced consumption. The overall savings might be as high as 13.0 quads in 1985. This would permit the annual U.S. consumption in that year to be reduced from about 102 quads (projected) to an actual 89 quads, or a reduction of 12 percent. This modest conservation scheme would require the institution of sufficient incentives and arrangements to achieve the objectives.

Table 20.15

A SET OF OBJECTIVES FOR ENERGY CONSERVATION
FOR THE U.S. IN 1985

Item	Annual energy saving (Quad = 10^{15} Btu)
1. Use of wastes for recycling and energy generation	2.0
2. Electric conversion and transmission efficiency improvement	0.5
3. Use of waste heat from electric plants for district heating	1.0
4. Efficiency labelling and efficiency improvement for appliances and room air conditioners	0.5
5. All new automobiles with at least 27 miles per gallon average	3.0
6. Industrial energy efficiency improvement	3.0
7. Conservation in building, heating and cooling	2.0
8. Revision of peak load pricing schemes for electric utilities	1.0
Total	13.0

REFERENCES

1. "Efficient use of energy," The Study Group on Technical Aspects of Efficient Energy Utilization of the American Physical Society, *Physics Today*, August 1975, pp. 23–33.

2. C. A. BERG, "A technical basis for energy conservation," *Mechanical Engineering*, May 1974, pp. 30–42.

3. E. T. Hayes, "Energy implications of materials processing," *Science*, February 20, 1976, pp. 661–665.

4. C. W. BEARDSLEY, "Let there be light, but just enough," *IEEE Spectrum*, Dec. 1975, pp. 28–34.

5. G. C. NEWTON, JR. "Energy and the refrigerator," *Technology Review*, January 1976, pp. 57–63.

6. G. DALLAIRE, "Designing energy-conserving buildings," *Civil Engineering*, April 1974, pp. 54–58.

7. C. A. BERG, "Energy conservation through effective utilization," *Science*, July 13, 1973, pp. 128–137.

8. R. S. BERRY AND H. MAKINO, "Energy thrift in packaging and marketing," *Technology Review*, February 1974, pp. 33–43.

9. W. HAUSZ AND C. F. MEYER, "Energy conservation: Is the heat storage well the key?," *Public Utilities Fortnightly*, April 24, 1975, pp. 34–38.

10. B. HANNON, "Energy conservation and the consumer," *Science*, July 11, 1975, pp. 95–102.

11. L. SCHIPPER, "Towards more productive energy utilization," Lawrence Berkeley Laboratory Report 3299, Oct., 1975.

12. J. E. SNELL *et al*, "Energy conservation in new housing design," *Science*, June 25, 1976, pp. 1305–1311.

13. F. VON HIPPEL AND R. H. WILLIAMS, "Energy waste and nuclear power growth," *Bulletin of Atomic Scientists*, Dec. 1976, pp. 18–21, 48–49.

14. R. H. WILLIAMS, *The Energy Conservation Papers*, Ballinger Publishing Co., Cambridge, Mass, 1975.

15. M. H. ROSS AND R. H. WILLIAMS, "Energy efficiency: Our most undervalued energy resource," *Bulletin of Atomic Scientists*, Nov. 1976, pp. 30–38.

16. P. SPORN, "*Energy in an Age of Limited Availability*," Pergamon Press, New York, 1977.

17. L. SCHIPPER AND A. J. LICHTENBERG, "Efficient energy use and well-being: The Swedish example," *Science*, Dec. 3, 1976, pp. 1001–1012.

18. E. HIRST, "Residential energy-use alternatives: 1976 to 2000," *Science*, Dec. 17, 1976, pp. 1247–1252.

19. "Will energy conservation throttle economic growth?", *Business Week*, April 25, 1977, pp. 66–80.

20. M. B. BEVER, "Recycling in the materials system," *Technology Review*, February 1977, pp. 23–31.

21. M. H. ROSS AND R. H. WILLIAMS, "The potential for fuel conservation," *Technology Review*, February 1977, pp. 48–57.

EXERCISES

*20.1** One strategy suggested for reducing the home heating energy requirement is to turn off the heater from 10 P.M. to 6 A.M. Calculate the percentage energy savings that this strategy accomplishes. Assume that the hourly temperature drop $d(T - T_0)/dt$ is proportional to the difference between the inside temperature T and the outside temperature T_0, so that

$$\frac{d(T - T_0)}{dt} = -\alpha(T - T_0).$$

Also, assume that $\alpha = 0.025$/hour and K, the thermal conductance of the house, is equal to 600 Btu/hr·°F. The initial difference between the indoor and outdoor temperature (at 10 P.M.) will be assumed to be 30°F.

20.2 Many beverages are sold in aluminum or steel cans. It takes about 1086 kWh to produce and use 1000 steel cans, accounting for energy consumption from ore mining to can disposal. [8] The comparable figure for aluminum cans is

---

* An asterisk indicates a relatively difficult or advanced exercise.

6893 kWh. These totals include filling, storage, collection, and disposal. Determine the energy cost per 1000 cans if they are recycled.

20.3 Explore the potential energy savings in one of the following industries: (1) oil refining, (2) ammonia synthesis, (3) iron and steel manufacture. Propose one or two approaches to energy conservation in that industry and discuss the advantages and disadvantages of that measure.

20.4 One possible energy-conservation scheme is based on instituting Daylight Savings Time throughout the whole year. One estimate is that the approach would save one percent of the nation's electric-power consumption. Explore the support for such an action in your community, and list the advantages and disadvantages of such a measure.

20.5 The U.S. military uses about 700,000 barrels of oil per day for ships, vehicles, airplanes, and installations. The military uses this energy to maintain and operate its fleet, Air Force, and Army, and keep them in a combat-ready status. Examine the potential for reducing this energy usage by 10% per year. What percentage of oil consumption of the U.S. does the military account for?

20.6 How many soft-drink cans can be manufactured from recycled aluminum with the energy needed to make a single can from aluminum ore?

20.7 The present technique for making paper involves the formation of a pulp slurry, the water content of which is reduced from 99.5% to a final, ideal value of 7%. The water is removed mechanically down to about 70%; from then on the removal is by evaporation at low efficiency. Even ideally, the current drying operation requires a minimum of 1.5 MWh/tonne of paper, but in practice at least three times this energy is consumed. Explore alternative methods of paper drying and other means of improving the efficiency of the present process. List several schemes for conserving energy in the paper-making process.

20.8 Estimate the total wattage of incandescent lamps and the total wattage of fluorescent lamps in your home or dormitory. Calculate the potential savings if all the lighting in your home was converted to fluorescent lighting.

20.9 Examine the potential for the use of bicycling or walking for short trips in your city. Propose two incentive schemes for attracting people to use bicycles or walking for short trips.

20.10 Contact a local heating-supply contractor about the possibility of using a heat pump for a local house. Determine the purchase price, the electric power requirement, and the C.O.P. Perform an analysis to determine whether you would purchase the heat pump for a house of 1800 ft^2 in your town.

20.11 Prepare a plan for your state which would require that the EER of room air conditioners be equal to 8 in two years from now, and 10 after four years from now. Estimate the savings in electric-energy consumption in your state if this action was adopted.

*20.12 Adobe brick houses are common in New Mexico and they often do not require air conditioning. If the brick is ten inches thick and the conductivity of the brick is 4 Btu/hr, calculate the heating load required if one assumes the walls and roof are adobe and the windows may be neglected. Assume the house is a 1500 ft^2 home (a one-story flat-roof rectangle of 30 by 50 feet). Repeat your calculations for a home constructed of wood with 3 inches of asbestos insulation. Assume in both cases that $\Delta T = 50°F$ and the house is 10 feet high.

*20.13 If there are 100 ft^2 of one-quarter-inch glass uncovered windows in the house of Exercise 20.12, calculate the heat loss due to the windows.

*20.14 If drapes cover the windows, reduce the conductivity, k, to 1 Btu/hr and increase the net thickness, d, to 1 inch. Calculate the loss due to the windows as considered in Exercise 20.13.

20.15 Calculate the percentage savings for the typical house in New York City if the point of operation of the house is changed from point A to a house with storm windows and ceiling and wall insulation $2\frac{1}{2}$ inches thick. Refer to Fig. 20.4.

20.16 If your city has one or more electric power plants, determine their power rating and the waste heat disposed of. Prepare a plan for utilizing this power for hot-water district heating in your city. Calculate the potential energy savings for your city by instituting your plan.

20.17 Consider the standard and target house described in Table 20.10. Assume that the insulation and infiltration measures for the target house would cost the owner of a standard home $3500 to install. Also, consider a home utilizing electric heating instead of gas heating; (a) calculate the savings of power capacity the utility achieves, and compare the capital cost of installing that capacity with the cost of improving the home. (b) Prepare an analysis for the utility which would enable it to decide whether to include the cost of the insulation measures in the homeowner's bill at 9% interest over the next four years. Assume that the home-owner pays 4¢/kWh for electric energy.

*20.18 Explore the advantages and disadvantages of several economic measures to obtain energy conservation. Consider at least two of the following: (1) rationing; (2) tax credit; (3) an energy tax; (4) subsidized low-interest loans for capital costs of conservation measures.

20.19 Prepare a set of objectives for energy conservation for the U.S. in 1990. List a set of items and the annual energy savings you expect to achieve.

20.20 About one-fourth of new dwellings in the U.S. are home trailers, which are inefficient users of heating and cooling energy. Determine the insulation standards and thermal loss rate for a typical house trailer in your area. Compare the house trailer with a regular house.

20.21 If the dollars saved by conserving energy are then used in new ways which consume even more energy, the net effect may be nonconserving. List several energy-conserving sequences which result in energy savings. Also, list several net-energy-consuming sequences. One such example is saving money by using a bicycle rather than a bus, but then using the money saved to take an added airplane trip.

20.22 The adoption of a four-day work week by U.S. business would reduce consumption of energy by commuters. While the business would maintain a five-day business week, each employee would work four (nine- or ten-hour) days. (a) List the advantages and disadvantages of the plan, and (b) determine whether an energy savings would be achieved. (This is a "think" problem. Some areas you should consider are: air pollution, parking, traffic congestion, police manpower; highway maintenance, replacement of automobile parts, etc.) Consider as many ramifications as you can think of, analyzing pros and cons of each factor; but in answering (a) and (b), use only those elements of the situation that would have an impact on energy consumed.

20.23 One version of the four-day week envisions a progressive sequence of days off for the individual employee (Monday one week, Tuesday the next, etc.). If you had to "sell" the idea to business and its employees, (a) what would be strong points of your presentation? (b) From what segments of the commercial world would you expect to receive opposition?

21

Energy and Economics

21.1 INTRODUCTION

The economics of energy concerns the allocation of the scarce resource, energy, which has various alternative uses. In part, prices are used to allocate this resource. Furthermore, governmental policies, taxation, and trade quotas are used to allocate the various alternative sources to potential consumers.

The economics of energy refers to its aspect as a commodity. Exchange takes place when energy is transferred from one party to another and some valuable good, usually money, is transferred in the opposite direction. Energy is principally used as an input to an economic process and as an intermediate good. When energy is exchanged, it has a price, in dollars per unit.

The affluence of a society is increased up to a point by the use of energy. Thus, there is a strong correlation between the Gross National Product (GNP) of a rising energy inputs from fossil fuels and alternative sources. However, increasing affluence itself results in increasing use of energy.

Another factor regarding energy use is its availability. The location of the source relative to the consumer results in a transportation problem. Also, there is a time factor; that is, when is this energy available? The transportation and sale of energy requires some appropriate form of organization. Therefore, the existence of markets, pertinent regulation, and adequate organization are important parts of the economics of energy.

A model of the interactions of the social process of an industrialized nation is shown in Fig. 21.1. This model portrays the important interaction of energy, economics, and the environment. Energy and economics interact directly through supply and demand, prices, and capital investments. The energy marketplace is a result of the population, investment capital, governmental policies, technology,

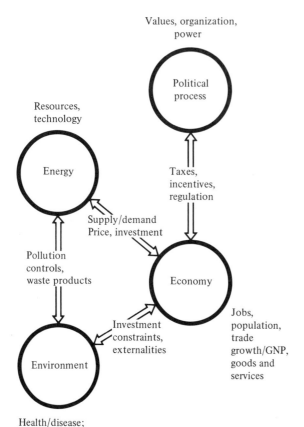

Values, organization, power

Resources, technology

Energy

Political process

Taxes, incentives, regulation

Supply/demand Price, investment

Pollution controls, waste products

Economy

Jobs, population, trade growth/GNP, goods and services

Investment constraints, externalities

Environment

Health/disease; Ecosystem/climate

Fig. 21.1 *A model of the inter-actions of the social processes in an industrialized nation.*

energy sources, and energy demand. In theory the marketplace balances all these factors and yields an equilibrium of supply and demand by means of prices and controls. As the world's energy marketplace changes, a redistribution of economic wealth occurs.

The question in the marketplace is how much energy customers will demand at a given price, and how much will it pay business or a government to supply energy at a given price. At some point these lines cross, and the market reaches a point of equilibrium. However, this market mechanism will be influenced by public policy or private monopoly, as in the case of OPEC. In the late 1960's, as a result of Federal Power Commission controls over prices at the wellhead for natural gas in the U.S., the country began to develop a systematic excess demand for gas.

The impact of energy and economics is felt in unemployment, GNP, inflation, and trade. As energy prices rise, the rate of growth of GNP in the U.S. may decline from its historic 3.8 percent per year to 3.0 percent per year or less. [1, 35]

21.2   GROSS NATIONAL PRODUCT AND EMPLOYMENT

Energy consumption and the Gross National Product of the U.S. have maintained a long-term relationship. From about 1880 to 1915 there was a steady rise in energy requirements per unit of GNP. From 1915 through 1965 the use of energy per unit of GNP has been steadily declining. The decline since 1915 resulted from an increased efficiency of energy production and utilization.

The energy used per unit of GNP rose during the period 1969–1970 and again during 1973–1974 due to the recessions of these periods when the real GNP declined slightly. A curve depicting the energy–GNP ratio is shown in Fig. 21.2.

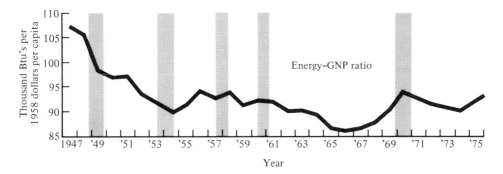

Fig. 21.2   *The energy–GNP ratio for the U.S. for the period 1947 to 1975.*

A quantitative study of the relationship of energy consumption to GNP was completed for thirty nations using the mathematical relation

$$\log E = a + b \log \text{GNP}, \qquad (21.1)$$

where E = energy consumption in millions of metric tons. [2] This study showed correlations between GNP and energy ranging from 0.84 to 0.99 for the period 1953 to 1965; the statistical reliability of the regression was uniformly high. However, there was a wide range in the factor b, often called the income elasticity. The value of b and the correlation coefficient R^2 are given in Table 21.1. By examining the table, one notes that the income elasticity is somewhat related to the nation's stage of economic development among other factors.

The energy–GNP ratio for selected nations in 1971 is given in Table 21.2. The range of this ratio is about three to one, with the U.S.S.R. using about 2.7 times as much energy to generate a unit of GNP as did France in 1971. The United States was achieving a reasonable average energy–GNP ratio and has essentially held that position since 1971 as a result of energy conservation schemes and reduced growth in energy consumption.

Table 21.1

RELATIONSHIP BETWEEN ENERGY
CONSUMPTION AND GNP

Nation	b	R^2
India	1.89	0.97
Italy	1.77	0.98
Canada	1.37	0.99
Brazil	1.19	0.98
Japan	0.96	0.99
U.S.A.	0.96	0.99
West Germany	0.66	0.95
United Kingdom	0.48	0.84

Table 21.2

THE ENERGY—GNP RATIO FOR SELECTED
COUNTRIES IN 1971

Nation	Energy per dollar of GNP (Kilograms of coal equivalent per 1971 dollars)
U.S.S.R.	3.30
Norway	2.69
Canada	2.63
United Kingdom	2.31
U.S.A.	2.24
Netherlands	1.96
W. Germany	1.65
Japan	1.62
Italy	1.53
France	1.23

The U.S. energy–GNP ratio has remained within the range 87 to 93 × 10^3 Btu on the basis of 1958 dollars, since 1952. Thus, the efficiency of utilization of energy has remained essentially constant. In 1974 dollars the ratio was 60 × 10^3 Btu per dollar (3.3 × 10^7 joules per dollar). [3] Thus, on an average basis, each dollar of goods or services requires an energy expenditure of 60 × 10^3 Btu or 17.5 kilowatt-hours (approximately $\frac{1}{2}$ gallon of gasoline).

Employment in the U.S. has been studied relative to energy consumption. [4] Energy use can be shown to be closely related to employment and when energy consumption rises employment also rises. However, it is difficult to ascertain whether energy is a causal factor or simply follows the trend of a recession resulting from other causes. For the period 1958 to 1972, energy consumption increased from 240 to 333 × 10^6 Btu's per person, while employment increased at a rate of 575 persons for each trillion Btu's consumed. During the period of the 1973–74 recession, GNP declined, unemployment increased, and energy consumption per capita declined.

Other causal factors besides the increases in energy prices contributed to the phenomenon of the recession. [29] Nevertheless, it is interesting and important to note the correlation between energy use and employment during earlier periods in the U.S.

21.3 ELASTICITY OF DEMAND AND SUPPLY

The change in the supply of a commodity in response to its market price is called *supply elasticity*. Similarly, a commodity has an elasticity of demand in response to price. In general we may write an equation for the elasticity e of demand q as

$$e = \frac{\Delta q/q_0}{\Delta p/p_0}, \tag{21.2}$$

where p is the price of the commodity. The values p_0 and q_0 are the initial values at which the percent changes $\Delta p/p_0$ and $\Delta q/q_0$ are evaluated. Therefore, we can state that the elasticity in the demand of a commodity is the ratio of the percent change in demand to the percent change of the price, where both are evaluated at specified values. As an example, consider the demand–price relationship shown in Fig. 21.3. Let us first evaluate the change in demand for a price of \$4 ($p_0 = 4$). Then we have

$$e = \frac{\Delta q/q_0}{\Delta p/p_0} = \frac{-(1000/6000)}{(\$1/\$4)}$$

$$= -0.667, \tag{21.3}$$

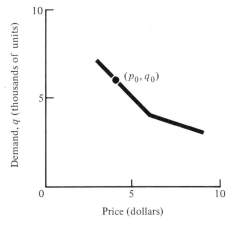

Fig. 21.3 *A demand-vs.-price curve for a commodity.*

where the elasticity is negative since the slope of the curve is negative. In other words, as the price increases, the demand declines.

If the price is set at $7, then the elasticity is

$$
\begin{aligned}
e &= \left(\frac{\Delta q}{\Delta p}\right)\left(\frac{p_0}{q_0}\right) \\
&= \left(\frac{-1000}{3}\right)\left(\frac{7}{3667}\right) \\
&= -0.636.
\end{aligned}
\tag{21.4}
$$

An inelastic commodity is one that occurs when $\Delta q/\Delta p = 0$ or $e = 0$.

The elasticity for the demand in electricity as a function of price, population, and income is given in Table 21.3. [5] For example, if the commercial electricity price increased 20 percent, it would cause the long-term demand to drop nearly 30 percent after seven years. Similarly, if the population increased by 10 percent, we would expect a 10 percent increase in commercial electricity demand and an 11 percent increase in industrial demand. The most important factors appears to be price, followed by population and per capita income.

Table 21.3
ELASTICITY OF DEMAND FOR ELECTRICITY

Factor	Residential	Consumer class Commercial	Industrial
Price	−1.3	−1.5	−1.7
Population	+0.9	+1.0	+1.1
Income per capita	+0.3	+0.9	+0.5
Percent of response in 1st year	10%	11%	11%
Years for 50% of total response	8	7	7

Thus, in the absence of other effects, increased prices of fossil fuels could result in a reduced demand. Of course, this has already occurred in the case of increased OPEC prices, causing declines in the use of oil in many countries. However, there is a long delay in responding to such price changes since countries must shift to alternative energy sources. For example, France has accelerated its nuclear power projects, but delays in starting plant operation are necessary for site relocation, approval, and construction.

The average response of the industrial sector to a 10 percent rise in energy prices is a 2.1 percent decrease in energy after one year, and 7 percent in the long run (one decade). [6] Therefore, one may estimate the elasticity to energy prices in the industrial sector as $e = -0.7$.

We can also express the elasticity of supply of a commodity as

$$e = \frac{\Delta S/S_0}{\Delta p/p_0},$$ (21.5)

where ΔS is the change in the supply and S_0 is the original supply. This elasticity will be positive, since increased prices will normally result in increased supply. However, if a cartel such as OPEC can in fact control the supply of oil, the supply can be artificially held down, and a "shortage" can be induced.

We may express the dependence of the Gross National Product (GNP) on the price of energy as follows: [7]

$$\frac{\Delta GNP}{GNP} = \left(\frac{\Delta p}{p}\right) \frac{p\beta}{1 - p\beta},$$ (21.6)

where p = price of energy and β is the energy–GNP ratio as recorded in Fig. 21.2. The value of β was 92×10^3 Btu/dollar of GNP in 1975. Therefore, if the price of energy increased by 20 percent ($\Delta p/p = 0.20$), then the change in GNP is

$$\frac{\Delta GNP}{GNP} = \left(\frac{\Delta p}{p}\right) \frac{0.092}{1 - 0.092} = (0.20)(0.101) = 0.020,$$ (21.7)

where it is assumed that the price of energy is \$1.00 per 1 million Btu in 1975. Thus a 20 percent increase in the price of energy would result in a 2 percent decrease in GNP.

Recently, an economist, Georgescu-Roegen, has noted that the concept of entropy should be integrated with the normal economic concepts of supply and demand. [8, 9] With the inclusion of the entropy law, the economic process is seen as irrevocably declining rather than cyclical. The standard view is that the price mechanism will offset scarcities. However, in his view, since entropy is increasing, scarcity of high-availability energy is irrevocable. Nevertheless, the time scale of this irreversible process is very long and depends upon the availability of alternative sources such as solar energy. Of course, eventually even solar energy will run out, but that time is infinite within the scale of the age of civilization.

Input–Output Analysis

When the price or supply of a fuel is changed, as occurred during the Arab oil embargo of late 1973, the question is not only whether the economy is impacted but also which industries and segments of society will change to the greatest degree. One method for answering questions about sectors or portions of the economic system is input–output analysis. Input–output analysis is the product of Professor Wassily Leontief, who received the Nobel Prize in economics for his research.

An input–output table gives a detailed picture of the flow of goods, services, or energy that individual sectors or industries buy and sell to each other in a particular year. Each horizontal row of the table gives the amounts that a particular industry sold to all sectors. Transactions are measured in dollars or in energy units (such as kilowatt-hours). Individual vertical columns indicate how much each sector purchased as inputs from other sectors. Input–output coefficients are used to form a system of linear equations connecting the outputs of all industries. [10] The central idea of input–output (I-O) analysis is the concept that there is a fundamental relationship between the volume of output of an industry and the size of the inputs going into it.

Only a small portion, about 30 percent, of the total energy usage in the U.S. is a result of direct purchases. The energy used indirectly to produce goods and services is therefore extremely important. For example, less than 10 percent of the energy needed to make an automobile is consumed directly by the manufacturer. Over 90 percent is required to produce the steel, glass, rubber, and other inputs. Therefore, to determine total energy cost, one must model the physical flow of materials and energy through the economic system. [11, 12]

Consider an economic system composed of N sectors, each producing a unique output. Assume an energy intensity E, measured in Btu per unit output associated with the output of sector j. The data base is an $N \times N$ matrix of interindustry transactions $\mathbf{T}$, a sector of total outputs $\mathbf{X}$, and a sector of energy extraction data $\mathbf{E}$. The total output x_j of a sector is the sum of its sales to other sectors plus its sales for final consumption y_j, or

$$x_j = \sum_{k=1}^{N} T_{jk} + y_j. \tag{21.8}$$

The energy-balance equations are

$$\sum_{i=1}^{N} \mathscr{E}_i T_{ij} + E_j = \mathscr{E}_j x_j. \tag{21.9}$$

In matrix notation we have

$$\mathscr{E}\mathbf{T} + \mathbf{E} = \mathscr{E}\mathbf{Ix}; \tag{21.10}$$

solving for $\mathscr{E}$ we obtain

$$\mathscr{E} = \mathbf{E(Ix - T)}^{-1}, \tag{21.11}$$

where $\mathbf{I}$ is the identity matrix. The values of the energy intensities are the important data to obtain and utilize in the analysis. For example, the energy intensity of aluminum and glass are 380 and 101×10^3 Btu per U.S. dollar, respectively.

For example, the actual fuel in the tank of an automobile accounts for only 57 percent of the total used by that sector. An additional 13 percent goes to the cost of extraction, refining, and other energy losses. Another 30 percent is attributable to manufacture of the car, parts maintenance, roads, etc. The same analysis can be applied to other sectors of the economy.

21.4 NET ENERGY

In examining how much energy results from a process such as the generation of electricity, it is useful to determine how much total energy goes into producing the output. Thus, for example, we need to account for the energy consumed in extracting and bringing the fuel to the generating plant, as well as the energy lost in the generation process.

Net energy is defined as the amount of energy that remains for consumer use after the energy costs of finding, producing, upgrading, and delivering the energy have been accounted for. Thus, we need to account not only for the energy used to power an automobile but also the energy required to find the oil, bring it to the wellhead, transport it, refine it, and get it to the customer. It is clear that obtaining energy from oil shale or tar sands may consume a large percentage of the energy held in the shale or sands, and thus result in a low amount of net energy. As wells are drilled deeper or it takes more energy to extract the coal from the ground, the net energy of the process will decline. [26]

The cost of constructing a pipeline to deliver gas from Alaska's North Slope would cost $6 billion, which is one-third of the U.S. investment in natural-gas pipelines. Yet the gas flowing through this pipe would supply only 5 percent of the present U.S. demand. Perhaps a company could earn a sufficient return on its financial investment to complete the project, but is it a wise investment of energy to construct and operate the pipeline? Another question is: What is the net energy of the nuclear-power reactor process which requires a large amount of energy to enrich the uranium used as fuel in the reactors? As new, more efficient technologies are introduced, net energy increases in the process.

With ample energy, all materials can be produced or substitute materials found. For example, sulfur can be removed from the coal before or during the combustion process in an electric generating plant. However, all of these processes require energy, and thus energy itself is an important limiting factor to increasing energy supply.

For all energy-production processes, energy subsidies are added in the form of energy used to extract, refine, or transport the energy. In summing the various types of energy subsidies, all energy measures must be of the same quality if they are equivalent in their ability to do work. For example, a calorie of electricity can do more work than a calorie of coal or oil, and both can do more work than a calorie of sunlight. Energy quality is calculated by evaluating the energy used in converting from one energy form to another—that is, by evaluating the amount of one type of energy required to develop another. [13,14]

One measure of net energy is the net-energy ratio, which is defined as

$$\text{NER} = \frac{\text{Delivered energy}}{\text{Energy sequestered}}, \tag{21.13}$$

where the energy sequestered is the energy subsidy for the process. The NER for several selected energy resources in the U.S. is given in Table 21.4. Thus, the NER for domestic gas is much greater than for coal gasification because of the large amount of energy required to operate the gasification process. The NER for electricity from oil is only 0.30; that is, the amount of energy sequestered to yield one unit of electric energy is 3.3 units of energy.

Table 21.4

NET ENERGY RATIOS FOR SELECTED
ENERGY PROCESSES IN THE U.S.

Energy process	NER
Light-water nuclear reactor	9
Domestic natural gas	60
Coal gasification	7
Oil from oil shale	2.8
Geothermal dry steam to electricity	12.6
Electricity from oil	0.30
Imported oil	6

21.5 THE PRICE OF ENERGY

In the past, economic growth has been closely associated with rising energy inputs from fossil fuels. As the price of these energy inputs rises, the rate of growth of a nation's economic product is expected to decrease. Also, a sharply rising price of fuels results in different impacts on various industries and sectors of the economy. Any change in the relative price structure of energy will result in substantial redistribution of income.

In the 1960's, as a result of Federal Power Commission (FPC) control over the wellhead price of natural gas produced in the U.S., this country developed an excess demand for inexpensive gas. As gas prices are deregulated or allowed to rise in the late 1970's, we can expect these changes to impact various segments of society in different ways. As oil prices increase in the U.S., a resulting transfer from the oil consumer to the oil producer occurs. The U.S. oil companies' profits increase in this process. During the period 1920 to 1970 there has been a declining trend in the cost of energy of about 2 percent per year. Since 1970, the price of energy has continually increased. However, the price of energy relative to the average cost of labor declined continually from 1900 to 1970 and has only levelled off since 1972. Nevertheless, the cost of energy in the U.S. is less than for other industrialized nations.

The average price for fossil fuels in the U.S. is given in Table 21.5. The price of petroleum to refineries rose by 144 percent over the period 1970–75, while the price of petroleum delivered to electric utilities rose by 235 percent during that same period. The price of natural gas and coal delivered to a utility for the period 1920–1975 rose by 126 percent and 134 percent, respectively. In 1975, the price per million Btu for utility fuel was $1.96, $0.82, and $0.52 for oil, coal, and gas, respectively. The price of oil is the highest since it has many uses in transportation in the chemical industry and is relatively easily transported. The price of coal is about one-half of that of oil, but the use of coal requires extensive investment in pollution-control equipment. Natural gas is the least expensive because Federal regulations hold the price low.

In late 1975, the U.S. Congress enacted a bill which rolled back the average price of domestic crude oil from $8.75 to $7.66 on January 1, 1976, and then

Table 21.5

AVERAGE PRICE FOR FOSSIL FUELS IN THE U.S.

	1970	1973	1974	1975	Percent increase 1970–75
1. *Domestic petroleum* (to refinery)					
Price ($/barrel)	3.58	4.00	8.00	8.75	144
Price ($/10⁶ Btu)	0.60	0.65	1.31	1.43	
2. *Petroleum* (all sources to electric utility)					
Price ($/barrel)	3.58	4.32	11.56	12.00	235
Price ($/10⁶ Btu)	0.60	0.71	1.88	1.96	
3. *Coal* (to utility)					
Price ($/ton)	7.70	9.00	14.50	18.00	134
Price ($/10⁶ Btu)	0.35	0.40	0.66	0.82	
4. *Natural gas* (interstate to utility)					
Price ($/1000 cu ft)	0.23	0.34	0.44	0.52	126
Price ($/10⁶ Btu)	0.23	0.34	0.44	0.52	

gradually allowed the price to increase over the 40 months beginning January 1, 1976, until price controls are phased out.

The price of natural gas to various sectors of the economy is shown in Table 21.6. The price to the industrial sector is relatively low, but the gas supply to industry usually is based on interruptible supply contracts. That is, with an impending shortage in a gas company's supply, it may cut off the supply to the industry temporarily. By contrast, the supplier must continue to supply residences, and the relatively higher price to the residential sector reflects this guarantee, as well as the cost of the distribution network.

Energy transmission costs, as shown in Table 21.7, must also be ultimately included in the final cost of the energy delivered to the consumer. For example,

Table 21.6

THE AVERAGE PRICE OF NATURAL GAS
IN 1971 IN THE U.S. FOR
VARIOUS SECTORS

	Price ($/thousand cubic feet)
Wellhead price	0.16
Industrial price	0.32
Commercial price	0.76
Residential price	1.03

Table 21.7
ENERGY TRANSMISSION COSTS

Form of energy	Transmission mode	Transmission cost ($¢/10^6$ btu/100 miles)
Natural gas	Pipeline	1.3
Hydrogen	Pipeline	3.0
Electricity	Overhead line, 500kV	4.2
Electricity	Overhead line, 200kV	7.8

the wellhead cost of gas in 1975 was about $0.40/10^6$ Btu while the transportation costs for delivering that gas through a 1000-mile pipeline was $0.13/10^6$ Btu.

The price of gasoline, a refined product of crude oil, has also increased significantly over the past several years. The average wholesale price of gasoline rose from 12.5¢/gallon in 1972 to 26¢/gallon in 1975.

The retail price of gasoline in the U.S., expressed in constant 1967 dollars, is given in Table 21.8, for the period 1920–1980. The price of gasoline declined from 1920 to 1970, with an interruption of the decline during World War II due to war-time shortages. After 1973, because of world-wide increases in the cost of oil, the cost of gasoline began to rise again.

Table 21.8
RETAIL PRICE OF GASOLINE IN THE
U.S. EXPRESSED (EXCLUSIVE OF TAX)
IN U.S. DOLLARS

Year	Price (Constant 1967 dollars)
1920	0.49
1930	0.39
1940	0.44
1950	0.37
1960	0.35
1970	0.30
1973	0.30
1975	0.40
1980 (Projected)	0.50

21.6 TRADE AND THE ECONOMY

The United States, Europe, and Japan have experienced a rising cost for the oil imports they depend upon for an important source of energy. The U.S. paid $25.1 billion in 1974 for its petroleum imports, up $17.6 billion from the year before. The cost of U.S. oil imports for the period 1950 to 1976 is given in Table 21.9. During that period the amount of imports increased steadily, as well as the cost per barrel of the imported oil.

Table 21.9

THE COST OF OIL IMPORTS IN THE U.S.

Year	U.S. oil imports (Millions of barrels)	Cost (Millions of dollars)	Average cost per barrel
1950	199	113	$0.57
1955	321	433	1.35
1960	590	1,140	1.93
1965	833	1,711	2.05
1970	1,154	3,060	2.65
1973	2,199	7,500	3.41
1974	2,400	25,100	10.46
1976	3,100	35,000	12.10

The United States balance of payments has held relatively steady over the recent years since, as the cost of oil has increased, the income obtained from exports of goods and resources has resulted in a relatively balanced trade account. In 1975 exports and imports essentially balanced at about $26 billion, with a resulting small trade surplus compared to a trade deficit of $3 billion in 1974. The U.S. has a strong export market in commercial aircraft, chemicals, computers, and agriculture.

The balance between agricultural exports and oil imports is a critical factor for the U.S. economy over the next five years. The U.S. has completed an agreement with the U.S.S.R. to deliver large shipments of grain and would like to import oil from Russia in return.

One possibility is that, if the price of domestic fuels in the U.S. are held by law to a relatively low level, a continuing stimulation of consumption will result. A stimulation of energy consumption over the next five years will result in increased oil imports. In order to meet this trade imbalance, the U.S. could be forced to increase its export of food, particularly grains, thus resulting in higher domestic prices for U.S. foods. Thus, the U.S. may be confronted with a near-term problem of living with either higher oil prices or higher food prices. Eventually, the substitution of an alternative energy source could replace the need to import oil in order to meet the nation's demand for energy.

21.7 CAPITAL COSTS OF PUBLIC UTILITIES

The cost of construction of electric generating plants, coal gasification plants or gas pipelines are examples of large investments of capital required, if new, modern facilities are to be available during the next decade.

A large portion of these expenditures will have to be borne by those companies licensed to serve a region with energy sources. These companies are called public utilities; representative utilities are Consolidated Edison of New York and the Los Angeles Department of Water and Power in California. Certain businesses, though privately owned, are generally considered to be "affected with a public interest" and customarily are classified as public utilities. Traditionally, such businesses have included those supplying water, artificial and natural gas, telegraph and telephone services, electricity, and street transportation. These utilities are said to supply a necessary or essential service and are given a monopoly of service to a region. Consumers' demands for the products of public utilities are not significantly elastic; that is, the quantity purchased is not significantly affected by small price changes. For this reason, these services, as a *public good*, are regulated by federal or state agencies.

The nature of electric-power generation makes the industry very capital-intensive. Technological development has increased capital intensity somewhat by substituting capital for fuel. In view of the extreme capital intensiveness of the utility industry, it is not surprising that these utilities invest a large amount of effort in the coordination of operations, load control, emergency procedures, and maintenance scheduling. Overall coordination leads to economies of scale on equipment, thus resulting in larger power plants and higher-voltage power transmission lines. Economies of scale both in equipment and in pooling of reserve capacity apply to transmission as well as to generating facilities.

Regulatory agencies permit utilities an income based on their capital investment. Before utilities can increase allowable income, they must increase the rate base upon which they reserve the "fair rate of return" permitted by the regulatory agency. Utilities have, in the past, achieved long-run economies of scale, thus receiving a further incentive to expand the service base. Utilities usually receive a long-term return on investment of about 7 to 10 percent of capital invested.

The consumption of electric energy grew at an annual rate of 7 percent over the period 1960 to 1973. However, it is expected this annual rate may decline to a 3 to 4 percent annual rate over the next decade. This decline in growth will result in a reduced need for new electric generating plants. Nevertheless, replacement for old plants and additional capacity will be required. About 50 percent of what consumers pay today for electricity is attributable to capital costs of utilities. The electric utilities spent about $17 billion in 1976 for capital investment, compared with $12 billion in 1971 and $5 billion in 1965.

The overall capital requirements for U.S. energy production over the period 1971 to 1985 are very large indeed. One estimate states that the total capital requirement for all forms of investment in capital expenditure for energy generation and extraction will equal in excess of $500 billion over the 15-year period.

In 1960 the total capitalization of investor-owned (private) electric utilities amounted to $40 billion. By 1970 it had grown 90 percent to $76 billion. By 1980 it may grow another 300 percent to $235 billion. [16] One estimate states that the electric utilities will need to raise $300 billion between 1975 and 1985 to pay for new capital investment.

The petroleum industry may require $350 billion over the decade 1975–85, and the coal industry may need $25 billion. Thus, the total capital investment for all energy needs may be as high as $700 billion over the period 1975 to 1985. Irrespective of the capital needs of the energy industry, there is real doubt whether these amounts of capital can be raised. Inflation, scarcity of material, high equipment costs, and high interest rates may make the projected level of capital investment unrealizable.

The need for reliable electric power forces the electric utility system to be highly capital-intensive. Electric utilities have a gross plant investment per dollar of sales revenue seven times higher than the average of all manufacturing corporations. The gross plant investment for the average electric utility was $4.42 per dollar of sales revenue. With their reduced ability to raise capital, utilities are forced to cut back expansion plans. These cutbacks may be fortunate, since the expected expansion of electric power use may never come to pass. However, some electric utilities, such as Consolidated Edison of New York, have encountered difficulty in borrowing capital and are unable to raise equity capital due to their low price-to-book-value ratio. The price–book ratio is the ratio of the current selling price of a company's common stock to its book (asset) value. The price–book ratio of many utilities dropped below 1 in 1974; thus the utilities hesitated to sell shares of their stock at an undervalued price.

The energy industry can only lower its capital needs as one or more of the following events occurs: (1) a lower inflation rate, (2) delayed installation of pollution-control equipment, (3) a lowered growth rate of energy consumption, (4) a decreased delay of construction, (5) an increased debt ratio, or (6) a decrease in the population growth rate.

The time period to construct a large new electric power plant is given in Table 21.10. [17] The cost per kilowatt of installed kilowatt capacity is also given in the table. Clearly, nuclear power plants cost the most and require the longest construction time. A 1000-megawatt power plant costs from $400 to $750 million dollars to construct. These are large capital costs for any utility to absorb without a ready ability to borrow capital or to raise large amounts of equity capital.

Table 21.10

CONSTRUCTION COSTS OF NEW ELECTRIC POWER PLANTS
FOR A PLANT LARGER THAN 800 MEGAWATTS

	Time to construct (Months)	Cost (Dollars per kilowatt)
Gas turbine	16	$200
Oil-fired steam	56	413
Coal-fired steam	58	526
Nuclear	96	800

The cost per kilowatt for new fossil plants declined from 1960 to 1966, when it reached a low of about $119 per kilowatt. However, it may cost $700 to $1000 per kilowatt for plants going into operation in 1984 or later. [18] Whether the capital investment will be available for these new plants is a real question for utilities to consider. If the growth rate of electric power consumption and the general use of energy drops below expectations in the U.S. over the next decade, this occurrence may well turn out to be an unexpected blessing.

In the U.S. most new oil or gas wells require a capital investment of $4.5 billion for every quad (10^{15} Btu) of energy produced annually. Synthetic oil and gas from coal would cost more than $10 billion per quad. New electric-power plants range from $45 billion per quad for coal to about $90 billion for nuclear. However, most investments for conservation measures cost less than $10 billion per quad saved and many cost less than $4 billion per quad. [35] Capital investment in conservation measures may be the nation's best buy in the 1980's.

21.8 PUBLIC-UTILITY RATES

A public utility's profit or return on investment is determined by multiplying the rate base (capital investment) times the rate of return. This rate of return, determined through a regulatory process, is normally 6 or 7 percent. Thus, a utility may increase its profit by increasing its capital investment or its rate of return. The value of the rate base falls between two extremes: original cost and cost of reproduction. In other words, when placing a value on an electric generating plant for purposes of determining the rate base, the regulatory commission will consider the original cost of building that plant and the cost of reproducing it today. Under inflationary conditions (what has been and is expected to be the normal condition for the future), the commission will favor the lower original cost, while the utilities will favor the higher cost of reproduction. A majority of the state commissions use original cost, one uses reproduction cost, and the rest use some intermediate value. During periods of high inflation, the use of original cost places a severe burden on the utility to raise rates.

A principal obligation of a public utility, gas or electric, has been to charge only reasonable prices or rates. Rates of return are fixed by a regulatory agency in each state. The determination of reasonable rates, in order to reach a fair rate of return, is the most difficult and controversial aspect of public utility regulation. The use of the original-cost methods for valuing the plant tends to provide more rate stability as well as lower rates during periods of rising prices. Reproduction cost, on the other hand, tends to reflect current economic conditions more accurately and to approximate more closely resource-allocation criteria used in private industry. [19] While private industry strives for a rate of return of 15 to 16 percent, utilities are usually limited to 6 to 7 percent. Thus, utilities cannot generate the necessary cash flow required to build new plants, and must use loans or equity financing.

The typical rate structure for public utilities today is called the *block rate structure* and results in an average cost per energy unit (kWh or therm) as consumption increases. The average cost declines with increased usage due principally to increased utilization of the distribution system and the fixed-cost portion of the system without proportionate increases in cost. An illustrative block-rate structure for electricity and gas supplied to residences is given in Table 21.11. Note that the price per unit of energy declines as consumption increases.

Table 21.11

A BLOCK-RATE STRUCTURE FOR GAS AND
ELECTRICITY SUPPLIED TO RESIDENCES

1. *Electricity*	*Charge per month*
Customer charge	65¢
First 100 kwh, per kwh	5¢
Next 100 kwh, per kwh	4¢
Next 100 kwh, per kwh	3¢
Next 700 kwh, per kwh	2¢
2. *Gas*	*Charge per month*
Customer charge	75¢
First 25 therms, per therm	14¢
Next 25 therms, per therm	12¢
Next 25 therms, per therm	10¢
Over 75 therms, per therm	9¢

The industrial and commercial customers are charged significantly lower rates. In addition, they often pay a demand or load charge—a charge on the energy consumed at specific periods of the day. Thus, if industries can use electricity at offpeak hours (night time), they are using idle capacity and are charged less than daytime rates.

Declining block rates are usually justified in terms of the fixed costs per unit declining with increased usage. An alternative rate structure designed to recover the fixed costs is called the *two-part tariff*. This rate structure charges a customer a fee to recover the fixed costs and then uses a flat rate for energy consumption, as shown in Table 21.12. This structure is set to yield the same price for some set energy usage (500 kWh in this case). However, the two-part structure does not reward increasing usage by affording a declining price.

Table 21.12

AN ILLUSTRATIVE TWO-PART RATE
SCHEME

	Charge per month
Customer charge	$4.00
Charge per kWh	$0.025

A three-part inverted rate structure is given in Table 21.13. This structure is a form of what is called an *inverted rate structure*, since rate per energy unit increases as consumption increases. [20] An inverted rate structure is useful if the aim is to discourage expansion of energy use. However, an inverted rate structure would provide a substantial economic incentive for large-volume industrial users to turn to alternative sources of energy rather than pay the progressively higher rates.

Table 21.13
A THREE-PART INVERTED RATE
STRUCTURE

	Charge per month
Customer charge	$2.00
First 300 kWh, per kWh	0.02
Over 300 kWh, per kWh	0.04

On balance, the two-part rate structure may be the most equitable approach for the future. This form of rate structure has been used for many years by the telephone utilities with reasonable equity and acceptance. Perhaps the two-part rate structure will achieve the objectives of providing a fair rate of return to the utility, a reasonable price to the consumer, and an incentive to the consumer to limit the consumption of energy.

21.9 THE STRUCTURE OF THE ENERGY INDUSTRY

The structure of the energy industry in the U.S. is a mixture of public and private ownership. This structure involves a complex combination of resource ownership, indirect taxation, tax subsidies, establishment of guaranteed markets, public-land mineral-disposal regulations, import quotas, and restrictions on output.

The U.S. has a unique mixture of private enterprise and public operation in the energy industry. There is, for the most part, a private-enterprise economy in oil, coal, and the mining of uranium. In gas transmission and distribution, we have a regulated utility organization. In the development of hydroelectric and nuclear power, the U.S. uses a combination of public and regulated private ownership. Nevertheless, the primary energy industry, the fuel industry, is largely in the private sector and operates unregulated.

Within recent years there has been a growing examination of the practices and alleged concentration of power in the hands of the major energy companies in the U.S. These companies are vertically integrated, in that they operate at four levels: production, refinement, transportation, and marketing. Furthermore, many of these energy companies have undertaken horizontal expansion into primary energy areas, other than oil or gas, such as coal or uranium.

By 1970, the top twenty U.S. petroleum companies controlled more than 70 percent of the U.S. domestic production. Nevertheless, the largest share of the market held by one company, Exxon, was only 8.5 percent in 1975. With the ten largest companies holding less than 10 percent of the market, competition is still able to thrive.

In addition, U.S. petroleum companies have experienced increasing profits during the recent years. The profits of several large U.S. oil companies are given in Table 21.14. The profit per share of common stock for Exxon increased by 83 percent in the period from 1972 to 1974. However, these oil companies had

Table 21.14
PROFITS OF SEVERAL LARGE U.S. OIL COMPANIES

| Company | Earnings per share of stock | | | 1974 price/book value |
	1972	1973	1974	
Exxon	$6.83	$10.00	$12.50	0.94
Mobil	5.65	7.50	11.00	0.53
Gulf	2.15	4.00	5.50	0.51

low price-to-book-value ratios due to investor concern regarding the future value of these companies. Exxon made a profit of $2.8 billion after taxes in 1974, and is the second most profitable company in the world. Exxon's total revenues in 1976 were $48.6 billion, which is about equal to the GNP of Belgium. [21] Of the ten biggest profitmakers in the U.S. in 1975, six are oil companies. These large oil companies also produce 21 percent of the nation's coal, 32 percent of the U.S. uranium, and 21 percent of the world's petrochemicals. The total assets of the petroleum industry are listed in Table 21.15. The oil industry is gargantuan, but the question is: Is it excessively powerful?

Table 21.15
FIXED ASSETS OF THE U.S. OIL INDUSTRY IN 1974

Segment	Assets (Billions of dollars)	Percent of total
Production	$71	44
Refining	32	20
Transportation of oil	16	10
Marketing	26	16
Chemicals	11	7
Other	5	3
Total	$161	100%

There is a significant movement towards the breakup of these large oil companies. Any change, however, could have a significantly larger impact than that of the 1911 breakup of Standard Oil. The Standard Oil Trust had, by 1905, managed to effectively dominate 90 percent of the entire U.S. petroleum industry by controlling the means of oil transport, as well as the refineries. Today, there are many oil companies, all with limited shares of the market. Nevertheless, there is a concern that the large energy companies are (1) using their size to extort excessive profits from their customers, and (2) they have more power than it is wise for a democracy to entrust to organizations only indirectly responsible to the citizens of the nation.

Any form of divestiture would be severe, but either horizontal or vertical restructuring is possible. Horizontal changes would be least drastic, and would result in the large energy companies divesting themselves of their non-oil business like coal or chemicals. Vertical restructuring could result in new firms being allowed to operate only in refining or marketing, for example. There have been several efforts in the Congress to introduce bills requiring a form of breakup of the large energy companies. It is not clear whether these efforts have sufficient popular support at this time.

Some critics have called for the establishment of a publicly controlled energy system. One proposal for a publicly controlled system calls for a new local government unit to establish and administer energy policy—the public energy district. [22] This district would produce, transport, and market energy in its region. The Tennessee Valley Authority (TVA) is an example of one form of a district, but the proposal would have the district responsible to the local electorate, not the federal government. Another recent proposal calls for the nationalization of all U.S. energy production, refining, and distribution systems. [23] Such a system exists in Great Britain and France.

21.10 METHODS FOR CONTROLLING GASOLINE CONSUMPTION

As a result of the 1973–74 gasoline shortage, there has been a serious discussion of the necessity for devising allocation schemes for gasoline and heating oil.

One means of limiting gasoline consumption is by rationing methods. If limiting the consumption of gasoline is necessary, the public appears to prefer rationing to the use of taxation schemes. However, all rationing schemes generate black markets and require large bureaucracies to administer the system.

Taxation schemes can be used to reallocate resources into energy-conserving areas. One report states that every one cent of added tax per gallon of gasoline will result in a decline of one percent in gasoline consumption within a five-year period. [24] Surtaxes of 15¢ to 45¢ a gallon could cut gasoline use by 16 percent to 40 percent within five years. This charge would result in a savings of 11 billion to 27 billion gallons a year. Other estimates of the elasticity of demand for gasoline are a short-run elasticity equal to -0.4 and a long-run elasticity equal to -0.7. It is not clear whether reallocation schemes such as these will be required in the near future, or whether alternative energy sources will substitute for oil, thus obviating the need for further controlling the consumption of oil by taxation or rationing methods.

Increased costs of fossil fuels over the next decade may result in a more rapid development of alternative sources of energy such as solar and geothermal.

REFERENCES

1. "The economics of the energy crisis," *Technology Review*, April 1974, pp. 49–59.

2. P. E. DE JANOSI AND L. E. GRAYSON, "Patterns of energy consumption and economic growth and structure," *Journal of Development Studies*, Jan. 1972, pp. 241–249.

3. J. H. KRENZ, "Energy per dollar value of consumer goods and services," *IEEE Trans. on Systems, Man and Cybernetics*, July 1974, pp. 386–388.

4. O. H. Zinke, *Energy in the Near Term*, Dept. of Commerce, Arkansas, 1973.

5. D. Chapman, *et al.*, "More resistance to electricity," *Environment*, Oct. 1973, pp. 18–36.

6. J. G. Meyers, "Energy conservation and economic growth—Are they incompatible?" Feb. 1975, pp. 27–32.

7. M. S. Macrakis, *Energy: Demand, Conservation, and Institutional Problems*, the MIT Press, Cambridge, Mass., 1974.

8. N. Georgescu-Roegen, "Energy and economic myths," *Ecologist*, Vol. 5, 1975, pp. 164–174. Also in the *Southern Economic Journal*, Jan. 1975.

9. N. Wade, "Nicholas Georgescu-Roegen: Entropy, the measure of economic man," *Science*, Oct. 31, 1975, pp. 447–450.

10. A. P. Carter, "Applications of input–output analysis to energy problems," *Science*, April 14, 1974, pp. 325–329.

11. C. W. Bullard and R. A. Herendeen, "Energy impact of consumption decisions," *Proceed of the IEEE*, Mar. 75, pp. 484–493.

12. R. A. Herendeen, "Affluence and energy demand," *Mechanical Engineering*, Oct. 1974, pp. 18–21.

13. M. W. Gilliland, "Energy analysis and public policy," *Science*, Sept. 26, 1975, pp. 1051–1056.

14. M. Slesser, "Accounting for energy," *Nature*, March 20, 1975, pp. 170–172.

15. "FPC issues May 1974 report on fuel cost, quality," *FPC News*, Vol. 7, No. 35, August 30, 1974, pp. 1–2.

16. "The big squeeze on U.S. companies," *Business Week*, Sept. 22, 1975, pp. 50–55.

17. R. R. Bennett, "Major impact of inflation on power-plant economics," *Ebasco News*, April, 1975, pp. 5–9.

18. J. Lessey, "Why we need an energy £," *New Scientist*, July 11, 1974, pp. 85–86.

19. F. C. Olds, "Power-plant capital costs going out of sight," *Power Engineering*, Aug. 1974, pp. 36–43.

20. M. T. Farris and R. J. Sampson, *Public Utilities: Regulation, Management and Ownership*, Houghton Mifflin, New York, 1973.

21. C. J. Cicchetti, "Electricity price regulation," *Public Utilities Fortnightly*, Aug. 29, 1974, pp. 13–18.

22. "The oil giants: An irresistible target," *Forbes*, Jan. 15, 1976, pp. 20–29.

23. J. Ridgway and B. Conner, "A publicly controlled energy system," *Current*, April 1975, pp. 11–18.

24. R. Lekachman, "The case for nationalization," *Christianity and Crisis*, Mar. 4, 1974, pp. 43–45.

25. S. Wildhorn, *et al.*, "How to save gasoline: Public policy alternatives for the automobile," RAND Corp., Report No. 1560, Oct. 1974, Santa Monica, California.

26. D. A. Huettner, "Net energy analysis: An economic assessment," *Science*, April 9, 1976, pp. 101–104.

27. S. Baron, "Energy cycles: Their cost relationship for power generation," *Mechanical Engineering*, June 1976, pp. 22–30.

28. B. COMMONER, *The Poverty of Power*, Knopf, Inc., New York, 1976.

29. *Energy Flow through the U.S. Economy*, Center for Advanced Computation, Univ. of Illinois Press, Urbana, 1977.

30. J. MADDOX, *Beyond the Energy Crisis: A Global Perspective*, McGraw-Hill Book Co, New York, 1975.

31. J. LESSEY, "Taxes for Conservation," *New Scientist*, Oct 7, 1976, p. 30.

32. P. SPORN, *Energy in an Age of Limited Availability*, Pergamon Press, New York, 1977.

33. A. W. SHURCLIFF, "The local economic impact of nuclear power," *Technology Review*, January 1977, pp. 40–47.

34. H. A. MERKLEIN, *Energy Economics*, Gulf Publishing Co., Houston, Texas, 1977.

35. "Will energy conservation throttle economic growth?", *Business Week*, April 25, 1977, pp. 66–80.

EXERCISES

21.1 In what form can you buy the most energy for a dollar today—gasoline, natural gas, or electricity? Assume that gasoline costs 60 cents per gallon, electricity costs 4 cents per kilowatt-hour, and natural gas costs 10 cents per therm (or 10^5 Btu).

21.2 A proposal has been made which would require an energy currency (*E*-currency) to be used alongside the dollar currency. The energy content of every product would be represented by the *E*-currency. [18] Therefore, for example, a television set might sell for $200 plus 10 *E*-units. Each individual would have a fixed annual amount of *E*-units to spend each year. Analyze the economics of the use of *E*-units and the wisdom of such a proposal.

21.3 The long-run elasticity of the demand for residential electricity as a function of the price of electric household appliances can be estimated. Tyrell has calculated the elasticity to the price of appliances to be $e = -0.42$. [7] Calculate the percent change in the use of electricity in residences in the U.S. as the average price of appliances increases by 15%.

21.4 The U.S. domestic demand for petroleum can be represented by an elasticity for an initial price of $7.00 and a consumption of 13 million barrels per day. This elasticity is $e = -1.0$ at the initial price and consumption. Calculate the expected consumption when the price of oil rises to $8.00.

21.5 Calculate the energy–GNP ratio for 1976 and add that data point to Fig. 21.2.

21.6 Estimate the net energy ratio for (1) the generation of electricity from natural gas, and (2) oil from tar sands.

21.7 Determine the average price of several fuels today and compare with Table 21.5 for earlier years. Calculate the average annual rate of increase of fuel prices for the most recent two years.

21.8 Determine the cost of oil imports to the U.S. for the immediately past year and estimate the cost for 1980.

21.9 What policies or occurrences do you support or desire that will lower the capital needs of public utilities? Discuss your willingness to forego an increasing availability and reliability of electricity and natural gas as a source of heat and light.

21.10 Discuss which public utility rate scheme you favor. Examine the implications for conservation of energy, cost to the low-income consumer, and fairness to the general consumer.

21.11 Do you favor the breakup of the vertically integrated oil companies? Explain the reasoning behind your position. What is the extent of the impact of such an action on the nation's economy?

*__21.12__ Contrast the advantages and disadvantages of rationing and taxation schemes for controlling the consumption of gasoline.

*__21.13__ Compile a table of retail prices for electricity, natural gas, and petroleum in the U.S. for the years 1940, 1960, 1970, and 1974. Adjust the price by the consumer price index so that the price index is 100 for 1970 for each fuel per a fixed unit of energy (joules or Btu). Use the *Statistical Abstract of the U.S.* to compile the tables.

*__21.14__ A typical home in the middle region of the U.S. might use about 6.5×10^{10} joules for space heating. Calculate the cost of heating this home using an 80%-efficient gas heater when gas costs $6.00 per 1000 cu ft. Compare this cost with electric heating when electric energy costs 3¢/kWh.

*__21.15__ Fuel prices have been governed by extraction costs, current demand, and market control. Prepare a position paper for setting the fuel price by the cost of replacement.

21.16 A barrel of oil saved may be more valuable than a new barrel of oil produced. For a model house, ceiling insulation may cost about $400 and save about seven barrels of oil each year for the lifetime of the house. Find the present value of the savings of oil and compare with the cost of the insulation.

21.17 Determine the return on equity for several oil companies for the latest fiscal year, and compare with the return for drug companies and soft-drink companies. Is the return of the oil companies out of reason or beyond permissible limits?

21.18 Should there be built into the taxation and pricing structures incentives designed to reduce the energy intensity of goods and services? Prepare a scheme for accomplishing this goal.

21.19 The average residential use of 500 kWh of electricity and 100 therms of natural gas resulted in a monthly (January) bill of $40 in the U.S. However, the

---

*An asterisk indicates a relatively difficult or advanced exercise.

bill for this energy would have been \$30/month and \$74, in San Francisco and New York, respectively. List several factors that account for this range of charges in the U.S. What would be the bill in your city for such a monthly use of energy?

21.20 Some groups, such as the Georgia Conservancy, advocate the enactment of a *severance tax* (or royalty) on all nonrenewable energy resources. Under this proposed federal law, all producers who remove and sell a fossil fuel would pay a severance tax. Examine the economic principles of a severance tax and determine whether this would be a wise law.

22

Energy and the Environment

22.1 ENVIRONMENTAL IMPACTS

Energy, the economy, and the environment are all interrelated, as we noted in Chapter 2 (see Fig. 2.1). The environmental consequences of energy consumption have become a matter of general concern only within the past two decades.

The demand for clean, nonpolluting energy has caused major shifts in the utilization of fuels. The electric utilities, under pressure to reduce sulfur-dioxide pollution, have turned from coal to oil, thus causing an increase in oil consumption and eventually in imports.

In this chapter we will consider some overriding issues of energy and the environment. In each preceding chapter on a specific fuel or energy source, we reflected upon the environmental impacts of that fuel. Here we will attempt to examine some of the issues common to all fuels.

The use of energy eventually results in waste heat and matter. Environmental concerns relate to air pollution, water pollution, thermal waste heat, and climatic changes of the earth, among others.

Most of today's environmental discussions revolve around specific pollutants that influence the biosphere, such as SO_2, carbon monoxide, the oxides of nitrogen, radioactive waste, and waste heat (causing temperature rise in water bodies). Also of importance are fundamental climatic processes, including local and global effects of air pollution resulting from energy consumption and the thermal balance of the earth in its setting in the universe. Unfortunately, the science of climatology is unable, at present, to predict the ultimate consequences for the earth's climate of man's production of energy. At what rate of energy production would the ice caps melt? Will the carbon dioxide or dust thrown into the atmosphere by the burning of fossil fuel threaten the stability of the weather system? How does the geography of man's energy production affect weather in various parts of the world? [1]

421

Environmental element affected	Type of impact	Pollutant	Areal extent: Global	Areal extent: Regional	Areal extent: Local	Cost of control: High	Cost of control: Intermediate	Cost of control: Low
Air	Chemical–physical	Carbon dioxide	■					
		Carbon monoxide			■	■		
		Sulfur oxides		■	■		■	
		Nitrogen oxides		■		■		
		Hydrocarbons		■				■
		Photochemical smog		■				
		Particulates	■	■	■		■	
		Water vapor	■		■			■
		Trace metals		■	■			■
	Radiological	Noble gases	■					■
		Particulates		■	■			
	Resource use	Oxygen						
		Helium		■				■
	Other	Thermal inputs		■	■			
		Electromagnetic emissions			■			■
		Noise		■	■	■		
Water	Chemical–physical	Oil spills		■	■			■
		Acid mine drainage		■	■	■		
	Radiological	Tritium	■					■
		Other effluents		■				■
		Uranium milling wastes			■			■
	Thermal	Thermal inputs		■	■		■	
Land	Chemical–physical	Acid fallout from the air		■	■			
		Mineral fallout from cooling towers			■			■
		Solid wastes			■		■	
	Radiological	High-level wastes		■	■		■	
	Resource use	Land subsidence			■		■	
		Strip-mining of coal		■	■	■		
		Land use for power production and transmission		■	■		■	
		Hydroelectric dams		■	■			

Fig. 22.1 *Environmental factors related to energy.*

The most important effect that a growing use of energy in industrialized societies has had on the environment has been to redirect the net production of ecosystems from natural consumers to mankind. The present global human population is dependent upon large inputs of energy to biological systems in order to provide enough food, since natural systems do not provide enough edible yield.

Environmental problems are created at each stage of the production, upgrading, transportation, and utilization of fuels. For example, acid-water drainage from coal mines and air and water pollution from electric generating plants present serious environmental impacts. Environmental factors related to energy are illustrated in Fig. 22.1.

22.2 SOCIAL IMPACTS

The noted author Ivan Illich has recently discussed the impact of energy use on society. [2] Illich views the increased use of energy as disruptive. He states that a low-energy policy allows for a wide choice of lifestyles and cultures while, if a society opts for high per capita energy consumption, its social relations are dictated by technology. In his view, below a threshold of per capita energy consumption, motors improve the conditions for social progress. Above this threshold, energy grows at the expense of equity and further energy affluence then means decreased distribution of control over that energy. In his view, beyond this threshold, energy can no longer be controlled by political processes, and social breakdown occurs.

Illich's viewpoint is challenging to the current energy ethic prevalent in the industrialized nations. Whether his concept will hold up under the light of historical examination is not clear. Yet, if there is a threshold of disruption, it is not clearly defined as yet. In a sense, Illich is raising again the important issue of man and his machines, and which is master. [3]

22.3 ENVIRONMENTAL AND SOCIAL CONTROLS

A model of the control of environmental and social impacts of an ecosystem is shown in Fig. 22.2. The impacts resulting from the interaction of energy use, man, and his environment are evaluated. These evaluations or estimates of the impacts result in attitudinal and technological corrective actions. This feedback process continues throughout time and, as people's attitudes are changed, they alter their institutions in order to improve the situation.

One method of institutional control is legislation. In the U.S., the Refuse Act of 1899 may have been one of the first environmental laws. The Refuse Act required a permit for discharge of refuse into navigable waters. The Water Pollution Control Act of 1948 and the subsequent Acts of 1956, 1965, and 1972 provided for the cleanup of rivers and lakes and established standards for water quality. The Clean Air Act of 1967 and its amendment in 1970 expanded the federal government's role in setting and overseeing standards for automobile emissions.

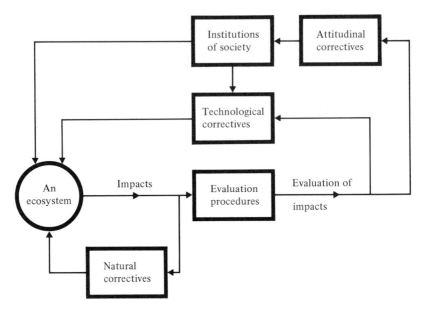

Fig. 22.2 *A model of the control of environmental and social impacts of an ecosystem.*

The Environmental Policy Act of 1969 requires every U.S. government agency that is planning a project to file an assessment of the impact on the environment. Environmental impact reports are now routinely required by most states and local planning agencies. The Act charged all planners with "a responsibility to contribute to the preservation and enhancement of the environment."

22.4 POLLUTION FACTORS

Pollution results from increasing population, population crowding, increasing affluence, and inefficiency in cleaning up the effluent from the energy-consuming process. The annual percentage increase in pollution concentration is approximately equal to the sum of the annual percentage increase in population, GNP per capita, and plant inefficiency. [4]

The pollution concentration is expressed as

$$C_P \propto \left(\frac{N}{A}\right) \times G_{CW} \times t_{1/2} \times \frac{1}{e}, \tag{22.1}$$

where (N/A) is the number of persons per area, G_{CW} is the per capita total consumption and wastage of goods, $t_{1/2}$ is the half-life of the pollutant in the environment in years, and e is the efficiency of cleaning up the effluent from the energy-consuming process. If the cleanup of the effluent is perfect, then $e = \infty$, and we have $C_P = 0$.

We can also express G_{CW} as equal to the Gross National Product per capita, since consumption is almost linearly related to GNP per capita. Also, we may write

$$G_{CW} \propto F, \tag{22.2}$$

where F is the amount of fuel consumed per capita to produce the goods. If we define $\eta = F_0/F$, where F_0 is the theoretical minimum energy per capita required to produce the goods, we get

$$G_{CW} \propto \frac{F_0}{\eta}. \tag{22.3}$$

Substituting Eq. (23.2) into Eq. (23.1), we have

$$C_P = \left(\frac{N}{A}\right) \times \frac{F_0}{\eta} \times t_{1/2} \times \frac{1}{e}. \tag{22.4}$$

Therefore, the pollutant concentration is proportional to the product of the concentration of people per sq km, the goods consumed per head, and how long the pollutant survives before the environment copes with it, divided by the fuel efficiency and the effluent-cleaning efficiency e. In order to reduce the pollution concentration, we could (1) reduce the population density, (2) reduce per capita consumption, (3) increase the efficiency of the energy process, and (4) increase the effectiveness of cleaning up the pollution.

Air Pollution

Roughly 85 percent of all air pollution in the U.S. is associated with the combustion of fossil fuels. Most of this comes from motor vehicles, but fossil-fuel power plants are also contributors. The principal pollutants are carbon monoxide, sulfur oxides, nitrogen oxides, hydrocarbons, and particulate matter. These pollutants have the potential of impairing public health, creating annoyance, and causing significant property damage.

Coal and oil produce the most pollutants when burned. Coal accounts for about two-thirds of the sulfur dioxides in the air. Therefore, systems for removing sulfur from the gas have been developed and utilized.

Cities are particularly affected by pollutants, since the atmosphere normally disperses them but, in congested areas, emissions accumulate more rapidly than they can be naturally dispersed.

The first control steps were to lessen pollution from autos, initially with engine modifications to cut down emissions and then to require further reductions by means of control devices.

A primary goal of environmental control is to reduce the damage due to sulfates. Sulfur dioxide is released in the exhaust from power plants burning oil or coal containing sulfur. The U.S. Environmental Protection Agency estimated that 33 million tons of SO_2 were emitted in 1972. Once in the air, the sulfur dioxide combines with oxygen to form sulfate, SO_4.

The conventional method of pollution control is a straightforward attack on sulfur: either prevent the sulfur from getting into the air (by burning low-sulfur fuel or by removing the sulfur from the stack gas) or dissipate the sulfur-dioxide concentration by building tall smokestacks that send the pollutant high in the air, where it diffuses before reaching lung level. However, the effectiveness of both these approaches is in doubt. The chemistry of the formation of sulfates in the air is still open to further research. The U.S. requirements for stack emissions from fossil-fuel burners, to become effective in 1977, are given in Table 22.1. Coal normally releases about 5 lbs of SO_2 per 10^6 Btu, in contrast to the requirement of 1.2 lbs/10^6 Btu.

Table 22.1
U.S. STACK EMISSION STANDARDS FOR
PLANTS BUILT AFTER 1971

| | Maximum permissible emissions in lbs/btu 10^6 Type of fuel | | |
Pollutant	Coal	Oil	Gas
Particulates	0.1	0.1	0.1
Sulfur dioxide	1.2	0.8	—
Nitrogen oxides	0.7	0.3	0.2

Flue-gas desulfurization systems have been implemented in many power plants. Other plants use tall stacks and low-sulfur fuel. It is not yet clear when all the desired air-quality standards can or will be met. Nevertheless, air-quality schemes are being pursued vigorously by both government and industry.

Thermal Pollution

Thermal pollution involves the releasing of waste heat into the environment. Steam-electric power generating plants release heat as a consequence of producing electricity. Also, automobile engines release large amounts of waste heat through the auto exhaust and cooling system.

The effects on the environment of thermal releases may be detrimental, beneficial, or insignificant, depending on many factors such as the manner in which the heated water used for cooling the power plant or engine is returned to its source or disposed of. The effect on the source of cooling water of a power plant depends upon the amount of water available, the ecology of the source water, and its desired use. The addition of the heated water to the source body of water does raise its temperature a few degrees in a mixing area, and this increased temperature can affect fish and other aquatic life. However, these thermal effects can be dissipated or minimized through suitable design measures. Recently, more utilities have chosen to utilize alternative cooling methods such as cooling towers and ponds.

The oxygen content of water is critical for most marine life and it is affected by temperature. The saturation level of dissolved oxygen decreases with increasing temperature and at the same time the demands for oxygen increase. The multiplication rate of plankton increases with increasing temperature. In addition, increasing water temperature can increase the toxic effect of various pollutants and alter the entire life food chain.

A recent study of the effects of a power plant are illuminating. [5] Professor Merriman studied the nuclear generating plant with a capacity of 590 megawatts located on the Connecticut River. This plant returned 828 cubic feet per second of water with an increase in temperature of 22.4°F. It was determined that the water discharge of this power plant "had no significant deleterious effect on the biology of the river." [5]

Many power plants use cooling towers. These towers are classified as wet or dry, and each of these may be forced-draft or natural-draft. In the wet tower, the cooling water is in contact with the air and cooling takes place by evaporation. Wet cooling towers operate well in cool, dry climates. Because of the water placed into the air, fogging or local rain can take place.

A dry tower is basically a large radiator. The cooling water flows through finned tubes and is cooled by air passing over them. Dry towers are more expensive to construct than wet towers, but they have no water loss. The costs associated with cooling towers, in comparison with other methods, is shown in Table 22.2. These costs can be compared to the cost of the plant, which might average 300 million dollars.

Table 22.2
COMPARATIVE COSTS OF COOLING SCHEMES FOR AN
800-MEGAWATT FOSSIL-FUEL POWER PLANT

	Cost (Millions of dollars)
Once-through cooling (river or lake)	5
Cooling pond	7
Wet tower:	
Forced draft	7
Natural draft	9
Dry tower:	
Forced draft	16
Natural draft	32

Heat emissions from power plants and energy-consuming devices, coupled with a surface changed from vegetation to buildings and roads, has an effect on the local climate of regions of the earth. For example, urban areas are noticeably warmer than their rural surroundings. However, global effects are insignificant since waste heat is insignificantly small compared to the natural energy balance of the earth.

Water is abundantly available on the earth, but it is not always located near the need. On the average day, some 3×10^{11} cubic meters of rain and snow fall on the land area of the world. Waste heat released to water or air fits into (1) the hydrologic cycle, in which great quantities of water circulate, and (2) the heat balance of the world's atmosphere. [6]

While the overall heat balance of the world is insignificantly affected, thermal-emission burdens are significant in cities. The heat releases in the Boston–Washington corridor equal 15 percent of the net solar radiation in summer and 50 percent in winter. The temperature rise over cities is as much as 15°F, resulting in the phenomenon called the *urban heat island.*

In 1975, about 11×10^{15} Btu of waste heat was discharged by power plants to condenser cooling waters in the U.S. It is proposed to construct clusters of 2 to 5 nuclear and coal-fired power plants in a large complex, often called an *energy park.* These clusters could be sources of large islands of heat, causing localized meteorological changes. Such a park may provide up to 10,000 megawatts and release up to 1.5×10^{14} Btu of waste heat. Whether it is wise to permit the construction of such large energy parks is not at all clear at this time. [12]

Radioactive Waste

Nuclear power plants do not produce the smoke and other air pollutants of fossil-fuel plants. However, there is the problem of the management of the relatively small amounts of radioactive wastes. These wastes fall into two categories, high- and low-level.

Low-level wastes are the very low levels of radioactivity such as those which occur in air, water, and solids outside the fuel elements in the routine operation of nuclear-power reactors. It now appears that low-level wastes can be controlled to a level consistent with the background level of radiation.

At the fuel processing plant, however, high-level wastes are produced during the reprocessing of spent fuel elements from nuclear reactors. Spent reactor fuel is removed from the reactor, securely packaged, and shipped to the reprocessing plant. It is during processing that high-level wastes are removed from the fuel elements and concentrated in liquid form for storage.

These liquid wastes are further concentrated into solids and stored at a federal depository. It is planned to use salt mines for storage of these wastes.

Other Environmental Effects

The actual and potential environmental effects of energy extraction, processing, storage, transportation, and consumption are numerous. One popular example of a large undesirable impact was the Santa Barbara Channel oil spill of 1969, which resulted in an oil slick 8.5 miles long.

Sometime before 1980, two or more deep-water terminals for the delivery of foreign crude oil will be established in the coastal waters of the U.S. One of these terminals may be necessary for the receipt of Alaskan oil. Thus, the environmental hazards associated with the supertanker and the potential for oil spoils are significant. As a result of locating one of these terminals, the associated refineries and land-use impact would also be important to consider.

As we noted earlier, the impact of mining, transporting, and burning coal is a significant part of the U.S. energy environment scheme. For example, what are the potential effects of mining coal in the northwest plains states? All these issues need to be considered as we continue to explore our energy sources.

22.5   THE COSTS OF ENVIRONMENTAL CONTROL

It is important to study the economic and energy costs of pollution-control devices and schemes. What is the energy cost of the U.S. program to clean up and maintain the environment?

The Council on Environmental Quality has estimated that the total cumulative cost for the period 1971–1980 for pollution control will be 287 billion dollars (constant 1971 dollars). It is planned to spend \$107 billion and \$87 billion on air and water pollution, respectively. [7] Another estimate states that the total costs may be as high as \$300 billion dollars. [8] The gross cost for environmental cleanup is made up of myriad small and large costs such as onetime capital costs and continuing operating costs.

In 1974, California Edison Company stated that 23 percent of its 714-million-dollar capital expenditure budget would go for environmental control. TVA stated that the installation of SO_2 scrubbers on all its plants would cost at least a billion dollars, and cost \$225 million a year to operate. To cover this, electric rates would go up 28 percent. Scrubber operation would take 6 percent of each station's production, would require mining 2.4 million tons more of coal each year, plus eight million tons of limestone. Sludge disposal would use 20,000 acres of land in 20 years.

The addition of emission controls to automobiles has added about \$200 to the price of the 1975 model. Auto makers estimate that meeting 1978 standards could cost \$450 per car.

The added increment of energy required to reduce the pollutants emitted from an automobile exhaust can be noted in the reduced gasoline mileage of the newer automobiles. This reduction, called the *fuel penalty*, is about 15 to 20 percent for the 1976 auto, relative to the 1968 automobile. Of course, the fuel utilization of automobiles can be reduced by redesign of the auto and the engine, and reduction of the vehicle weight. [9]

Eric Hirst estimates that waste-water treatment in the U.S. for all residential and industrial waste water would have been 25 billion kWh per year, or 1.8 percent of the total 1970 electric energy use. [9]

The removal of particulates from stationary sources is by means of electrostatic precipitators, which require less than 0.1 percent of the power output. Limestone scrubbers to remove 80 percent of the sulfur from stack gases consumes about 4 percent of the power-plant output.

The use of forced-draft wet cooling towers for power plants results in a power penalty of about 2 percent for a nuclear plant. The power loss for dry towers is about 6 percent of the plant output. Thus, one may estimate that the energy requirements for environmental improvement may average 4 to 6 percent of the total energy consumed over the next several years. As new, efficient, control technologies are developed, we can expect an improvement in this penalty factor. [13]

REFERENCES

1. C. A. S. HALL, "Look what's happening to our earth," *Bulletin of Atomic Scientists*, March 1975, pp. 11–21.

2. I. ILLICH, "Energy and social disruption," *Ecologist*, Feb. 1974, pp. 48–52.

3. R. C. DORF, *Technology, Society, and Man*, Boyd and Fraser Publishing Co., San Francisco, 1974.

4. M. THRING, "The equations of survival," *New Scientist*, March 1, 1973, pp. 482–483.

5. D. MERRIMAN, "Calefaction of the Connecticut River, U.S.A.," *Transactions of the N.Y. Academy of Sciences*, Jan. 1973, pp. 59–65.

6. F. C. OLDS, "Water resources and waste heat," *Power Engineering*, June 1973, pp. 26–33.

7. E. HIRST, "The energy cost of pollution control," *Environment*, Oct. 1973, pp. 37–44.

8. F. C. OLDS, "Environmental cleanup 1975–1985," *Power Engineering*, Sept. 1975, pp. 38–45.

9. E. HIRST, "Pollution-control energy," *Mechanical Engineering*, Sept. 1974, pp. 28–35.

10. A. D. WATT, "Placing atmospheric CO_2 in perspective," *IEEE Spectrum*, Nov. 1971, pp. 59–72.

11. "Does pollution control waste too much energy?", *Business Week*, March 29, 1976, p. 72.

12. M. TERRY AND P. WITT, *Energy and Order*, Friends of the Earth, San Francisco, 1977.

13. H. ASHLEY *et al.*, *Energy and the Environment: A Risk–Benefit Approach*, Pergamon Press, New York, 1977.

14. A. P. CARTER, *Energy and the Environment: A Structural Analysis*, Univ. Press of New England, Hanover, New Hampshire, 1977.

15. G. L. TUVE, *Energy, Environment, Population, and Food*, Wiley and Sons, Inc., New York, 1977.

16. E. DRAKE AND R. C. REID, "The importation of liquefied natural gas," *Scientific American*, April 1977, pp. 22–29.

EXERCISES

22.1 In the use of coal to produce electricity, list each stage of the process and describe the environmental impact of each step. Estimate which step in the process has the greatest impact on the environment.

22.2 Illich maintains that, if a society passes a certain threshold in per capita consumption of energy, it loses its ability to control that energy. [2] Examine this theory of a threshold associated with the control of the uses of energy, and provide your own analysis of the social interrelationships with energy use.

22.3 Prepare a schematic diagram similar to the model shown in Fig. 22.2 for the control of the environmental and social impacts of nuclear power plants in the U.S. List the institutions active in this area and describe the attitudinal changes that have occurred over the past decade.

22.4 If, for a certain energy process, one is able to increase the fuel efficiency by 10% and increase the effluent-cleaning efficiency by 15%, determine the percent change in the pollution concentration.

22.5 Contact one of the auto manufacturers and obtain data on their system of emission control. Obtain data on the fuel penalty imposed by that control device.

22.6 Contact your local electric utility or steel mill and obtain data on the fuels they have used over the past year and the content of sulfur in these fuels.

22.7 Obtain the total cost of a relatively new electric power plant in your state. Also obtain data regarding the cost of the pollution control devices installed on that plant and the effectiveness of that equipment.

22.8 Obtain information on the cooling-water system used in the system discussed in Exercise E22.7. Obtain data on its costs, its size, and the amount of water consumed per year by the plant for cooling purposes.

22.9 If there is a nuclear power plant operating in your state or nearby, contact the power company for pertinent data regarding the measured radioactivity levels. Compare these data with the estimated background radiation for your state.

*__**22.10**__ The energy and environmental features of recycling materials are interesting. Recycling saves processing energy. For example, the energy requirement for recycling aluminum is as low as 5% of the requirement for originally processing bauxite ore. Explore the energy and environmental relationship for (1) aluminum, (2) paper, or (3) steel. List the advantages of recycling the material in terms of both energy and environment.

22.11 The environmental effects of a plant with the purpose of turning coal into a gas or synthetic oil may be large. Consider the effects of such a plant located in Wyoming or Montana, which would use 4 million tons of coal a year.

22.12 The use of throwaway beverage containers is a waste of energy. Determine the net savings in requiring returnable bottles for soft drinks and other beverages rather than using nonreturnable bottles. Assume that a bottle can be returned an average of eight times. Also explore the Oregon bottle law.

22.13 Obtain the total capital costs of the pollution-control equipment used by your local utility. Compare that cost to the total asset value of the utility.

*__**22.14**__ The burning of fossil fuels causes the addition of carbon dioxide (CO_2) to the atmosphere. The carbon dioxide molecules, along with water vapor and ozone,

* An asterisk indicates a relatively difficult or advanced exercise.

have the property of absorbing in the infrared or long-wavelength region. [10] Thus the short-wavelength radiation from the sun passes through the earth's atmosphere, but the long-wavelength radiation emitted by the earth is absorbed by CO_2, H_2O, and O_3 molecules in the atmosphere and then reradiated in all directions. These molecules inhibit the escape of the long-wavelength radiation and are responsible for additional heating of the earth's atmosphere. This phenomenon is called the "greenhouse effect" because the same principle works in a greenhouse, where the glass roof and sides transmit short-wavelength radiation from outside but inhibit the loss of long-wavelength radiation from inside. Explore the greenhouse effect and determine whether it is presently an important environmental factor. At what point do you estimate that the greenhouse effect may have a profound impact on the earth's ecology?

*22.15 Assume that one-half of the hot-water heating of your home could be provided by solar heating and the other half by electric energy. This system, incorporating solar and electric heat, would cost $600 to purchase, in contrast to $120 for a comparable total electric heating unit. Determine the power rating of a total electric heater and assume it operates one-tenth of the time, to obtain the average power required. (a) Add the capital cost of the electric power plant and compare the total capital cost to society of each approach. (b) Assume that the cost of electric energy is 4¢/kWh and a total electric water heating system requires 10,000 kWh per year. Calculate the number of years for the combined solar–electric system to break even for the homeowner.

*22.16 Several recent studies indicate that energy conservation and pollution control goals may be in direct conflict. [11] The environmental controls required in the iron and steel industry, for example, may require an increase of 10% in energy consumption. Other industries estimate that, to meet pollution control goals, an increase of energy consumption by that industry as high as 10% may be required. Prepare an analysis of the energy costs of pollution control of one industry. Obtain information from the Environmental Protection Agency, the industrial trade association, and representative companies, in order to prepare your analysis. Provide your view of whether these environmental goals should be pursued over the next decade if the energy costs are as high as 10% of total consumption by that industry.

*22.17 Consider an electric power plant using 2900 metric tons of coal to produce 400 MW average and 800 MW peak power. Using once-through cooling water, estimate the thermal pollution and the exhaust pollutants. Assume the plant has an efficiency equal to 40%.

23

Energy Policy

23.1 POLICY ISSUES

The national security, financial stability, and standard of living of a nation are interwined with its energy consumption. Recent international events have brought many nations to an abrupt recognition of the fact that domestic inflation and related economic factors, adjustments in lifestyle, and national security can be challenged and impacted by the reduced availability of energy supplies. Thus, many nations are finding it necessary to inquire what energy and environmental policies are required in order to balance supply and energy consistent with acceptable economic, social, and environmental goals.

Policy issues are numerous for the U.S., and include those listed in Table 23.1. Action on many of these policy issues will be required over the next decade in order to avoid significant unforeseen impacts on the national economy and standard of living. It is critical to reconcile environmental policies with energy policies, so that they are coordinated, not counterproductive. Also, decisions on the mix of foreign and domestic oil to be used over the next decade is an important issue. [8] The control and regulation of energy prices is an issue of large magnitude. Should large energy companies be allowed to grow and possibly dominate the energy marketplace?

Energy policy issues can be classified as near-term (between now and 1985), intermediate-term (1985 to the end of the century), and long-range. The most difficult problems to resolve may be the short-term issues, since time may not be available to develop new technology or implement new policies. For example, one primary immediate issue is the provision of enough electricity with a coal-based utility industry while meeting environmental constraints, either through improving technology or by converting to other fuels. It should be noted that the direction of a policy initiated for the short term may have profound implications for long-term events and dramatically affect the types of decisions that will have to be made in the future.

Table 23.1

ENERGY POLICY ISSUES FOR THE U.S.

I. *Supply and demand of energy*
   A. Should the U.S. adopt a policy of self-sufficiency in energy resource development and production? Or should the U.S. pursue a policy of maximum importation of energy sources in order to conserve domestic supplies?
   B. How dependent can the U.S. become on imported petroleum? Should the nation pursue a policy designed to increase the development of its urban, onshore, and offshore oil reserves?
   C. Should the U.S. seek ways to negotiate contracts that guarantee delivery of imported natural gas?
   D. Should tax and fiscal policies be designed to provide increased incentives for the development of potential energy reserves? To increase oil refining capacities?
   E. Can synthetic fuels close the energy gap? What are the prospects for solar power, geothermal, magnetohydrodynamics, fusion or other alternative sources?
   F. What are the air quality implications of increased use of low sulfur oil, high sulfur oil, and coal?
   G. Should the nation adopt policies to alter existing building codes that would reduce energy demand? How effective would they be?
   H. How can the energy demand of the transportation market be lowered?

II. *Policy implementation*
   A. Are national security, clean energy and a low-cost national energy base compatible?
   B. Should the price of energy be allowed to be set by a free-market system or an alternative system?
   C. What are the implications of decisions for siting refineries and associated industries in connection with deep-water terminals to receive imported energy supplies?
   D. Should there be established a "one-step" approval procedure for energy-related projects (i.e., power plant siting, urban and offshore oil drilling, geothermal development, etc.)?
   E. Should a coordinated federal–state energy research and development effort be developed?

Perhaps the primary objectives of the U.S. energy policy can be summarized as: (1) the development of an adequate supply of energy at reasonable prices to permit the nation to enjoy a good standard of living, (2) the achievement of relative self-sufficiency of energy supply, (3) the maintenance of a safe and healthy environment, (4) the attainment of maximum efficiency in the production, distribution, and utilization of all forms of energy, and (5) the conservation of energy resources and the reduction of demand for energy.

23.2 PROJECT INDEPENDENCE

On November 7, 1973, former President Nixon responded to the Arab oil embargo by a call for American self-sufficiency: "Let us pledge [that] by 1980, under Project Independence, we shall be able to meet America's energy needs from

America's own energy resources." By November, 1974, only a year later, the government concluded that the goal was not attainable by 1980, perhaps not even by 1985. [1] The U.S. Federal Energy Administration discerned four strategies available for Project Independence: (1) continuation of present policies, (2) accelerating domestic supply, (3) government-mandated conservation, and (4) creation of a domestic stockpile. One or a combination of these strategies can be adopted. The quickest path toward balancing demand and domestic supply is conservation and the longer-term development of alternative energy sources. [2] Project Independence, as originally conceived, included rapid development of oil and gas in Alaska and the continental shelf, greater utilization of coal, and further development of nuclear reactors.

However, the achievement of complete energy independence by 1980 was found to require measures beyond political or economic desirability. [3] Probably, the U.S. can never be completely independent of the rest of the world in energy or any other commodity. In a world of interdependence, the achievement of total independence by any nation is likely to be beyond reach. However, the Arab oil embargo and the response of Project Independence probably presented a challenge which was healthy in that the nation came awake to needs of the future and the inevitability of imposing limits on unbridled energy growth. One wise response was the formation of a single federal agency for research and development of energy resources and technology. [10]

23.3 THE U.S. FEDERAL ENERGY ORGANIZATION

The U.S. federal government's involvement in the nation's energy policy has been developing for over 125 years. The Organic Act of 1849 established the Department of the Interior to manage the public lands and the resources of these lands. In 1866 the Mining Act declared these public lands to be "free and open" to mining. The U.S. Geological Survey was created in 1879 to classify and examine the geological structure and the mineral resources of the nation. Under the Coal Land Act of 1873, lands known to contain valuable coal deposits were sold in small tracts for mining purposes. The U.S. Department of Commerce was founded in 1903 with the objective, in part, of fostering, promoting, and developing domestic commerce and mining of the U.S. The Bureau of Mines Act of 1910 created the Bureau to conduct investigations concerning mining and mineral substances. The Federal Water Power Act of 1920 established the Federal Power Commission, later reorganized as an independent agency in 1930. In 1935, the FPC was given authority over interstate transmission of electric energy and the investigatory and advisory functions were extended to the whole electric industry. The 1938 Natural Gas Act extended the FPC controls to include interstate transmission and sales of natural gas.

Two months after the explosion of the atomic bomb at Hiroshima, President Truman urged Congress to establish the Atomic Energy Commission (AEC). The Atomic Energy Act of 1946 established a civilian Atomic Energy Commission with control over all policy aspects of atomic-energy development. The Atomic

Energy Act of 1954 gave the ACE authority over the development, licensing, and regulation of private power reactors. In 1970, the President's Reorganization Plan created the Environmental Protection Agency (EPA).

After the energy crisis of 1973–74, it became clear that it was necessary to reorganize the federal government's tangled energy bureaucracy. In 1974 a bill was passed by Congress that split the Atomic Energy Commission into two agencies—an Energy Research and Development Administration (ERDA) and a new regulatory body, the Nuclear Regulatory Commission (NRC). The objectives of this organization were to (1) reduce the inflexibility and lack of coordination among federal agencies and regulators in the energy field, (2) improve the completeness, reliability, and accuracy of energy data, and (3) to work towards a permanent means of settling federal–state energy conflicts. The Energy Reorganization Act of 1974 satisfied these objectives, at least in part.

The Energy Research and Development Administration was activated on January 19, 1975. The agency's mission, as defined by Congress, is to develop all energy sources to meet the needs of present and future generations; to increase the productivity of the national economy and make the nation self-sufficient in energy; to restore, protect and enhance the environment; and to assure public health and safety. ERDA has six divisions for energy research and development; (a) fossil energy; (b) nuclear energy; (c) environment and safety; (d) solar, geothermal, and advanced energy systems; (e) national security; and (f) conservation. ERDA inherited a staff of 7200 persons and a budget of $3.6 billion transferred from AEC, Department of the Interior, the National Science Foundation, and the Environmental Protection Agency.

The Nuclear Regulatory Commission (NRC) was also established in January, 1975, as a separate agency from the AEC, to perform the licensing and regulation of nuclear-power plants and the safeguarding of nuclear materials and facilities.

One additional federal agency involved in energy policy is the Federal Energy Administration (FEA), created in May, 1974 as a reaction to the Arab oil embargo of 1973–74. This agency has about 3000 employees and a budget of $190 million. Its primary mission is the implementation of petroleum regulatory and allocation programs. It also generates and maintains data on energy supplies.

U.S. Policy Decisions in 1975–1977

ERDA released a national plan for energy in 1975 which called for major changes in U.S. energy policy. In this plan, the pursuit of commercially feasible solar-energy systems was elevated to the highest priority. This was a change from a former priority commitment to nuclear power as the sole source of new energy alternatives. Now, nuclear fission, solar energy, and fusion power are all called possible sources of energy in the long term. Also, ERDA has expanded its support for research and development into fossil-fuel programs. The principal message of the ERDA plan is that there isn't any single energy technology that can be guaranteed to meet long-term energy needs, so as many options as possible should be nurtured. However, ERDA recognizes that it may take up to 60 years for a full transition from a dependence on oil to a new alternative energy source such as solar or fusion power. The transition from coal to oil took about 60 years (1910 to 1970), as it did for wood to oil (1850 to 1910). Nevertheless, a shift of one-fourth of an energy source could be achieved in ten years and that would relieve the U.S. dependence on oil imports.

The Office of Technology Assessment (OTA) of the Congress criticized the ERDA plan as providing insufficient attention to the near-term potential for enhancement of oil and gas recovery, as well as the production of synthetic oil and gas from coal and oil shale. [4, 7] The OTA report also called for increased emphasis on energy conservation and the nonhardware aspects of the energy problem. The goals, as proposed in the 1975 ERDA plan, are summarized in Table 23.2. The enhanced recovery of oil and gas is a goal that may be reached in the near term, if sufficient research and development funds are invested.

Table 23.2
A SUMMARY OF THE ERDA PLAN GOALS

Energy source	U.S. energy conversion goals (10^{15} Btu)	
	1985	2000
Coal gasification and liquefaction	0	9
Direct coal utilization	6	9
Oil shale	2.0	4
Enhanced oil and gas recovery	6	9

Energy policy in the Congress and the executive branch was under thorough review during 1975. The Congress was insisting on volume restriction of oil imports by means of allocation schemes, import quotas, and regulation. The executive branch supported the use of increased prices to limit demand. The President's plan in 1975 was the imposition of a tariff of $1 per barrel on imported oil, eventually rising to $3 per barrel. Congress, however, was unable to accept the imposition of the import tax, and voted a suspension of this tariff.

President Ford's goal was to reduce imports of oil by one million barrels a day; he proposed an excise tax on domestic oil and an import fee on crude oil and refined products, in order to lower consumption. Congressional leaders consented to an allocation proposal and an increase in the gasoline tax. One alternative approach from some Congressional members called for a rationing scheme that would give each motorist only enough coupons to buy 36 gallons of gasoline a month. Currently, 125 million U.S. drivers buy an average of 50 gallons per month. This alternative, however, would require a bureaucracy of up to 20,000 employees to administer it, at a cost of $2 billion per year.

Finally, a compromise bill was signed by the President on December, 1975, and became effective in 1976. First, the price for domestic oil was rolled back from the January 1976 price of $8.75 a barrel to $7.66 on February 1, 1976, and then would be allowed to rise at an annual rate of about 10 percent over a period of 40 months. A strategic oil reserve will be established, with a goal of 150 million barrels by 1979 and 400 million barrels by 1983.

In the area of conservation, the Energy Policy and Conservation Act required that the auto manufacturers must achieve specified fuel economies. An average goal of 20 miles/gallon in 1980 and 27.5 miles/gallon in 1985 will be required. Also, states will get federal grants to develop energy-conservation programs. Appliance makers will be required to label their products with energy-efficiency information.

During the week of April 18, 1977 President Jimmy Carter placed before the Congress and the public his energy policy. He called for personal sacrifice, environmental protection, and a continued economic growth coupled with a reduced demand for energy. His plan is based on the increased exploitation of coal resources, a significant reduction in the importation of petroleum, and significant conservation measures. Specifically, his goals are, by 1985, to:

1. reduce annual energy growth to 2%;
2. cut gasoline consumption by 10%;
3. cut oil importation to one-half of the 1977 rate; and
4. increase annual U.S. production of coal to one billion tons.

The President proposed that Congress authorize the establishment of a cabinet-level Department of Energy, incorporating FEA, ERDA, and the energy roles of several federal agencies. In addition he proposed that by 1985 a strategic oil reserve of one billion barrels be developed.*

President Carter's proposal included measures concerned with:

1. transportation,
2. building heating and cooling,
3. cogeneration,
4. utility-rate reform,
5. oil and natural gas pricing,
6. coal use and environmental policy,
7. solar power, and
8. geothermal energy.

Mr. Carter requested authority to raise the tax on gasoline by 5 cents each year that consumption exceeded target levels.

Also Mr. Carter proposed a tax on automobiles which achieve less than a specified mileage per gallon, and provide rebates to owners of new cars that get more than a specified mileage per gallon. Mr. Carter also asked Congress to delay indefinitely construction of the fast breeder nuclear reactor at Clinch River, Tennessee. Finally, Mr. Carter requested the approval of a tax-credit program for the installation of solar heating and cooling equipment in homes.

---

* The Department of Energy commenced in October 1977 with 818 employees and a budget of $10,500,000,000.

The issues remaining unresolved in 1977 were the deregulation of natural gas prices, federal standards for thermal efficiency in new building construction, the transportation and refining of Alaskan oil, and the use of oil from naval reserves for the storage program.

23.4 ENERGY RESEARCH AND DEVELOPMENT

The pursuit of new, alternative sources of energy and the conservation of energy is the task of research and development over the next several decades.

During the past century, wood was replaced as the primary fuel by coal. Then, in the 20th century, oil replaced coal as the leading fuel. By 1990 or 2000, it is likely that a new source of energy will be replacing oil as the primary fuel. It is the role of research and development (R & D) to develop these alternative sources. In R & D, one sets the task, the financial resources, and the time to reach the goal. If one has a clear and well defined R & D goal and is assured of certain resources, one still cannot be sure of how long it will take to reach the goal. Thus, for example, we are clearly aware of the goal of nuclear fusion and have invested heavily over the past decade, yet it may still be 20 or more years before we reach the goal of controlled fusion power.

Over the next five-year period, the U.S. may spend up to $10 billion for R & D. Of this amount, about $3.5 billion may go for nuclear fission, $1.5 for nuclear fusion, $3 billion for coal gasification and liquefaction, and $1 billion for solar, geothermal, and wind power. In addition, over $1.2 may go for research on environmental effects.

The ERDA budget for research and development is given in Table 23.3. The expenditures for solar R & D have grown rapidly over recent years, but still are

Table 23.3
THE ERDA BUDGET* (MILLIONS OF DOLLARS)

Program area	Fiscal year			
	1973	*1974*	*1975*	*1976*
Conservation	32.2	65.0	71	91
Oil, gas, and shale	18.7	19.1	43	44
Coal	85.1	164.4	418	442
Environmental control	38.4	65.5	185	199
Nuclear fission	406.5	530.5	625	709
Nuclear fusion	74.8	101.1	264	304
Solar	4.0	13.8	89	116
Geothermal	4.4	10.9	32	50
Systems studies	7.2	17.3	50	60
Other	0.9	11.5	203	398
Total	672.2	999.1	1,980	2,413

* Some line items are not fully comparable because of changes in categories for reporting.

lower than the expenditures for nuclear fusion or fission. However, fission reactors are commercially available, whereas solar conversion systems may not be commercially available for 10 to 20 years.

One potential scenario for the next 40 years calls for a near-term dependence on coal, with nuclear fission as an intermediate-term source, and solar or fusion energy as a long-term solution. Based on this prospect, R & D expenditures should emphasize: (1) the development of proven methods of the gasification, liquefaction, and desulfurization of coal in the near term, (2) the improvement of the safety and reliability of nuclear-fission reactors, (3) research into methods of exploiting the potential of solar and fusion energy, and (4) the development of effective and economic conservation methods.

23.5 STATE ENERGY POLICY

In matters of energy policy in the U.S., focus usually turns to the White House and Congress. However, one must not overlook the 50 state governments and the critical contribution they can make to energy-policy formulation and implementation. State policy towards population growth and population density have an impact on the required growth of energy sources within that state.

Recently, Florida reviewed its consumption of energy and its projected needs for the next several decades. The Legislature and the governor appointed and funded an Energy Policy Committee in 1973, and recently received an interim report. [5] Florida is growing by the migration of 300,000 persons each year to its sunny climate. Also, some 25 million tourists visit Florida each year. Florida's consumption of electricity, of which almost three-fourths goes for residential and commercial use, has been increasing at an average annual rate of 11 percent. Heavy use of air conditioning, in both private residences and tourist accommodations, is a major factor in this large increase.

The Florida Energy Committee analyzed high-, slowed-, and low-growth cases. The intermediate or slowed-growth case would require 50 new 1000-megawatt plants, 1150 miles of transmission line, and 13 large oil refineries to be built by the year 2000, according to the report. In order to limit and control the growth of energy consumption, Florida may adopt (1) land-use regulation, (2) impact fees to be levied on new development projects, (3) conservation through better building design, (4) inverted rate schedules for utilities, and (5) greater energy efficiency in transportation. [5] In addition, Florida, as is the case in many states, is considering the issue of the siting of energy facilities such as refineries and nuclear plants.

In California, since 1973, many energy bills have been submitted to the legislature. These bills cover the issues of energy policy and propose several means of dealing with them. One such bill, passed in 1974, prohibited the sale or installation of a new residential-type gas appliance equipped with a pilot light 24 months after the development of an intermittent-ignition device. California also passed an important law in 1974 establishing the California Energy Resources and

Development Commission (AB 1575). Powers and duties of the Commission include: (1) development of energy-conservation policy and measures, (2) administration of energy-research and development programs, (3) independent analysis of energy supply and demand, (4) development of safety and environmental guidelines for siting electrical power facilities, (5) certification of electrical power facilities, (6) monitoring and enforcement powers, and (7) emergency powers to relieve energy shortages. The functions provided in the Act, including research and development, are funded by a surcharge of 0.1 mils ($0.001) per kilowatt-hour on electricity sold in California.

The following energy issues are illustrative of those analyzed by the Commission: (a) effects of massive movement of oil from the West Coast to the Midwest, including the effects of increased tanker traffic upon California's coastal waters and coastline; (b) need for and character of refinery expansion and modification; (c) environmental, economic, and technical trade-offs between alternative means for generating electricity; (d) energy-system siting, including especially the vulnerability of systems to seismic activity; (e) relationship of the California transportation system, including especially air-pollution aspects, to energy use.

Several states are considering the enactment of legislation establishing a state energy commission similar to the Califronia Energy Commission. Energy policy is as important to the future economic and social well-being of states as it is to the nation.

23.6 ENERGY POLICY IN EUROPE

Most, if not all, of the industrialized nations in Europe have taken actions to further centralize and integrate national energy policy. In Great Britain, a new Department of Energy was established in 1974. Mr. Eric Varley thus became the Secretary of State for Energy, a unique role in Europe. The British Department of Energy has concentrated its policy development into five main areas: [6] the coal industry; nuclear policy; North Sea oil and gas; international aspects; and energy conservation. A major task in Britain will be to maintain and increase the deep-mine coal production of 120 million tons a year over the next decade, while the North Sea oil and gas is being fully developed. In Britain, coal is produced and managed by the National Coal Board, an arm of government. A British National Oil Corporation is being set up, through which the government will exercise its participation rights in North Sea oil.

The Commission of the European Economic Community (EEC) is attempting to develop an integrated energy policy for the EEC. The Commission has proposed that the share of the total energy consumed in the EEC that is provided by nuclear reactors increase from 1.4 percent in 1973 to 17 percent in 1985. This increase will permit, according to the Commission's plan, the dependence on oil to drop from 61 percent in 1973 to 41 percent of the total energy consumed in the EEC in 1985. The plan maintains that by the year 2000, the EEC could meet as much as 50 percent of its energy need with nuclear power. One aim of this energy policy is to reduce the EEC's dependence on imported energy supplies.

The policy issues for Europe, Japan, and the U.S. include the rates of increase of energy consumption and the social and environmental costs of mining, converting, transporting, and using fossil fuels and nuclear power. [12, 15, 16] The dependence of a nation on imported oil and the development of new alternative energy sources are critical issues. One of the key issues is the allocation of research and development support for various alternative energy sources of the future. Choosing the right energy source for the future is an element of policy that is difficult to obtain agreement upon in any nation. These decisions must be taken in the face of great uncertainty and the risk of investment in an inadequate substitute technology that could have serious consequences to any nation.

REFERENCES

1. S. E. ROLFE, "Whatever happened to Project Independence?", *Saturday Review*, January 25, 1975, pp. 25–28.

2. H. A. KISSINGER, "The Washington energy conference," *Vital Speeches of the Day*, March 15, 1974, pp. 322–325.

3. D. J. ROSE, "Energy policy in the U.S.," *Scientific American*, January 1974, pp. 20–29.

4. J. GREY, "Energy since the oil war," *Astronautics and Aeronautics*, November 1975, pp. 16–17.

5. L. J. CARTER, "Florida: An energy policy emerges in a growth state," *Science*, April 19, 1974, pp. 302–305.

6. M. KENWARD, "Government of energy," *New Scientist*, April 11, 1974, pp. 70–72.

7. *An Analysis of the ERDA Plan and Program*, Office of Technology Assessment, U.S. Government Printing Office, 1975.

8. D. KASH, *Our Energy Future*, Univ. of Oklahoma Press, Norman, 1977.

9. R. J. KALTER AND W. A. VOGELY, *Energy Supply and Government Policy*, Cornell Univ. Press, Ithaca, New York, 1977.

10. T. H. TIETENBERG, *Energy Planning and Policy*, Lexington Books, Lexington, Mass, 1976.

11. A. B. LOVINS AND J. H. PRICE, *Non-Nuclear Futures: The Case for an Ethical Energy Strategy*, Ballinger Publishing Co, Cambridge, Mass, 1975.

12. A. B. LOVINS, *Soft Energy Paths: Toward a Durable Peace*, Ballinger Publishing Co, Cambridge, Mass, 1977

13. J. J. DI CERTO, *The Electric Wishing Well*, MacMillan Publishing Co, New York, 1976.

14. J. C. SAWHILL, "Paralysis on the Potomac," *Saturday Review*, Jan. 22, 1977, pp. 19–20.

15. A. B. LOVINS, "Energy strategy: The road not taken?", *Foreign Affairs*, October 1976, pp. 65–96.

16. A. WEINBERG, *Economic and Environmental Implications of a U.S. Nuclear Moratorium*, Institute for Energy Analysis, Oak Ridge, Tenn., 1976.

17. J. M. BLAIR, *The Control of Oil*. Pantheon, Inc., New York, 1977.

EXERCISES

23.1 Several policy issues are listed in Table 23.1. What process should be used to prepare national policies for these issues? Should the energy experts prepare the policy or should Congress and other bodies be involved? Prepare a position paper on the process of energy policy development.

23.2 Enumerate several short-term policy issues (between now and 1985) and indicate those relatively difficult to solve in the short term.

23.3 Discuss the wisdom of developing a large domestic stockpile of fossil fuels in the U.S. in order to be essentially independent of any embargo of oil. Estimate the costs of such a stockpile and the advantages of having the stockpile.

23.4 Investigate the organization of agencies in your state responsible for energy policy. Prepare a list of the agencies and their duties. Would you recommend that these agencies be merged into one energy department?

23.5 Obtain a summary of the annual updated ERDA plan for research and development. (ERDA, Wash., D.C. 20545.) Analyze the shifts, if any, in the plan as the years progress from its original 1975 plan.

23.6 Obtain a summary of the latest fiscal-year budget for ERDA (Wash., D.C. 20545) and compare it with the budget for fiscal 1976. Analyze any significant shifts in budget priorities.

23.7 Prepare an energy-policy scenario for the U.S. for the near term, intermediate term, and long term.

23.8 Briefly summarize the recent energy policy of the EEC, Great Britain, or France. How does this policy differ from the current energy policy in the U.S.?

23.9 Should the policy of the United States in regard to world energy and other resources be a "spaceship" or a "lifeboat" policy? To share or to husband? The question is especially pertinent in regard to the U.S. production of food, the most valuable energy resource in the world. Prepare a policy analysis of the U.S. energy approach to sharing its energy and food resources.

*23.10 Prepare a policy analysis of one of the following issues and provide a series of recommendations on that issue:

a) Should environmental controls on energy exploitation be relaxed in order to allow the use of domestic instead of foreign fuels, or of more abundant in place of scarcer fuels?

b) Should emphasis be placed more on maintaining or increasing energy supply, rather than on conserving energy through improved technical and social efficiencies?

c) If we assume that energy imports will continue to be needed, how can the nation move toward greater assurance of foreign supply at tolerable prices, at

---

* An asterisk indicates a relatively difficult or advanced exercise.

the same time developing some form of insurance against interdiction of this supply?

23.11 President Carter proposed to the Congress in April, 1977, that homeowners be entitled, under the proposed law, to a tax credit of 25% of the first $800 and 15% of the next $1400 spent on approved conservation measures such as the addition of insulation on solar heating. Examine the potential effectiveness of these incentives and indicate whether you recommend that they be implemented.

23.12 The President's proposed tax-rebate schedule for *new* automobiles in 1979 is as follows (where mpg = miles per gallon):

$$12.0 \leqslant mpg < 13.0 \qquad \text{a penalty of \$553}$$
$$14.0 \leqslant mpg < 15.0 \qquad \text{a penalty of \$339}$$
$$16.0 \leqslant mpg < 17.0 \qquad \text{a penalty of \$176}$$
$$21.0 \leqslant mpg \leqslant 22.0 \qquad \text{a rebate of \$90}$$
$$23.0 \leqslant mpg \leqslant 24.0 \qquad \text{a rebate of \$165}$$

Examine (a) the equity and (b) the effectiveness of this conservation proposal.

24

International Factors

The rapid increase in demand for energy has caused great concern for the adequacy of long-range supply, not only in the U.S. but in the world as a whole. In this chapter we will consider the use and supply of energy throughout the world.

24.1 WORLD ENERGY CONSUMPTION

The growth in the use of energy in the world has increased at a rate of 5 percent per year since 1860, with a brief interruption during the depression years of 1930 to 1935. In Table 24.1, the world consumption of energy over the period 1860 to 1970 is given. The projections for the years 1980 and 2000 are based on a continuing world growth rate of 4.5 percent per year.

The annual world energy production of individual energy sources is shown in Fig. 24.1. Currently, coal, oil, and natural gas provide the predominant shares of

Table 24.1
WORLD ANNUAL
CONSUMPTION OF ENERGY

Year	Consumption ($\times 10^{15}$ Btu)
1860	4.03
1880	9.79
1900	22.59
1920	41.89
1940	67.86
1960	122.72
1970	202.43
1980 (Projected)	310
2000 (Projected)	722

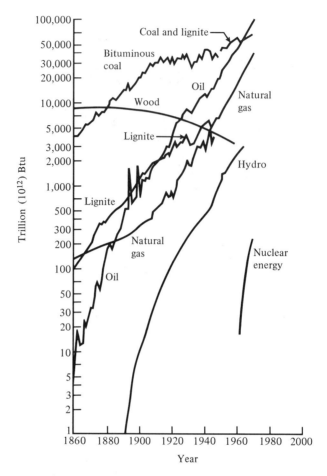

Fig. 24.1 *Annual world energy production of individual energy sources.*

the world's supply of energy; wood and hydropower supply about equal shares of the world's energy supply. Petroleum is the leading source of energy in the world today, supplying about 46 percent of the world's energy consumed in the form of fossil fuels, hydropower, or wood. The world's use of petroleum may continue to grow over the next two decades while other alternatives are developed. After the year 2000, we might expect that the dependence upon oil will decline as a shift to solar and nuclear energy occurs. The world's individual energy sources in 1970 are given in Table 24.2.

During the decade 1961 to 1970, world consumption of natural gas rose 113 percent compared to a 103 percent rise for petroleum. During that period, consumption of petroleum tripled in Western Europe. Over the decade the Middle

Table 24.2

WORLD INDIVIDUAL ENERGY SOURCES IN 1970

Source	Supply ($\times 10^{15}$ Btu)	Percent of total
Coal and lignite	69.4	34.3
Oil	84.8	41.9
Natural gas	36.8	18.2
Nuclear/hydro	5.8	2.8
Wood	5.6	2.8
Total	202.4	100%

Eastern and African share in total world oil production rose from 27 percent to 43 percent. Petroleum is such an important trade commodity that it accounted for 55 percent (by weight) of all international shipping in 1970. Nuclear-fission energy has grown in importance as an energy source in the world, and now accounts for about 1.6 percent of the total energy production.

Per capita consumption of energy in 1970 is listed in Table 24.3. Note that the U.S. had a per capita use five times as great as the world average use. However, many of the world's nations that possess energy resources are rapidly increasing their per capita use of energy; for example, note the growth rate for Saudi Arabia

Table 24.3

PER CAPITA USE OF ENERGY IN 1970

Country	Per capita use of energy (kg of coal equivalent)	Per capita use of energy (10^6 Btu)	Percent increase 1965 to 1970
U.S.A.	11,128	294	21
Canada	8,997	238	25
Sweden	6,304	166	41
E. Germany	5,944	157	9
United Kingdom	5,358	148	5
W. Germany	5,151	136	21
Australia	5,374	142	18
U.S.S.R.	4,436	117	24
France	3,799	100	28
Japan	3,215	85	80
Italy	2,685	71	49
World average	1,883	52	
Mexico	1,203	32	25
Iran	887	23	128
Saudi Arabia	827	22	126
China	522	14	14
India	189	5	8

and Iran. These nations reached the world's average per capita use of energy by
1976.

One measure of industrialization is the use of electric energy by a nation. In
Table 24.4, the electric-energy production is given for several nations. Fossil
fuel use for generating electric power is a significant factor in many nations.

Table 24.4

ELECTRIC ENERGY
PRODUCTION IN 1969 IN
SELECTED NATIONS

Country	Production $(10^9\ kWh)$
U.S.A.	1,552
U.S.S.R.	689
Japan	316
United Kingdom	238
West Germany	221
France	132
Italy	110
East Germany	65

24.2 A TRANSITION TO THE FUTURE

Over the next several decades, the world may double or triple its consumption of
energy. During this period it is possible to assist the less developed countries
(LDC) to attain some reasonable measure of energy use and associated quality
of life. During the next 30 years, several new alternative sources of energy may be
developed. These new sources, such as solar energy, may enable the LDC to raise
their standard of living. Also, as the growth of the per capita use of energy levels
off in the industrialized nations, the LDC may increase their per capita use of
energy. By increasing the overall efficiency of world energy use by a factor of two,
the LDC could double their use of energy. [17, 18]

In Table 24.5, a comparison of energy-consumption patterns in five industrialized nations is given. [1] It is interesting to note that Sweden can afford a high
quality of life while using one-half the energy per capita consumed in the U.S. Also,
Sweden required about one-half of the energy per million dollars of GNP compared
to the U.S., as indicated in the table. [16] Therefore, one might infer that it may be
possible to maintain a well-developed, affluent society at an energy-per-capita
rate considerably less than that of the U.S. If the U.S. were to stop increasing its
energy-per-capita usage, other nations could then strive to achieve a ratio similar
to that of Japan, about 80 million Btu per capita.

Another way of examining the possibility for a transition of energy usage
throughout the world is to divide the world into two groups of nations. We will
denote the relatively well-off nations as wealthy, and the poorer, less well-off or

Table 24.5

A COMPARISON OF ENERGY CONSUMPTION PATTERNS
FOR FIVE COUNTRIES IN 1972

	Sweden	Britain	U.S.A.	West Germany	Japan
Population (millions)	8.1	55.8	208.8	61.7	107.0
Gross National Product ($1 million U.S.)	43.8	157.0	1,159.3	174.2	251.4
Energy consumption (10^{15} Btu)	1.3	8.4	67.3	9.2	9.6
Energy consumption per capita (10^6 Btu)	152	143	307	143	86
Energy consumption per million of dollars of GNP (10^7 Btu per 10^6)	3.0	5.4	5.8	5.3	3.8

less developed countries (LDC) as the other group. The 2.5 billion people living in countries with per capita gross national product less than $650 (1972 U.S. dollars) will be denoted as the less developed countries (LDC). The remaining 1.2 billion people living in nations with a per capita GNP greater than $650 will be denoted as developed countries. The global distribution of energy use for this oversimplified division is given in Table 24.6. [2] In this table, the developed nations are divided into the U.S. and other nations.

In order to permit an acceleration of energy use by the less developed nations, a slowdown in the growth of use by industralized nations would be required. This is a possibility, since population growth in the developed nations has slowed down

Table 24.6

GLOBAL DISTRIBUTION OF ENERGY USE IN 1972
AND RECENT RATES OF CHANGE

	Less developed countries	All developed countries	
		Other developed countries	U.S.A.
Population (billions)	2.5	1.0	0.2
Per capita energy use (10^6 Btu/person)	13.7	100	307
Total energy use (10^{15} Btu)	36.5	103.8	67.3
Percent of total energy use	18%	50%	32%
Recent annual growth of:			
Population	2.5%	1.0%	0.7%
Total energy use	5.5%	5.0%	4.0%
Per capita energy use	3.0%	4.0%	4.0%

to less than 1 percent per year. Maintaining a constant per capita use of energy in the developed nations would enable the less developed nations to significantly expand their use of energy. One optimistic scenario for redistribution of the growth of energy use is given in Table 24.7. This scenario calls for a reduced rate of growth of population of the LDC over the next three decades while the per capita use of energy increases. Also, this projection assumes a zero rate of population growth of developed nations by the year 2000 and a leveling off in the rate of growth of per capita energy use. Then, with this transition in the rates of energy growth, the per capita energy use would increase by 41 percent in the developed nations over the 30-year period while increasing by 140 percent in the less developed nations. Using this projection the energy consumption in the year 2000 would be 305×10^{15} Btu and 155×10^{15} Btu in the developed and less developed countries, respectively. Then, the total world consumption in 2000 would be 460×10^{15} Btu compared with 230×10^{15} Btu in 1975. Whether this redistribution of energy consumption in the world can be attained will depend, in part, on the availability of energy resources in the world and the political will of individual countries.

Table 24.7
AN OPTIMISTIC SCENARIO FOR THE GROWTH OF ENERGY
USE FOR THE PERIOD 1975 TO 2005

	Developed countries		*Less developed countries*	
	Population	*Per capita energy use*	*Population*	*Per capita energy use*
Growth rates				
1975–1985	0.8%	3.0%	2.0%	2.5%
1985–1995	0.3%	1.0%	1.5%	3.5%
1995–2005	0.0%	0.2%	1.0%	4.5%
Values at specified times (billions)		$(10^6$ btu)	(billions)	$(10^6$ btu)
1975	1.25	150	2.75	15
1985	1.35	192	3.41	20
2000	1.44	212	4.31	36

24.3 WORLD ENERGY RESOURCES

Estimates of world energy resources vary widely. Energy *resources* means those quantities of mineral fuels that have been identified and are considered, on the basis of engineering and geological knowledge, to be recoverable or producible under current economic conditions with existing technology. The term *resources* includes reserves, but also encompasses deposits already identified but not presently recoverable, as well as undiscovered deposits which may or may not be recoverable. Through discovery and improvements in technology, undiscovered and uneconomical resources are being moved into the reserve category. Table 24.8 shows the world's known recoverable reserves and an estimate of the total re-

Table 24.8

ESTIMATED WORLD RESERVES AND
RESOURCES OF FOSSIL FUELS

	Known recoverable reserves	Total resource
Coal (10^9 tons)	665	10,000
Petroleum (10^9 barrels)	556	2,000
Natural gas (10^{12} cubic feet)	2230	13,000
Shale oil (10^9 barrels)	—	2,000
Uranium ore (10^3 tons)	1085	4,000

Table 24.9

ESTIMATED WORLD
RESERVE OF FOSSIL
FUELS EXPRESSED IN
QUADRILLIONS OF
Btu'S (10^{15} Btu)

Coal	15,000 Q
Petroleum	3,500 Q
Natural gas	2,000 Q
Uranium ore	800 Q
Total	21,300 Q

source of the fuel. [3] The world's estimated reserves of fossil fuels expressed in quadrillions of btu's (10^{15} Btu) are given in Table 24.9. The world consumption of energy in 1975 was about 230 quads and the estimated known reserve is 21,300 quads. Thus, the reserve-to-production ratio is approximately 100.

The known reserves of fuels are not evenly distributed throughout the world. The distribution of the world's petroleum reserves are listed in Table 7.6. The Middle East and Africa hold about 69 percent of the world's proven reserves. The distribution of the world's coal reserves are given in Table 24.10. Note that the U.S. holds about 32 percent of the world's proven coal reserves.

Table 24.10

THE WORLD'S PROVEN COAL RESERVES
(IN BILLIONS OF TONS)

U.S.S.R. and other Communist nations	305
United States	216
Western Europe	72
Far East and Oceania	46
Africa	17
Other	9
Total	665

Perhaps one of the best areas of potential development of energy for less developed nations lies in the construction of hydroelectric projects. Asia and Africa hold vast resources of potentially developable hydropower projects. For example, India and Zaire have 41 gigawatts and 132 gigawatts of potential hydropower, respectively. However, hydroelectric projects are more capital-intensive than many other kinds of power plant, and require long construction times. Therefore, less developed nations often are able only slowly to develop their hydroelectric potential. The estimate of the energy that could be potentially generated from hydroelectric power plants is given in Table 24.11. [4] If one-quarter of the world's hydroelectric resource could be exploited by the year 2000, then approximately 2400×10^9 kWh (8.2 quads) could be obtained annually. This, unfortunately, might be only 3 or 4 percent of the world's total energy consumption.

Table 24.11
THE WORLD'S HYDROELECTRIC
RESOURCE ENERGY POTENTIAL

Country	Annual energy output (10^9 kilowatt-hours)
China	1320
U.S.S.R.	1095
U.S.A.	701
Zaire	660
Canada	535
Brazil	519
India	205
World total	9802

Canada

An abundance of natural resources makes Canada almost self-sufficient in the supply of energy. Yet the country's vast reserves of coal, oil, and tar sands and its large potential for hydroelectric development do not insulate the nation from the possibility that it will experience an energy shortage. [5] Canada is self-sufficient in petroleum, but the consumers and producers are a long way apart and there is no pipeline connecting them. Canadian producers are actually closer to many U.S. consumers. In the past, Canada has exported oil in the western part of the country to pay for the oil imported in the more densely populated and industrialized east. As shown in Table 24.12, the imports exceeded exports in 1975. [6]

At the end of 1974, the Canadian government decided to reduce the amount of oil exported to the U.S. with the intention of stopping the export altogether by 1982. It is planned to reduce the export to the U.S. by 400,000 barrels daily in 1977. By comparison, oil exports to the U.S. were over 1 million barrels per day at their peak in 1973.

Table 24.12

CANADIAN OIL EXPORTS AND IMPORTS (in thousands of barrels per day)

Year	1970	1971	1972	1973	1974	1975
Exports	680	750	950	1120	900	700
Imports	570	670	790	850	900	950

One proposal is to exchange surplus U.S. oil from the Alaskan slope for the continued flow of Canadian oil to U.S. refineries. The proposal calls for an interchange of Alaskan oil in the west for Canadian oil in the U.S. midwest. This would save considerable transportation costs.

Of the estimated Canadian resources of 121 billion barrels, only nine billion barrels are proven reserves, as shown in Table 24.13. Canada also holds an estimated 55×10^{12} cubic feet of proven reserves of natural gas. The western province of Alberta produces 85 percent of Canada's oil and 88 percent of its natural gas. In addition, the untapped tar sand resource lies in Alberta.

Table 24.13

CANADIAN RESOURCES

	Resource	Proven reserve
Petroleum (billions of barrels)	121	9
Natural gas (10^{12} cubic feet)	780	55
Alberta's tar sands (billions of barrels)	300	

Hydroelectric power is another important part of Canada's energy resource. The total installed hydroelectric capacity exceeds 34 gigawatts, representing 63 percent of the nation's total electric-power capacity. Hydroelectricity provides about 23 percent of Canada's primary energy supply. This is in contrast to the U.S., where hydroelectric energy supplies less than 2 percent of the total energy consumed in the nation. Over the next decade, Canada is planning to further develop hydroelectric projects and explore storage schemes.

Western Europe

Energy consumption and resources in Western Europe can be examined for the European Common Market, which has many common policies in energy development and use. In later sections, we will consider the resources and consumption patterns of several specific member nations of the Common Market (European Economic Community, EEC). [19]

About 50 years ago, Europe was a net exporter of fuel, in the form of coal, while today it is the largest importer of primary energy in the world. As late as 1940 Europe was dependent on imported sources for only 5 percent of its needs.

Table 24.14
PETROLEUM IN WESTERN EUROPE*
(EEC) IN 1975

Annual production	0.18 billion barrels
Annual consumption	4.97
Annual imports	4.79
Total proven reserves	20 billion barrels

* European Economic Community nations are:
Belgium, Denmark, France, Great Britain, Ireland,
Italy, Luxemburg, Netherlands, and West Germany.

Today Europe's imported energy is in the form of oil and represents approximately 60 percent of the world trade in oil. The use and importation of oil in western Europe is shown in Table 24.14. The EEC nations were dependent upon imports for 96 percent of their petroleum in 1975.

The EEC is dependent upon the importation of oil for 60 percent of their total energy needs, whereas they imported only 31 percent of their total energy consumed in 1960. As western Europe's energy use has grown at an annual rate of 5 percent over the past two decades, the dependence upon imported oil has grown proportionately. Petroleum supplies 60 percent of the total energy consumed in western Europe as shown in Table 24.15.

Table 24.15
THE ENERGY SOURCES IN
WESTERN EUROPE

Source	Percent of total
Coal	29
Oil	60
Natural gas	7
Hydroelectric	3
Nuclear	1
Total	100

Natural gas is also used in western Europe and supplies 7 percent of the total energy consumed. About 10 percent of this natural gas is imported from Algeria and the U.S.S.R. (via Czechoslovakia). The primary source of natural gas in the EEC is the Netherlands.

The total proved reserves of coal in western Europe are estimated to be 3.5×10^{18} Btu (1.5×10^{11} tons), which is sufficient to supply Europe for many decades at the present rate of consumption.

Western Europe has three organizations responsible for energy policy and planning. Euratom is responsible for nuclear energy, coal comes under the European Coal and Steel Community, and the European Economic Community itself deals with oil and natural gas. The EEC policy calls for a reduction in Europe's reliance on imported energy; the keystones of the proposed plan formulated by the European Commission are that oil should in future be regarded less as a fuel for burning and more as a raw material for the chemical industry, and that by 1985 the nine member countries' appetite for energy should be 10 percent below the present estimates for 1985 consumption, this saving to be brought about by emphasizing conservation and instituting economic incentives to limit energy consumption.

Euratom and the member nations are aggressively pursuing the development of nuclear-power plants. It is proposed to increase the percent of total energy supplied by nuclear-power plants from the present 2 percent to between 13 and 16 percent by 1985. At that time, nuclear power plants would supply one-half of the electricity in the EEC. [7] The cost of shifting to nuclear power plants in the EEC may be more than $30 billion over the next decade.

The dependence of western Europe on imported oil has profound economic and political consequences. [8] The 1974 balance-of-payments deficit of Britain, France, and Italy combined exceeded $25 billion. This deficit has subsequently been somewhat reduced, but the European nations are striving to work out a policy for independence in the future. This policy will depend upon the exploitation of North Sea oil and gas and the rapid development of nuclear power plants.

The Less Developed Countries

The recent rise in oil prices has had a severe impact on the developing nations. These nations have attempted to develop an industrial base and a mechanized, modern agricultural system, only to be faced with a rise in energy prices. The recent jump in oil prices and natural gas prices has caused a related increase in fertilizer costs, which is an important factor in the building of an expanded agricultural output. It is estimated that the cost of oil for developing nations increased by $10 billion in 1974, and a deficit in the balance of payments may have risen to $30 billion a year in 1976.

Developing nations depend to a great extent on firewood for a primary fuel source. More than 60 percent of the energy needs of the Central American countries are satisfied through the use of firewood and charcoal. Nine-tenths of the people in most poor countries rely on firewood for their energy. [11, 12] For perhaps a third of the world's people, the real energy crisis is a daily effort to find the wood for cooking dinner and heating their home. However, in many nations, the growth in the population is outpacing the growth of trees; the demand on woodlands and forests has led to denudation of the land in certain areas. One prime cause for the recent floods in the Indian subcontinent was the removal of tree cover in the catchment areas for fuel wood. Also, the continual cutting of fuel wood in the Southern Sahara/Sahel region of West Africa has facilitated the rapid southward encroachment of the desert, and has made living in this area even more precarious.

The LDC are estimated to use 2000 million tons (air-dry) of wood annually. This wood is equivalent to burning 1300 million tons of coal (31.2×10^{15} Btu).

The developing nations, such as Brazil, have the potential to plan the use of fuel-wood plantations. Brazil is contemplating the use of a plant—cassava (manioc)—to yield ethyl alcohol through a fermentation process. [13] Also, Brazil is pursuing the development of a significant nuclear-energy scheme and plans to add 8 additional large nuclear-power reactors by 1985. Developing nations may find that the potential for harnessing wind, solar, or geothermal energy may be the necessary ingredient of the future in order to achieve a reasonable measure of self-sufficiency, industrialization, and affluence.

24.4 ENERGY AND U.S. FOREIGN POLICY

During the past five years, the U.S. has learned that energy trade is a critical element in the determination of its foreign policy. The U.S. foreign trade of oil, natural gas, and coal is shown in Table 24.16.

Table 24.16
U.S. FOREIGN NET TRADE FOR 1950 TO 1973

Year	Exported coal (Million tons)	Imported oil and refined products (Million barrels)	Imported natural gas (10^9 cubic feet)
1950	29.0	198.9	(25.7)*
1960	37.7	590.2	144.3
1970	72.4	1153.5	751.0
1973	52.8	2158.0	950.0

* 25.7×10^9 cu ft of gas were exported in 1950.

In 1925, the U.S. exported 3 percent of its aggregate energy production. By 1950, net imports amounted to $1\frac{1}{2}$ percent of U.S. energy consumption. The overall energy net import share rose to 15 percent by 1973 due to increasing imports of oil and refined products. In 1973 the U.S. imported 2158 million barrels of petroleum (12.5×10^{15} Btu) and 950×10^9 cubic feet (9.8×10^{12} Btu) of natural gas. In 1973, the U.S. exported only 52.8 million tons (1.4×10^{15} Btu) of coal. Thus, the net energy trade in 1973 was about 11.1×10^{15} Btu or approximately 14.7 percent of the total energy consumption. The U.S. expended 1.5 billion dollars for this net importation of fuels in 1973. By 1976, this net energy expenditure had risen to more than five billion dollars.

The impact of a dependence on imported fuels has brought to the fore the following issues: (1) the reliability, (2) the impact of balance of payments, (3) national security, (4) worldwide interdependence and cooperation in energy, and (5) the use of oil as an international political weapon. (The latter issue became clearly demonstrated in the aftermath of the Arab-Israeli conflict in the fall of 1973.) In addition there is the economic dimension of energy supply and the readiness of oil-exporting states to fulfill the energy demands of the oil-importing nations.

In the matter of energy imports, foreign policy and domestic energy policies regulate and restrict the action of the economic marketplace. The OPEC nations have become powerful enough to control production and raise prices to levels far beyond the cost of extracting the oil. The U.S. has chosen to work with other importing nations in an effort to develop an energy policy rather than attempt to break the power of the oil cartel and restore competition. [23] The U.S. is pursuing a policy of attempting to develop long-run energy source alternatives while maintaining a sizeable dependence on oil imports. Foreign-policy alternatives will have to be developed over the next decade so that energy can be made continuously available without overly disarranging the international relationships.

REFERENCES

1. J. McHALE AND M. C. McHALE, "Human requirements, supply levels, and outer bounds," The Aspen Institute for Humanistic Studies, 1975.

2. J. P. HOLDREN, "Energy and prosperity," *Bulletin of Atomic Scientists*, Jan. 1975, pp. 26–28.

3. V. E. McKELVY, "World energy reserves and resources," *Public Utilities Fortnightly*, Sept. 25, 1975, pp. 27–33.

4. A. PARKER, "World energy resources," *Energy Policy*, March 1975, pp. 58–63.

5. "Anxieties and opportunities," *New Scientist*, April 17, 1975, pp. 140–141.

6. R. A. DUBIN, "The feud that keeps Canada's oil underground," *Business Week*, Feb. 10, 1975, p. 46.

7. "Energy: Building atom power in Europe and Japan," *Business Week*, March 17, 1975, pp. 36–37.

8. W. HAFELE, "Energy choices that Europe faces," *Science*, April 19, 1974, pp. 360–366.

9. W. MARSHALL, "Stretching energy independence," *New Scientist*, March 20, 1975, pp. 695–697.

10. B. SORENSEN, "Energy and resources," *Science*, July 25, 1975, pp. 255–260.

11. M. KENWARD, "Firewood—A growing energy crisis," *New Scientist*, Nov. 27, 1975, p. 519.

12. K. OPENSHAW, "Wood fuels the developing world," *New Scientist*, Jan, 31, 1974, pp. 271–272.

13. P. H. ABELSON, "Energy alternatives for Brazil" *Science*, Aug. 8, 1975, p. 14.

14. H. J. FRANK, "Economic strategy for import–export controls on energy materials," *Science*, April 19, 1974, pp. 316–321.

15. M. WILLRICH, *Energy and World Politics*, Free Press, New York, 1975.

16. L. SCHIPPER AND A. J. LICHTENBERG, "Efficient energy use and well-being: The Swedish example," *Science*, Dec. 3, 1976, pp. 1001–1012.

17. C. WILSON, *Energy Demand Studies: Major Consuming Countries*, MIT Press, Cambridge, Mass., 1977.

18. G. J. MANGONE, *Energy Policies of the World*, Elsevier Publishing Co, New York, 1976.

19. R. BAILEY, "Heading for an EEC Common Energy Policy," *Energy Policy*, Dec. 1976, pp. 308–321.

20. J. RUSSELL, *Energy as a Factor in Soviety Foreign Policy*, Lexington Books, Lexington, Mass., 1976.

21. D. HAYES, *Rays of Hope: A Global Energy Strategy*, W. W. Norton Co., New York, 1977.

EXERCISES

24.1 By the year 2000, the world's average per capita energy consumption might be 75×10^6 Btu (7.9×10^{10} joules) per person. At that time the range of per capita use might be 30 to 200×10^6 Btu. Do you find this a real possibility? What estimates would you propose? Estimate the impact of your proposal on the total consumption of energy in the world.

24.2 Coal production has decreased over the decade 1962 to 1972 in West Germany. [8] Examine the causes for this decline, and determine whether the production of coal could be increased to 30% above its all-time high. What coal exploitation policy would you recommend for West Germany as a member of the EEC?

24.3 Electric-energy consumption in Poland grew at a rate of 8% during the period 1960 to 1970. Poland's electric-energy consumption was about 100 billion kilowatt-hours and the installed capacity was 20 gigawatts. Determine the population of Poland and compare the consumption per capita with that of the U.S. Coal is the major fuel in Poland. Obtain estimates of the proven coal reserves in Poland, and estimate how many years they could last at the present rate of consumption, which is about 150 million tons (consumed in 1972).

*24.4 Prepare an alternative scenario in a format similar to Table 24.7. For this estimate, assume that the growth rate of per capita energy use in the less developed nations is 5.0% for the period 1985 to 2005 while the growth rate of the per capita energy use for the developed nations is 2% for the period 1975–1985 and 0.5% after 1985. Assume all other growth rates hold as specified in Table 24.7.

24.5 What are the impediments to the development of hydroelectric power projects in the less developed nations? Compare the development of hydropower in contrast to nuclear power, and give several factors in favor of each approach.

24.6 Determine the energy resources and proven reserves of Mexico. What is the domestic energy policy of Mexico? What is the country's foreign policy regarding energy?

24.7 Contrast the energy policies of Great Britain and France. Both nations are dependent upon imported oil, yet each nation has taken somewhat different

---

* An asterisk indicates a relatively difficult or advanced exercise.

actions and formulated different long-range plans. Examine these plans and analyze their implications for the rest of the EEC.

24.8 Determine the dependence of Brazil on imported oil, and explore alternative energy sources for that nation. Formulate an energy policy for that nation that will assist in economic development.

24.9 Consider the possible effects of a significant development of the nuclear breeder reactor and its use in many nations. What are the implications for foreign policy with regards to safeguards for the plutonium generated by the breeder?

*24.10 Compile statistics for the cost of gasoline per liter as a multiple of the cost of gasoline in the U.S., for several countries. Also, calculate the cost relative to GNP, as a multiple of U.S. cost relative to GNP. Include at least France, West Germany, Japan, and Italy.

24.11 Determine the reserves of fossil fuels of the U.S.S.R. and compare with U.S. proven reserves.

24.12 Determine the amount of petroleum imported by Japan. Also, determine what percentage of Japan's total energy supply is provided by imports.

25

Summary, Conclusions, and the Future

In this chapter we shall summarize the discussions of the preceding chapters and draw some conclusions about the consumption and supply of energy in the U.S. and the world in the future. Any attempt to envision the future is fraught with risk, but if we are to control our future we must try to discern it with some confidence and sensitivity to unforeseen consequences.

25.1 A SUMMARY

Energy consumption and world population have grown steadily over the last one hundred years. Furthermore, the energy consumed per capita has increased, particularly since 1940. Thus, the nations of the world have, more or less, become dependent upon energy as an important factor in this standard of living, their industrial growth, and their uses of leisure. Industrialized nations have become heavily dependent upon the uses of energy resources, and their economies are often dependent upon the importation of energy resources. The economic growth of nations has been linked, in the past, with the increased consumption of energy. The U.S. and Canada have developed an energy-intensive economic system which uses about twice as much energy per capita as do the nations of the European Economic Community. Yet the growth of the energy intensiveness in the economy of the U.S. and Canada may now be leveling off, while other nations continue to increase their energy use per capita. As the world population stabilizes, the standard of living of the people of the less developed nations will increase, particularly if effective and appropriate new energy technologies become available for their use.

The U.S. may consume about 90 quads (95 $\times$ 10^{18} joules) in 1985 and approximately 125 quads (132 $\times$ 10^{18} joules) in the year 2000. Whether these projections are achieved is dependent upon the supply of fossil fuels, the expansion of nuclear-reactor construction, the availability of imported oil and gas, and the development

461

of commercially economic solar heating and cooling systems. In addition, the U.S. and other industrialized nations are counting on significant energy savings through the implementation of effective conservation schemes.

The domestic U.S. supply of oil and natural gas over the next twenty-five years is anticipated to remain constant or decline somewhat. Thus, the U.S. is counting on an increased exploitation of its extensive coal resources. The expanded use of coal in the U.S. will depend upon the development of economic coal gasification methods as well as effective environmental-control devices for coal-fired electric plants.

The expanded utilization of electricity as an energy carrier will depend upon the increased availability of hydroelectric, nuclear, and coal-fired plants. With electric power becoming an increasingly important means of transport energy, electric autos, trains, and streetcars may become an important part of our transportation system. Electric power is particularly important in an industrialized economy and is also important for convenient use in the home.

The transportation sector currently accounts for about 25 percent of the total energy consumed in the U.S. If the movement of passengers and goods can be shifted to less energy-intensive modes of travel, the net result can be a significant energy savings without any loss of convenience or lowering of the standard of living. Also, a more efficient automobile (less energy-intensive) would be a major boon to the U.S. and other nations heavily committed to auto travel.

Nuclear-fission reactors are commercially available, and the electric utilities plan to increase the number of nuclear-power plants significantly. Nuclear-power plants may supply ten to fifteen percent of the total energy in the U.S. by 1985, if the utilities' plans are achieved. However, the concern of many citizens about the safe operation of these plants may limit their development. Also, there are the unresolved issues of nuclear wastes and proliferation of nuclear materials, that may constrict an increase in the number of nuclear plants.

Nuclear fusion power and energy from the oceans and winds may contribute significantly to energy resources by the end of this century. Nuclear fusion is still in a research-and-development stage; it will require an engineering breakthrough to bring fusion power to commercial use by the year 2000. Wind power may contribute to the energy supply by the late 1980's, especially in a decentralized manner. Thus, many small wind machines may contribute to the electric power needs of neighborhoods and small towns.

Solar energy methods will be a significant contributor to energy needs by 1985 and could become an important part of our energy supply by the year 2000. Solar heating and cooling systems could supply up to 10 percent of the nation's energy needs by the turn of the century, if sufficient incentives for expansion of this industry are provided. In addition, commercial methods of converting solar energy to electricity may be available by 1990.

The utilization of the energy of waste materials and the recycling of materials will increasingly become a significant contribution to the energy needs of industrialized nations. Total energy systems, which are conserving of energy, will become used more widely as the price of fossil fuels continues to increase. Also, the storage of energy, generated at offpeak periods, will become important in the future. As the U.S. and other nations become more heavily dependent upon electric power and more geographically remote generating plants, a shift to hydrogen as an energy carrier may occur. This is a possibility, particularly if large wind machines, ocean thermal schemes, or solar farms are implemented.

Conservation of energy in all nations of the world will become important as the price of fossil fuels continues to increase. In addition, the cost of constructing a power plant and the shortage of capital have resulted in increased emphasis on conservation and efficient use of fuels.

The industrialized nations are increasingly emphasizing the necessity of reducing the social and environmental impacts of energy consumption. Maintenance of the quality of the air, water, and land in the U.S. and other nations will require an effective environmental policy and a complex system of controls. The need for environmental controls will often conflict with the demand for energy and will result in a need for continuing compromise.

Energy policy in the U.S. and other nations will undergo several stages. The goals of many nations may shift from energy independence to conservation to cooperative research in the hunt for new sources of energy. The U.S. will, in all probability, spend twenty-five billion dollars on energy research and development during the period 1977 to 1987. In addition, many state governments will formulate their own energy policies in order to protect their own economies and environments.

As the per capita use of energy reaches equilibrium, the less developed nations will increase their use of energy. By the year 2000, the less developed nations may have achieved a per capita energy use equal to one-fourth of the rate of the developed nations. By the mid-21st century the per capita energy use of industrialized nations may be only twice that of less developed nations.

25.2   GROWTH AND SOCIETY

The industrialized nations of the world have been shifting from unlimited growth as a primary objective, to a mixed set of objectives which include a clean environment, reasonable leisure, employment for those who desire it, and a stable world. These nations have turned their attention from energy-consuming items, like cars and electric gadgets, to the quality of life instead of quantity.

Just as the goals of an individual change with each stage of development and the circumstances of life, so do the goals of a nation as its population levels off, its economy matures, its technology develops, and its resource constraints evolve. A mature society seeks to make maximum energy available at the least economic and social cost, and furthermore to make energy equitably available to all. Thus a nation seeks reasonable per capita energy availability, broad public access to that energy, reasonable energy prices, convenient conversion devices, and little or no adverse environmental effects.

The evolution of the social and cultural development of a nation is related to that nation's wise use of energy and technology. The cultural development, C, of a nation may be related to the amount of energy per capita consumed, E, and the efficiency of the technological means with which it is put to work, T. This relationship may be expressed succinctly, in one observer's view, by the formula: [1]

$$C = E \times T. \tag{25.1}$$

The cultural and social well-being of a nation may increase as its technological use of energy increases effectively.

The growth of energy demand in the U.S. has been very large. During the period 1950 to 1975, total energy use in the U.S. has grown by 123 percent. Energy consumption for automobiles for that period has grown by 214 percent. We must now work on increasing the technological efficiency of our use of energy. In place of prodigalities of consumption must come new frugal attitudes.

New styles of living may emerge where people gather together to share automobiles, homes, and tools. This possibility can be seen in the growth of new living arrangements such as communes.

Some proponents of the "limits to growth" argue that resources have already been depleted to a point beyond which we cannot recover. Nor, in the view of these critics, can there be a move towards an equitable distribution of energy resources in the world without revolution and dislocation. The energy crisis, if we may label it this, is a severe test of world stability and the economic system.

25.3 THE FUTURE OF ENERGY AND ECONOMICS

An unlimited desire to increase energy consumption per capita is unwise as well as unrealistic. As industrialized nations increase their per capita energy use, they reach a saturation point beyond which the rate of economic growth may level off or even decrease. Thus, the marginal value of increasing the energy consumption per capita in the U.S. or Canada may have become negligible.

Several nations such as Sweden and Switzerland may have already passed the U.S. with respect to GNP per capita. One possible reason is that these countries do not have to divert so much capital away from investment in growth-promoting activities toward dealing with the consequences of excessive energy use and low energy prices, as in the U.S.

Nations with extensive hydroelectric power available have developed a system of electric streetcars, and electric trains. This form of transportation militates against urban sprawl and wasteful use of energy by automobiles. This difference is illustrated by Table 25.1.

In the past, business managers have striven to replace expensive labor with cheap energy by using automation. This has resulted in higher productivity per worker, but also in some measure of unemployment. The unemployment rate in the period 1974 to 1976 in the U.S. remained above six percent. One means of promoting employment is to increase the price of energy more rapidly than the

Table 25.1
ECONOMIC ACTIVITY AND ENERGY IN 1972
FOR SELECTED COUNTRIES

Country	Energy consumption (lbs of coal equiv./dollar)	Automobile use autos (10^3 passenger-km)
U.S.	4.64	7.17
New Zealand	2.54	1.88
Norway	2.71	0.34
Japan	2.60	0.04

price of labor over the next decade. However, where the price of energy will undoubtedly increase during that period, it may still lag behind the increase in the cost of labor. The ratio of the cost of labor to the cost of energy has risen by a factor of 3.5 during the period 1950 to 1975. This constant rise continued during recessions and periods of expansion, and it is doubtful that it will level off over the next few years. Cheap energy is an assumption basic to our economy, while cheap labor is not.

In the economies of industrialized nations, economic growth is counted on to provide new jobs and capital. If new increased productivity is difficult to obtain, then new jobs may not be available.

25.4 DIVERSITY AND SIMPLICITY

The balance between economic stability and economic efficiency can also be viewed as a balance between diversity and simplicity. Diversity within a society implies a good deal of stability. A diverse economy also implies many individual firms or agencies and small units within a large society. Then, however, economies of scale cannot be exploited.

Simplicity, on the other hand, implies a small number of large units, making possible economies of scale. For example, a large 1000-MW nuclear reactor is more efficient than ten 100-MW reactors. This system tends towards centralization, in contrast to the decentralized economy of the diverse society. The diverse society depends upon windmills, solar collectors, geothermal plants, and small electric generators, while the centralized society depends upon energy plants providing upwards of 1000 MW.

In the recent past, our culture has placed a large value on economies of scale and economic efficiency and a lower value on diversity, decentralization, and the avoidance of instability.

Some recent reports call for a centralized, regulated, planned system of energy allocation and consumption. [2] Others point towards an optimal balance between energy use and environmental effects achieved by a free-market economy. [3]

A decentralized society operating on the basis of appropriate or intermediate technologies might slow the pace of urban sprawl, crime, and energy consumption. [4] Appropriate or alternative technologies are seen as relatively nonpolluting, cheap, and labor-intensive, nonexploitative of natural resources, compatible with local cultures, functional in a noncentralist sense and nonalienating. If we define the levels of technology in terms of equipment per workplace, we can call the indigenous technology of a typical underdeveloped country as, symbolically speaking, a $1 technology, while that of an industrialized nation is a $1000 technology. Then, an intermediate technology might be a $100 technology. The achievement of intermediate technology still rests upon scientific knowledge, but strives to achieve a good standard of living based on a reasonable energy consumption and a moderate technology suitable to the culture and life-style of the inhabitants.

A commitment to efficient use of energy and rapid development of renewable energy sources matched in scale and energy quality to end-use needs may result in a reduced rate of energy growth without harming the nation's or the world's economy as measured by GNP or by employment. [6] A mix of solar energy heating and cooling, conversion of agricultural, forestry and urban wastes, total systems, and efficient industrial processes may lead to a zero-growth rate for energy consumption by the 21st century.

In the future, we might anticipate that the quality of life may rest on more than energy consumption and expensive high-level technology. A balanced, decentralized, environmentally sound economy may be achieved without the prodigious use of energy in the future. However, it will take several decades for the world to work out a new social contract in order to achieve this new balanced economy.

REFERENCES

1. L. A. White, *A Science of Culture*, Farrar, Straus, and Co., New York, 1949.

2. *A Time to Choose*, Report of the Energy Policy Project of the Ford Foundation, Ballinger Publishing Co., Cambridge, Mass. 1974.

3. *No Time to Confuse*, Institute for Contemporary Studies, San Francisco, 1975.

4. E. F. Schumacher, *Small is Beautiful*, Harper and Row, New York. 1973.

5. R. C. Dorf, *Technology, Society, and Man*, Boyd and Fraser Publishing Co., San Francisco, 1974, Chapter 23.

6. A. B. Lovins, "Energy strategy: The road not taken?," *Foreign Affairs*, Oct. 1976, pp. 65–96.

Appendix A

Energy-Related Terms

Alternating current (ac): An electric current that reverses its direction of flow periodically (see frequency) as contrasted to direct current (dc).

Anthracite: A hard, black, lustrous coal that burns efficiently and is therefore valued for its heating quality.

Atomic Energy Commission (AEC): A five-member commission established after World War II to supervise and promote use of nuclear energy. The commission was abolished in 1975, and its functions were transferred to the Energy Research and Development Administration (ERDA) and the Nuclear Regulatory Commission (NRC).

Barrel: A liquid-volume measure equal to 42 U.S. gallons, commonly used in expressing quantities of petroleum and petroleum products (bbl).

Baseload: The minimum load of a utility (electric or gas) over a given period of time. Units in baseload service are operated in full capacity all the time.

Binary cycle: An energy-recovery system which results in heat exchange between two separate fluid-circulation systems. The purpose of a binary cycle is to obtain higher efficiencies from the energy source.

Bioconversion: A general term describing the conversion of one form of energy into another by plants or microorganisms. Synthesis of organic compounds from carbon dioxide by plants is bioconversion of solar energy into stored chemical energy. Similarly, digestion of solid wastes or sewage sludge by microorganisms to form methane is bioconversion of one form of stored chemical energy into another, more useful form.

Bituminous coal: Soft coal; coal that is high in carbonaceous and volatile matter. When volatile matter is removed from bituminous coal by heating in the absence of air, the coal becomes coke.

Breeder: A nuclear reactor that produces more fuel than it consumes. Breeding is possible because of two facts of nuclear physics: (1) Fission of atomic nuclei produces on the average more than two neutrons for each nucleus undergoing reaction. In simplified terms, then, one neutron can be used to sustain the fission chain reaction and the excess neutrons can be used to create more fuel. (2) Some nonfissionable nuclei can be converted into fissionable nuclei by capture of a neutron of proper energy. Nonfissionable uranium-238, for example, can thus be bred into fissionable plutonium-239 upon irradiation with high-speed neutrons.

British thermal unit (Btu): The quantity of heat necessary to raise the temperature of one pound of water one degree Fahrenheit. One Btu equals 252 calories, 778 foot-pounds, 1055 joules, and 0.293 watt-hours.

Calorie: Originally, the amount of heat energy required to raise the temperature of 1 gram of water 1 degree Centigrade. Because this quantity varies with the temperature of the water, the calorie has been redefined in terms of other energy units. One calorie is equal to 4.2 joules. (When capitalized, Calorie means 1000 calories (or 1 kcal). Therefore, one Calorie is equal to 4184 joules.

Central station power: Production of power—usually electrical—in large quantities at a generation plant as opposed to production at the point of consumption.

Coal: A solid, combustible organic material formed by the decomposition of vegetable material without free access to air. Chemically, coal is composed chiefly of condensed aromatic ring structures of high molecular weight. It thus has a higher ratio of carbon to hydrogen content than does petroleum.

Coal gasification: The conversion of coal to a gas suitable for use as a fuel.

Coal slurry pipelines: A pipeline which transports coal in pulverized form suspended in water.

Conduction: The process by which energy is transferred directly from molecule to molecule. It is the way in which electricity travels through a wire or heat moves from a warm body to a cool one when the two bodies are placed in contact.

Continental shelf: The submerged shelf of land sloping gradually from the exposed edge of a continent. It is usually defined as those areas where the water is less than 200 meters (600 feet) deep.

Controlled thermonuclear reactor (CTR): Controlled fusion, that is, fusion produced under research conditions, or for production of useful power.

Convection: The transfer of heat by the circulation of a liquid or gas.

Core: The central part of a nuclear reactor which contains the nuclear fuel.

Critical mass: The minimum amount and arrangement of a fissionable material, such as uranium-235 or plutonium-239, that is required to sustain fission in a nuclear reactor.

Crude oil: Petroleum liquids as they come from the ground. Also called simply "crude."

Curie: The unit of radioactivity. One curie is 3.7×10^{10} disintegrations per second. One gram of radium has 1 curie of radioactivity.

Decay, radioactive: The process whereby atoms of radioactive substances experience transformation into atoms of other elements with attendant emission of penetrating radiations (gamma ray) and some nuclear particles. Each radioactive substance has a unique decay rate, which may range from a fraction of a second to hundreds of years or more.

Depletion allowance: A tax allowance extended to the owner of exhaustible resources based on an estimate of the permanent reduction in value caused by the removal of the resource.

Deuterium: A nonradioactive isotope of hydrogen whose nucleus contains one neutron and one proton and is therefore about twice as heavy as normal hydrogen. Often referred to as heavy hydrogen.

Diesel oil: The oil fraction left after petroleum and kerosene have been distilled from crude oil.

Direct current (dc): An electric current, such as that produced by a battery, in which the electrical potential does not change its sign, and the voltage is often invariant with time. In a direct current, therefore, energy is carried by a continuous, unidirectional flow of electrons through a conductor.

Electric energy: The energy associated with electric charges and their movements. Measured in watt-hours or kilowatt-hours. One watt-hour equals 860 calories.

Electron: An elementary particle with a rest mass of 9.1×10^{-28} grams, bearing a negative electric charge. Electrons orbit the atomic nucleus; their transfer or rearrangement between atoms underlies all chemical reactions. Either negative or positive electrons (sometimes called positrons) may be emitted from atomic nuclei during nuclear reactions: they are then called beta particles.

Energy: The capacity to do work. A quantity which is conserved, although it may be exchanged among bodies and transformed from one form to another, converted between heat and work, or interconverted with mass.

Energy conversion: The transformation of energy from one form to another.

Enrichment: The process of increasing the concentration of fissionable uranium-235 in uranium from the naturally occurring level of about 0.7 percent to the concentration required to sustain fission in a nuclear reactor, generally about three percent. The principal method of enrichment is gaseous diffusion, but gaseous centrifugation is also receiving much attention, particularly abroad.

Entropy: The change in entropy, ΔS, of any system is equal to the increment of heat added divided by its absolute temperature, so that:

$$dS = \frac{dQ}{T} \frac{\text{joules}}{\text{kelvin}}.$$

The total entropy of the universe is continually increasing.

Environmental impact statements: The analytical statements that balance costs and benefits of a federal decision. Required by the National Environmental Policy Act (NEPA), section (102(2) (c).

Exponential growth: Type of growth illustrated by the compound interest law. A useful characteristic of an exponential growth rate is that the "doubling time," or the length of time required for the growing thing to double in size, is constant.

Fast breeder reactor: A nuclear reactor that operates with neutrons at the fast speed of their initial emission from the fission process, and that produces more fissionable material than it consumes.

Fast neutron: High energy neutron. Fast neutrons are utilized in the fast breeder actor both to produce nuclear fissions and to transform fertile material (e.g., ^{238}U) into fissionable nuclear fuel.

Fast reactor: A nuclear reactor in which the fission chain reaction is sustained primarily by fast neutrons. Fast reactors contain no moderator and inherently require enriched fuel. They are of interest because of favorable neutron economy which makes them suitable for breeding.

Feedstock: Fossil fuels used for their chemical properties, rather than their value as fuel, e.g., oil used to produce plastics and synthetic fabrics.

Fertile material: A material, not itself readily fissionable by thermal neutrons, which can be converted into a fissionable material by irradiation in a nuclear reactor. The two basic fertile materials are uranium-238 and thorium-232.

Fission: The splitting of a heavy nucleus into two approximately equal parts, accompanied by the release of a relatively large amount of energy and generally one or more neutrons. Fission can occur spontaneously, but usually is caused by nuclear absorption of neutrons or other particles.

Fissile materials: Any material fissionable by slow neutrons. The three basic ones are uranium-235, plutonium-239, and uranium-233.

Fly ash: The fine, solid particles of noncombustible material residue carried from a bed of solid fuel by the gaseous products of combustion.

Fossil fuel: Any naturally occurring fuel of an organic nature, such as coal, oil shale, natural gas, or crude oil. Fossil fuels are organically formed from living matter.

Fuel: A substance used to produce heat energy, chemical energy by combustion or nuclear energy by nuclear fission.

Fuel cell: A device for converting the energy released in a chemical reaction directly into electrical energy.

Fuel cycle: The series of steps involved in supplying fuel for nuclear power reactors. It includes mining, refining of uranium, fabrication of fuel elements, their use in a nuclear reactor, chemical processing to recover remaining fissionable material, reenrichment of the fuel, refabrication into new fuel elements, and waste storage. Fuel cycle is sometimes used to refer to a similar series of steps for fossil fuels.

Fusion: The combining of atomic nuclei of very light elements by collision at high speed to form new and heavier elements, resulting in the release of energy.

Gallon: A unit of measure. A U.S. gallon contains 231 cubic inches, 0.133 cubic feet, or 3.785 liters. It is 0.83 times the imperial gallon.

Gas-cooled fast breeder reactor (GCBR): A fast breeder reactor which is cooled by a gas, usually helium, under pressure.

Gas, manufactured: A gas obtained by destructive distillation of coal, or by the thermal decomposition of oil, or by the reaction of steam passing through a bed of heated coal or coke. Examples are coal gases, coke-oven gases, producer gas, blast-furnace gas, blue (water) gas, carbureted water gas. Btu content varies widely.

Gas, natural: A naturally occurring mixture of hydrocarbon gases found in porous geologic formations beneath the earth's surface, often in association with petroleum. The principal constituent is methane.

Gas turbine: An engine which converts chemical energy of liquid fuel into mechanical energy by combustion. Gases resulting are expanded through a turbine.

Gaseous diffusion: The principal process for enrichment of uranium; that is, for increasing the concentration of fissionable uranium-235 in a mixture of uranium isotopes to the level required to sustain fission in a nuclear reactor.

Gasification: In the most commonly used sense, gasification refers to the conversion of coal to a high-Btu synthetic natural gas under conditions of high temperatures and pressures; in a more general sense, conversion of coal into a usable gas.

Gasoline: A petroleum fraction composed primarily of small branched-chain, cyclic, and aromatic hydrocarbons.

Generator (electric): A machine which converts mechanical energy into electrical energy.

Geothermal energy: The heat energy available in the rocks, hot water, and steam in the earth's subsurface.

Geothermal steam: Steam drawn from deep within the earth. There are about 90 known places in the continental United States where geothermal steam might be harnessed for power. These are in California, Idaho, Nevada, and Oregon.

Gross National Product (GNP): The nation's total national output of goods and services at current market prices.

Half-life, radioactive: Time required for a radioactive substance to lose 50 percent of its activity by decay. Each radionuclide has a unique half-life.

Heat: A form of kinetic energy, whose effects are produced by the vibration, rotation, and general motions of molecules.

Heat exchanger: Any device that transfers heat from one fluid (liquid or gas) to another or to the environment.

Heavy water: Deuterium oxide; that is, water in which all hydrogen atoms have been replaced by deuterium atoms.

Heat pump: A refrigeration machine that is used for heating buildings rather than cooling them. Expanding refrigeration fluid removes heat from the outside air; the fluid is then compressed, and the heat resulting from compression is discharged to a heat exchanger in the surroundings to be heated.

Heat sink: The medium or location to which waste heat is discharged.

High-temperature gas-cooled reactor (HTGR): A reactor using blocks of graphite containing fissile and fertile material and cooled with helium. HTGR's are operated at a high temperature which permits conversion of heat to electricity with improved efficiency.

Hydraulic fracturing: A general term, for which there are numerous trade or service names, for the fracturing of rock in an oil or gas reservoir by pumping a fluid under high pressure into the well. The purpose is to produce artificial openings in the rock in order to increase permeability.

Hydrocarbon: A compound containing only carbon and hydrogen. The fossil fuels are predominantly hydrocarbons, with varying amounts of organic compounds of sulfur, nitrogen, and oxygen, and some inorganic materials.

Hydroelectric plant: An electric power plant in which energy of falling water is converted into electricity by turning a turbine generator.

Joule: A unit of energy or work which is equivalent to one watt-second or 0.239 calories.

Kerosene: The petroleum fraction containing hydrocarbons that are slightly heavier than those found in gasoline and naphtha.

Kilowatt (kW): 1,000 watts. A unit of power equal to 1,000 watts or to energy consumption at a rate of 1,000 joules per second. It is usually used for electrical power. An electric motor rated at one horsepower uses electrical energy at a rate of about $\frac{3}{4}$ kilowatt.

Kilowatt-hour (kWh): A unit of work or energy equal to one kilowatt in one hour. It is equivalent to 3.6 M joules.

Kinetic energy: The energy of motion; the ability of an object to do work because of its motion.

Light-water reactor (LWR): A nuclear reactor in which ordinary water or light water is the primary coolant/moderator with slightly enriched uranium fuel. There are two commercial light-water reactor types—the boiling water reactor (BWR) and the pressurized water reactor (PWR).

Lignite: A low-grade coal of a variety intermediate between peat and bituminous coal.

Liquefaction (of coal): The conversion of coal into liquid hydrocarbons and related compounds by hydrogenation.

Liquefied natural gas (LNG): Natural gas that has been liquefied by cooling to about $-140°C$. In this form, it occupies a relatively small volume and can be transported economically by ocean tanker.

Liquid-metal fast breeder (LMFBR): A nuclear breeder reactor cooled by molten sodium in which fission is accompanied by fast neutron irradiation of U-238 giving rise to Pu-239, a fissile material.

L thium: Element No. 3 (symbol, Li; atomic weight 6.94). As found in nature, li hium consists of a mixture of two stable isotopes—lithium-6 (7.5 percent) and lithium-7 (92.4 percent). Lithium-6 is of interest as a possible fuel or source thereof for the generation of power from a controlled thermonuclear reaction.

Load: The amount of power needed to be delivered at a given point on an electric system.

Load growth: The growth in energy and power demands by a utility's customers.

Lurgi Process: The chief commercially available process for coal gasification. Having originated in Germany, this process has limited application in the United States because of problems of scaling up the size of operations and characteristics of U.S. coal. The Office of Coal Research and American Gas Association are jointly funding further development.

Margin: The difference between the net system generating capability and system maximum load requirements including net schedule transfers with other systems.

Megawatt (MW): 1,000 kilowatts, 1 million watts.

Metallurgical coal: Coal with strong or moderately strong coking properties that contains no more than 8.0 percent ash and 1.25 percent sulfur, as mined or after conventional cleaning.

Methane (CH_4): The lightest in the paraffinic series of hydrocarbons. It is colorless, odorless and flammable. It forms the major portion of marsh gas and natural gas.

Middle distillate: One of the distillates obtained between kerosene and lubricating oil fractions in the refining processes. These include light fuel oils and diesel fuel.

Mine-mouth plant: A steam-electric plant or coal gasification plant built close to a coal mine. It is usually associated with delivery of output via transmission lines or pipelines over long distances, as contrasted with plants located nearer load centers and at some distance from sources of fuel supply.

Moderator: A material, for example, graphite or water, used to absorb a large portion of the energy of fission neutrons in a nuclear reactor. By slowing down neutrons to speeds at which they tend not to be absorbed by nonfissionable nuclei, the moderator prevents quenching of the chain reaction, and permits the reactor to operate at low fuel enrichments.

National Environmental Policy Act (NEPA): An act passed in 1970 requiring that the environmental impact of most large projects and programs be considered. Among its important provisions is one requiring a detailed statement of environmental impact of and alternatives to a project to be submitted to the government before the project can begin.

Natural gas: Naturally occurring mixtures of hydrocarbon gases and vapors, the more important of which are methane, ethane, propane, butane, pentane, and hexane. The energy content of natural gas is usually taken as 1032 Btu/cu ft.

Natural uranium: Uranium as found in nature, containing 0.7 percent uranium-235, 99.3 percent of uranium-238, and a trace of uranium-234. It is also called normal uranium.

Net reserves: The recoverable quantity of an energy resource that can be produced and delivered.

Neutron: An elementary particle with approximately the mass of a proton but without any electric charge. It is one of the constituents of the atomic nucleus. It is frequently released during nuclear reactions and, on entering a nucleus, can cause nuclear reactions including nuclear fission.

Nitrogen oxides (NO_x): Chemical compounds of nitrogen (N) and oxygen (O). A product of combustion of fossil fuels whose production increases with the temperature of the process. It can become an air pollutant if concentrations are excessive.

Nuclear fission: The splitting of large atomic nuclei into two or more new nuclear species, with the release of large amounts of energy.

Nuclear fuel cycle: The various steps which involve the production, processing, use, and reprocessing of nuclear fuels.

Nuclear fusion: The process by which small atomic nuclei join together with the release of large amounts of energy.

Nuclear power plant: Any device, machine, or assembly that converts fission nuclear energy into some form of useful power, such as electrical power.

Nuclear reactor: A device in which a fission chain reaction can be initiated, maintained, and controlled. Its essential component is a core with fissionable fuel. It usually has a moderator, reflector, shielding, coolant, and control mechanisms. It is the basic machine of nuclear power.

Oil shale: A sedimentary rock containing solid organic matter from which oil can be obtained when the rock is heated to a sufficiently high temperature.

Organization of Petroleum Exporting Countries (OPEC): Founded in 1960 to unify and coordinate petroleum policies of the members. The members and the date of membership are: Abu Dhabi (1967); Algeria (1969); Indonesia (1962); Iran (1960); Iraq (1960); Kuwait (1960); Libya (1962); Nigeria (1971); Qatar (1961); Saudi Arabia (1960); and Venezuela (1960). OPEC headquarters are in Vienna, Austria.

Particulate matter: Solid particles, such as the ash, which are released from combustion processes in exhaust gases at fossil-fuel plants.

Peaking capacity: That part of a system's equipment which is operated only during the hours of highest power demand.

Peaking load: The greatest amount of all of the power loads on a system, or part thereof, which has occurred at one specified period of time.

Petroleum: An oily flammable bituminous liquid that may vary from almost colorless to black, occurs in many places in the upper strata of the earth, is a complex mixture of hydrocarbons with small amounts of other substances, and is prepared for use as gasoline, naphtha, or other products by various refining processes.

Petroleum refinery: A plant that converts crude petroleum into the many petroleum fractions (asphalt, fuel oil, gasoline, etc.). Usually this conversion is accomplished by fractional distillation.

Photon: A quantum of electromagnetic energy having properties of both a wave and a particle, but without mass or electric charge.

Photovoltaic conversion: Transformation of solar radiation directly into electricity by means of a solid-state device such as the single-crystal silicon solar cell.

Plasma: An electrically neutral, partially ionized gas in which the motion of the constituent particles is dominated by electromagnetic interactions. The study of plasma motions is called magnetohydrodynamics (MHD).

Plutonium (Pu): A heavy, fissionable, radioactive, metallic element with atomic number 94. Plutonium-239 occurs in nature in trace amounts only. However, it can be produced as a byproduct of the fission reaction in a uranium-fueled nuclear reactor and can be recovered for future use.

Pollution: The accumulation of wastes or byproducts of human activity. Pollution occurs when wastes are discharged in excess of the rate at which they can be degraded, assimilated, or dispersed by natural processes. Sometimes noxious environmental effects not caused by human activity are also called pollution.

Potential energy: Energy which is not associated with motion—thus, that which is stored in chemical bonds and in water at high elevations are forms of potential energy.

Power: The rate at which work is done or energy is transformed. Power is measured in units of work per unit time; typical units are the horsepower and the watt.

Pressurized-water reactor: A power reactor in which heat is transformed from the core to a heat exchanger by water kept under high pressure to prevent it from boiling. Steam is generated in a secondary circulation system.

Primary fuel: Fuel consumed in original production of energy, as contrasted to a conversion of energy from one form to another.

Proton: A positively charged elementary particle having a mass of about 1.7×10^{-24} grams, roughly 1840 times as great as the mass of an electron. Protons are constituents of atomic nuclei and are emitted in some nuclear reactions.

Proved reserves: The estimated quantity of crude oil, natural gas, natural gas liquids, or coal, which analysis or geological and engineering data demonstrates with reasonable certainty to be recoverable from known oil, coal, or gas fields under existing economic and operating conditions.

Pumped storage: An arrangement in which water is pumped into a storage reservoir at a higher elevation when a surplus of electricity is being generated; during times of peak demand for electricity, this water is then used for the generation of electricity as in a hydroelectric power plant.

Radiation: The process by which energy in the form of electromagnetic radiation is emitted from matter. Also, the electromagnetic or particulate rays that are emitted from atoms or molecules as they undergo internal change.

Radioactivity: The spontaneous decomposition of an atom accompanied by the release of energy.

Reactor: Any device in which a chemical or nuclear reaction is sustained in a self-supporting chain.

Refinery: An industrial complex for processing crude oil by distillation and chemical reactions so as to produce a separate petroleum product. Typical crude fractions, from top to bottom or simple to complex, are: ether, methane, and ethane (the gasolines); propane and butane; kerosene, fuel oil, and lubricants; jelly paraffin, asphalt, and tar.

Reserve generating capacity: Extra capacity maintained to generate power in the event of unusually high demand or a loss or scheduled outage of regular generating capacity.

Residual fuel oil: A high-viscosity fuel oil that must be heated before it can be pumped and handled conveniently. Residual fuel oil is the petroleum fraction that is collected after all lower-boiling fractions have been distilled away. It is used primarily in industry, in large commercial buildings, and for the generation of electricity.

Secondary recovery: Oil and gas obtained by the augmentation of reservoir energy; often by the injection of air, gas, or water into a production formation.

Slow neutron: A low-energy neutron, sometimes called a thermal neutron. The energy of a slow neutron is about 0.025 electron volts, in contrast to the energy of a fast neutron, which may exceed 1,000 electron volts. Slow neutrons are very efficient in causing fission of uranium-235. Nuclear-power plants now in operation are designed to use slow neutrons to sustain the fission reaction.

Solar cell: A device which converts solar radiation to a current of electricity.

Solar constant: The average intensity of solar radiation striking the atmosphere. The solar constant is measured on a plane perpendicular to the path of the radiation. Its value is 1.36 kilowatts per square meter.

Solar furnace: An optical device with large mirrors that focuses the rays from the sun upon a small focal point to produce very high temperatures.

Solar spectrum: The total distribution of electromagnetic radiation emitted from the sun, minus those wavelengths that are absorbed by the solar atmosphere.

Stack-gas desulfurization (scrubber): Treating of stack gases to remove sulfur compounds.

Steam power plant: A plant in which the prime movers (turbines) connected to the generators are driven by steam.

Strip mining: The mining of coal by removing covering material (the overburden) and "stripping" away the entire underlying coal seam. Other forms of coal mining are underground, and auger mining, in which coal is drilled out of seams exposed along the side of a mountain.

Synthetic natural gas (SNG): A manufactured gaseous fuel generally produced from naphtha or coal. It contains 95 to 98 percent methane, and has an energy content of 980 to 1035 Btu per standard cubic foot, about the same as that of natural gas.

Tar sands: Sedimentary rocks which contain viscous, heavy petroleum that cannot be recovered by conventional methods of petroleum production.

Tertiary recovery: Use of heat and other methods other than fluid injection to augment oil recovery (presumably occurring after secondary recovery).

Thermal pollution: An increase in the temperature of water resulting from waste heat released by a thermal electric plant, for example, added to the cooling water.

Thermal power plant: Any electric power plant which operates by generating heat and converting the heat to electricity.

Thermal reactor: A nuclear reactor in which the fission process is propagated mainly by thermal neutrons, i.e., by neutrons that have been slowed down until they are in thermal equilibrium with the atoms of the moderator.

Thermodynamics, laws of: The First Law of Thermodynamics states that energy can neither be created nor destroyed. The Second Law of Thermodynamics states that when a free exchange of heat takes place between two bodies, the heat is always transferred from the warmer to the cooler body.

Thorium (Th): A naturally radioactive element with atomic number 90 and, as found in nature, an atomic weight of approximately 232. The fertile thorium-232 isotope is abundant and can be transmuted to fissionable uranium-233 by neutron irradiation and in turn used as nuclear reactor fuel.

Ton: A unit of weight equal to 2,000 pounds in the United States, Canada, and the Union of South Africa, and to 2,240 pounds in Great Britain. The American ton is often called the short ton, while the British ton is called the long ton. The metric ton, or 1,000 kilograms, equals 2,204.62 pounds.

Tritium (T): A manmade radioactive isotope of hydrogen with two neutrons and one proton in the nucleus.

Turbine: An engine, the shaft of which is rotated by a stream of water, steam, air, or fluid from a nozzle forced against blades of a wheel.

Uranium (U): A radioactive element with the atomic number 92 and, as found in natural ores, an average atomic weight of approximately 238. The two principal natural isotopes are uranium-235 (0.7 percent of natural uranium), which is fissionable (capable of being split and thereby releasing energy), and uranium-238 (99.3 percent of natural uranium), which is fertile (having the property of being convertible to a fissionable material). Natural uranium also includes a minute amount of uranium-234.

Waste heat: Heat which is at temperatures very close to the ambient and hence is not valuable for production of power and is discharged to the environment.

Wastes, radioactive: Equipment and materials, from nuclear operations, which are radioactive and for which there is no further use. Wastes are generally classified as high-level (having radioactivity concentrations of hundreds to thousands of curies per gallon or cubic foot), low-level (in the range of 1 microcurie per gallon or cubic foot), or intermediate.

Watt: A unit of power. It is the rate of energy use or conversion when one joule of energy is used or converted per second. (A joule is about 0.25 calories.)

Watt-hour: The total amount of energy used in one hour by a device that uses one watt of power for continuous operation. Electrical energy is commonly sold by the kilowatt hour (1,000 watt-hours).

Wellhead: Oil or gas brought to the surface, ready for transportation to refinery or ship or pipeline. Wellhead costs usually refer to the cost to bring the oil or gas to the surface and do not include costs of transportation, refining, distribution, or profit.

Work: The transfer of energy from one body to another; or the energy itself, in the process of transfer. Work and energy are measured in the same units.

Appendix B
Selected Data and Conversion Factors

Temperature

Absolute Zero: $0°K = 0°R = -273.15°C = -459.67°F$
Kelvin $T = T(°C) + 273 = T(°R)/1.80$
Rankine $T = T(°F) + 460$

Water

Density $= 1 \text{ g/cm}^3 = 1 \text{ kg/liter}$
Specific heat $= 1 \text{ cal/g·°C}$
$\qquad\qquad = 1 \text{ Btu/lb·°F}$

Conversion Factors

Length

1 inch $= 2.54$ cm
1 foot $= 0.3048$ m
1 mile $= 1609$ m

Area

1 ft^2 $= 0.0929$ m^2
1 acre $= 4047$ m^2

Volume

1 gallon $= 3.785$ liters
1 acre-foot $= 1233$ m^3

Speed

1 mile/hour = 0.4470 m/sec

Mass

1 lb = 0.4536 kg
1 ton = 2000 lb
1 tonne = 1 metric ton = 1000 kg

Energy

1 calorie = 4.184 J
1 Btu = 1054 J
1 kcal = 1 Calorie = 4184 J
1 kWh = 3.60×10^6 J
1 therm = 10^5 Btu
1 quad = 10^{15} Btu

Power

1 hp = 746 watts
1 Btu/sec = 1054 W

Insulation U-Value

1 Btu/ft²-hr-°F = 5.674 W/m²·°C

Index